Aquatic Ecosystem Management

Aquatic Ecosystem Management

Edited by **Jeremy Harper**

SYRAWOOD
PUBLISHING HOUSE

New York

Published by Syrawood Publishing House,
750 Third Avenue, 9th Floor,
New York, NY 10017, USA
www.syrawoodpublishinghouse.com

Aquatic Ecosystem Management
Edited by Jeremy Harper

© 2016 Syrawood Publishing House

International Standard Book Number: 978-1-68286-039-7 (Hardback)

Printed in the United States of America.

Contents

Preface

The purpose of the book is to provide a glimpse into the dynamics and to present opinions and studies of some of the scientists engaged in the development of new ideas in the field from very different standpoints. This book will prove useful to students and researchers owing to its high content quality.

Aquatic ecosystems are primarily divided into two types - marine and fresh water ecosystems; based on the percentage of dissolved components. This book compiles researches from across the globe to provide a comprehensive overview of aquatic ecosystems. Some of the diverse topics covered in this text are predation behavior, solar salt production, crude oil biodegradation, etc. The aim of this book is to bring forth new horizons of research and aid students in better understanding the concepts of this field.

At the end, I would like to appreciate all the efforts made by the authors in completing their chapters professionally. I express my deepest gratitude to all of them for contributing to this book by sharing their valuable works. A special thanks to my family and friends for their constant support in this journey.

Editor

Prey items and predation behavior of killer whales (*Orcinus orca*) in Nunavut, Canada based on Inuit hunter interviews

Steven H Ferguson[1,2]*, Jeff W Higdon[3] and Kristin H Westdal[4]

Background: Killer whales (*Orcinus orca*) are the most widely distributed cetacean, occurring in all oceans worldwide, and within ocean regions different ecotypes are defined based on prey preferences. Prey items are largely unknown in the eastern Canadian Arctic and therefore we conducted a survey of Inuit Traditional Ecological Knowledge (TEK) to provide information on the feeding ecology of killer whales. We compiled Inuit observations on killer whales and their prey items via 105 semi-directed interviews conducted in 11 eastern Nunavut communities (Kivalliq and Qikiqtaaluk regions) from 2007-2010.

Results: Results detail local knowledge of killer whale prey items, hunting behaviour, prey responses, distribution of predation events, and prey capture techniques. Inuit TEK and published literature agree that killer whales at times eat only certain parts of prey, particularly of large whales, that attacks on large whales entail relatively small groups of killer whales, and that they hunt cooperatively. Inuit observations suggest that there is little prey specialization beyond marine mammals and there are no definitive observations of fish in the diet. Inuit hunters and elders also documented the use of sea ice and shallow water as prey refugia.

Conclusions: By combining TEK and scientific approaches we provide a more holistic view of killer whale predation in the eastern Canadian Arctic relevant to management and policy. Continuing the long-term relationship between scientists and hunters will provide for successful knowledge integration and has resulted in considerable improvement in understanding of killer whale ecology relevant to management of prey species. Combining scientists and Inuit knowledge will assist in northerners adapting to the restructuring of the Arctic marine ecosystem associated with warming and loss of sea ice.

Keywords: beluga whales, bowhead whales, group size, hunting behaviour, narwhal whales, predator-prey relations, prey capture techniques, Traditional Ecological Knowledge, seals, walrus

Background

In recent years there has been significant interest in the role of killer whale (*Orcinus orca*) predation in shaping marine ecosystems and regulating prey populations [1]. Killer whales are widespread in world oceans and are the top predator in all regions where they occur [2-4]. The species consumes a wide variety of prey items, ranging from small schooling fish to large baleen whales [5], and there has been considerable debate over the role of killer whale predation in trophic cascades and prey species dynamics [6-8]. Killer whale predation can limit small prey populations [9-11], but more information on the species and number of prey consumed is needed to better address this issue [12]. Researchers have determined that in many areas killer whales with different and largely non-overlapping foraging specializations can co-exist. Examples include the coastal temperate northeast Pacific, where the transient ecotype feeds primarily if not exclusively on marine mammals, and the resident ecotype eats fish [13,14]. Four to five different ecotypes with different prey preferences have been identified in Antarctic waters [15-17]. Similar patterns have

* Correspondence: Steve.Ferguson@dfo-mpo.gc.ca
[1]Fisheries and Oceans Canada, Central and Arctic Region, 501 University Crescent, Winnipeg, Manitoba R3T 2N6 Canada
Full list of author information is available at the end of the article

been identified for the northeast Atlantic [18], and to some extent for the eastern Tropical Pacific and southern Indian Oceans [19].

In most other regions of the world, little is known about the ecology, life-history, and population status of killer whales [4]. One of these areas is the eastern Canadian Arctic (Figure 1), where there has been little directed research on killer whales until recently [20,21]. Killer whales were historically present in Davis Strait and Baffin Bay during the ice-free season, where they were occasionally reported in whaling logbooks in the

1800s [22]. The species appears to have recently colonized the Hudson Bay region [12], where sightings are increasing, possibly in response to declining summer sea ice distribution [23]. Killer whales are seasonal visitors to high latitude regions and sightings peak during the summer months (August and September) [24]. They have been observed killing and consuming other marine mammals, including both cetaceans (beluga (*Delphinapterus leucas*), narwhal (*Monodon monoceros*), bowhead (*Balaena mysticetus*)) and phocid seals (e.g., ringed (*Pusa hispida*) and bearded (*Erignathus barbatus*) seals

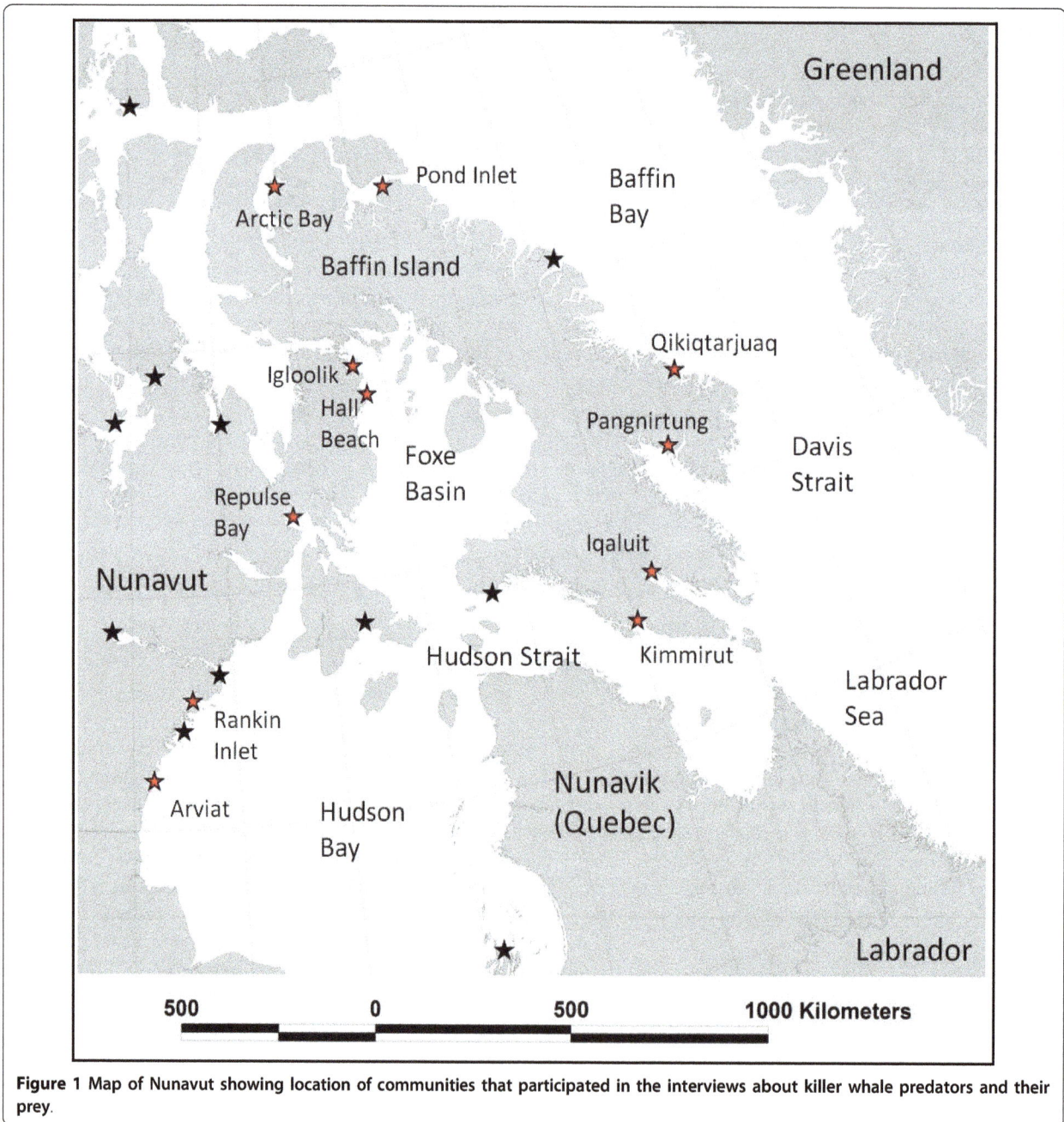

Figure 1 Map of Nunavut showing location of communities that participated in the interviews about killer whale predators and their prey.

[12,22,24,25]). These species are important to Inuit cultural and socio-economic well-being, and local hunters have expressed concern over increasing predation pressure on narwhal, beluga, and other Arctic marine mammals [12,26-28].

Traditional Ecological Knowledge (TEK) is being increasingly used in Nunavut [29-32], providing a long temporal record of rare species. Both Inuit and research observations of killer whale hunting and feeding can be scarce and hunting behaviour cryptic and hard to interpret. Observations of predation are rarely acquired by scientists, but Inuit hunters spend significant amounts of time on these waters hunting, fishing, boating, and observing marine mammals. We interviewed Inuit hunters and elders in Nunavut to develop an extensive baseline knowledge of Arctic killer whales. We collected information via 105 semi-directed interviews [33,34], including questions on killer whale prey items, hunting techniques, prey avoidance behaviour, and consumption patterns. Interviewees also provided information on killer whale abundance, distribution, and movements, which is summarized elsewhere [12,24].

Methods

We used semi-directed interviews with Inuit hunters to document spatial and textual information on killer whales and their prey. This method provides a way to collect TEK in an open and flexible manner and avoids the rigidity of questionnaires [33,34]. A list of questions was developed in advance (Appendix 1), based on knowledge gaps of killer whale information (e.g., prey items, hunting behaviour), but interviews remained open-ended and each interviewee was given the option to elaborate on matters that were important to them. Interviewees do not always address every topic, but the approach also allows the interviewee to provide important information not anticipated by the researcher. Note that quotes are usually via the interpreter and not a direct quotation from the interviewee although some were in English. The research protocol was approved by the Office of Research Services, University of Manitoba (i.e., ethics approval) and the Nunavut Research Institute (NRI), and interviews were conducted with the consent and assistance of local Hunters and Trappers Organizations/Associations (HTO/HTA) in each community.

Communities (Table 1 Figure 1) were chosen to provide a wide representation throughout the Hudson Bay and Baffin Island regions of Nunavut and with consideration of historic and recent killer whale observations [24,27] and logistical constraints. Interviews were conducted with the aid of a local interpreter, with most conducted in Inuktitut (some in English). Spatial information (hunting and traveling locations, outpost camps, killer whale sightings, migration routes, etc.) was

Table 1 Summary of eleven Nunavut communities visited to collect Inuit traditional ecological knowledge (TEK) of killer whales predation, with dates of visit, number of semi-directed interviews conducted, and region used for analyses.

Community	Date visited	Region	No. interviews
Repulse Bay	July-August 2007	Hudson Bay	17
Igloolik	February-March 2008	Foxe Basin	16
Hall Beach	February-March 2008	Foxe Basin	7
Rankin Inlet	March 2008	Hudson Bay	10
Arviat	March 2008	Hudson Bay	5
Pangnirtung	January 2009	South Baffin Island	11
Kimmirut	February 2009	South Baffin Island	5
Arctic Bay	April 2009	North Baffin Island	11
Iqaluit	April-May 2009	South Baffin Island	7
Pond Inlet	March 2010	North Baffin Island	8
Qikiqtarjuaq	March 2010	North Baffin Island	8
Total	July 2007-March 2010		105

recorded on maps copied from the Nunavut Land Use Study [35]. Interviewees were identified using a reputational sampling method [36] where the local HTO in each community provided a list of potential interviewees, augmented with other participants identified by the interpreters or other interviewees (i.e., snowball technique [37]). Our research protocol allowed participants to remain anonymous or have their names acknowledged as a source of information, but stipulated that their names not be attached to any specific comments. All participants, that chose to be acknowledged, are listed in Appendix 2.

A mix of qualitative and quantitative survey data on killer whales was collected and analyzed. Here we focus on the predation information only, without any specific focus on the information related to abundance, group sizes, spatiotemporal distribution, migrations and movements, etc. (data on file), although we give brief mention of these subjects where appropriate. Qualitative data were analyzed using an interpretive approach to connect ideas and categorize results [38], where results are grouped into related categories (e.g., specific prey items) and patterns summarized. Results were summarized within and between communities and four different regions (Table 1), and across Nunavut as a whole. All individual sighting reports were added to a larger killer whale sightings database [27], and observations from nine communities (pre-2010 interviews) were included in a recent analysis of the database, including predation observations [24] (also see [11,12] regarding communities in the Hudson Bay region).

Results

We conducted 105 semi-directed interviews in 11 Nunavut communities between July 2007 and March 2010 (Table 1, Figure 1) (5-17 interviews per community: mean 9.5). Most interviewees were male (91%, one female interviewee in Arctic Bay, Arviat, and Rankin Inlet, two in Hall Beach, Pangnirtung, and Pond Inlet). We made an effort to interview older hunters, and 71 of 89 interviewees who provided age information were born in the 1950s or earlier. Younger hunters were interviewed on occasion, but only one interviewee was under 30. Most participants were active hunters or formerly active hunters (full-time harvesters) when they were younger, and most interviewees have spent a considerable amount of time hunting, fishing and boating throughout their lives (data on file). Nearly all participants (97%) had seen killer whales at least once, although not necessarily in the vicinity of their current community. Informants provided information on the seasonal and spatial distribution of killer whales in Nunavut, in addition to information on movements, migrations, and relative abundance. This information is presented in detail elsewhere [24].

Marine mammal prey items

Participants in all communities provided extensive information on killer whale prey items. Inuit throughout Nunavut reported that marine mammals are the main prey items, and that all species of the phocid seals and the most commonly occurring cetaceans are consumed. Most interviewees reported multiple prey items, and five (two in Repulse Bay, three in Igloolik) referred to killer whales as the "wolves of the sea", or being "like wolves". Killer whales were reported to "eat whatever they can catch", "eat anything", and similar remarks by 14 interviewees (seven in Foxe Basin, three each in Repulse Bay and south Baffin, and one in north Baffin). Phocid seals (mainly ringed seals, but also harp (*Pagophilus groenlandicus*), bearded, and hooded (*Crystophora cristata*) seals, in descending order of frequency) were most commonly reported as prey (73 interviewees, 70% of total), followed by narwhal (63 interviewees, 60%), beluga (55, 52%) and bowhead whales (48, 46%). All communities and regions indicated similar marine mammal prey items, but the proportions of interviewees indicating each species varied by region (Figure 2).

Reports of bowhead whales as a prey item were non-randomly distributed across regions (chi-square test, $\chi 2$ = 11.00, df = 3, P = 0.012), and concentrated in Foxe

Basin (21 of 23 interviewees, 44% of the total interviewees that reported this species as prey). Bowhead predation was reported in the other three regions, but by fewer interviewees. Reports of narwhal as prey were also non-randomly distributed ($\chi 2$ = 11.22, df = 3, P = 0.011) and concentrated in the northern Baffin Island communities, with few south Baffin interviewees reporting this species as prey. Conversely, few north Baffin interviewees reported beluga as prey in comparison to the other three regions ($\chi 2$ = 7.63, df = 3, P = 0.055). Narwhal was also commonly reported as prey for the Hudson Bay region, particularly in Repulse Bay (16 of 17 interviewees). Reports of seals as prey items were evenly distributed across all four regions ($\chi 2$ = 2.89, df = 3, P = 0.409). One north Baffin interviewee (from Qikiqtarjuaq) also identified minke whales (*Balaenoptera acutorostrata*) as killer whale prey. Phocid seals were identified as prey by 73 (70%) interviewees, in all four regions (Figure 2). Most of these (70, 96%) identified ringed seals as a prey item, and this was the most commonly identified seal species in all four regions (Figure 3, ranging from 83% of the 12 Hudson Bay interviewees who listed "seals" to 100% of those in Foxe Basin and north Baffin). Harp seals were the second most commonly reported phocid prey, with 22 of the 73 interviewees (30%) identifying this species. Harp seals were identified as killer whale prey in all four regions, particularly in Foxe Basin and the south Baffin communities. Bearded seals were also mostly reported as a prey item by Foxe Basin interviewees (11 of 14 total reports). Hooded seals were reported as killer whale prey by one interviewee in Qikiqtarjuaq.

Predation on fish

There is little evidence that Canadian Arctic killer whales eat fish [24], and interviewee responses suggested that predation on fish, if it does occur, is extremely rare (Table 2). Most interviewees (n = 79, 75%) did not mention or discuss fish as killer whale prey at all. Five interviewees noted that they did not know, or were not sure, if killer whales ate fish, and nine suggested they probably did, or might. Only four interviewees specifically stated that killer whales did not eat fish or that they had never heard of it occurring. Seven interviewees stated that killer whales did eat fish (five from Foxe Basin, two of which were in reference to the Pond Inlet area; and two from southern Baffin Island). However, none of these seven interviewees indicated that they had observed predation on fish first-hand.

Killer whale hunting techniques

Interviewees provided extensive discourse of the hunting behaviour of killer whales, and described the methods used to hunt bowhead, narwhal, beluga, and seals.

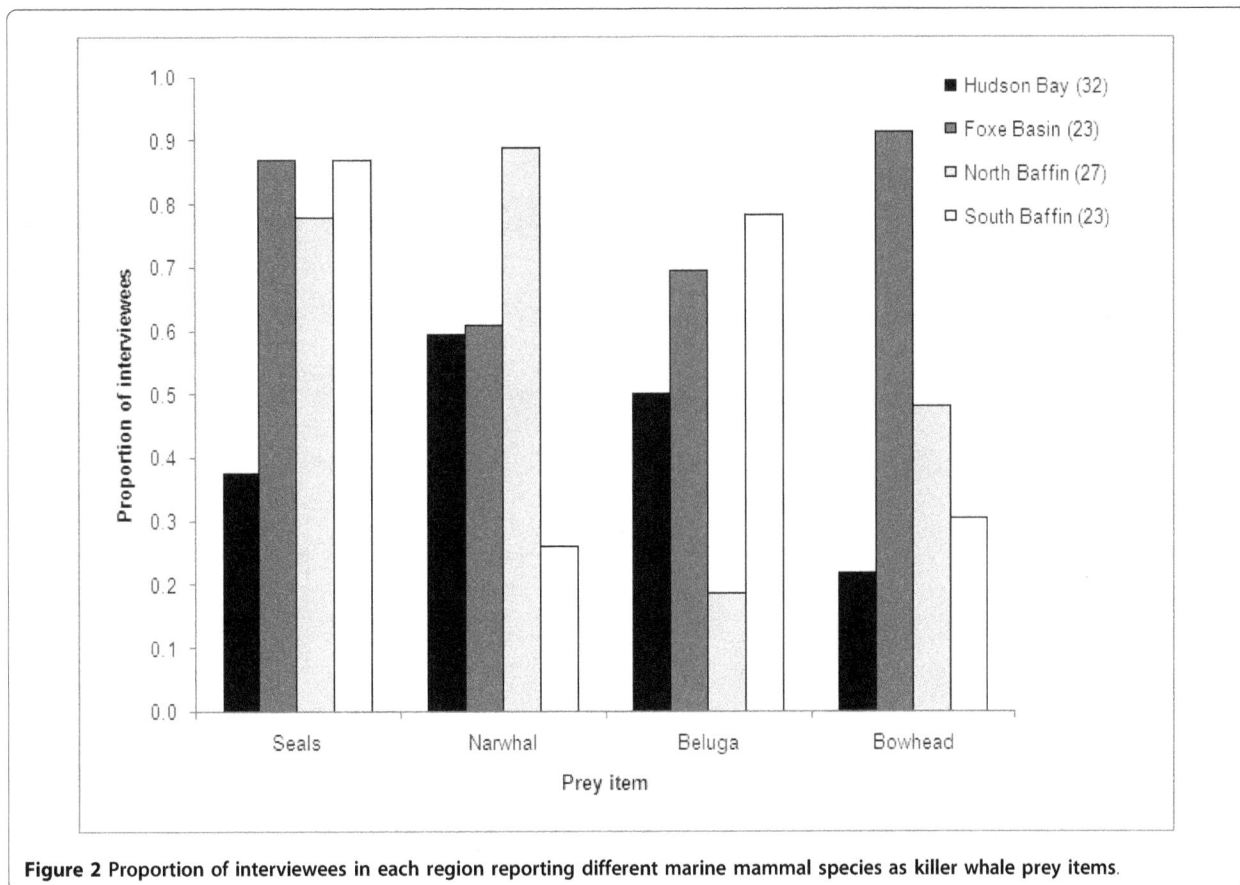

Figure 2 Proportion of interviewees in each region reporting different marine mammal species as killer whale prey items.

Bowhead whales

In total, 35 interviewees provided information on the methods killer whales use to attack and kill bowhead whales, including 17 who described first-hand observations (most from Igloolik, 10). The other 18 interviewees described killer whale attack behaviour from stories they had heard from others (Table 3). Foxe Basin interviewees recounted either direct observations or second-hand stories of 13 different killer whale attacks on bowhead whales (data on file), and provided most of the information on this species as a killer whale prey item. Eight Foxe Basin interviewees estimated the number of bowhead whales killed there each year ranged from 3 to 10. Killer whale attack behaviour is summarized in Table 3. Most interviewees provided similar descriptions of the techniques used when killer whales attack bowhead whales that included: (1) to drown prey by holding the whale underwater and/or covering the blowhole and (2) immobilize the prey prior to suffocation by biting and holding on to the front flippers and tail, ramming the whale to cause internal damage such as breaking ribs, and tearing chunks of flesh out of the living whale. Many interviewees described the cooperative nature of killer whale attacks, such that some whales would be on top of the bowhead, others biting the flippers and tail, while others rammed it. Two interviewees also described observations where several killer whales continued to circle the bowhead, to keep it from escaping, while the others attacked. One interviewee noted that smaller groups of killer whales occur in Foxe Basin, where they concentrate on smaller bowheads; but larger groups of killer whales occur in Arctic Bay where they are able to kill adult bowhead whales. There were 14 bowhead attack observations that included an estimate of killer whale group size, ranging from 1-30 (mean 6.2, median 3.5).

Inuit hunters occasionally found dead bowhead whales that were killed by killer whales, and this was noted by 22 interviewees. Reports were again mostly from Foxe Basin (Table 3), but dead bowhead whales were also reported from Hudson Bay (Repulse Bay) and northern Baffin Island (all three communities). Inuit assumed killer whales were responsible based on the external condition of the carcass (bite marks, chunks removed, evidence of internal injury such as broken ribs). Fifteen Foxe Basin interviewees identified locations of dead bowhead whales that had been killed by killer whales. Foxe Basin hunters reported finding from 3 to 5 dead

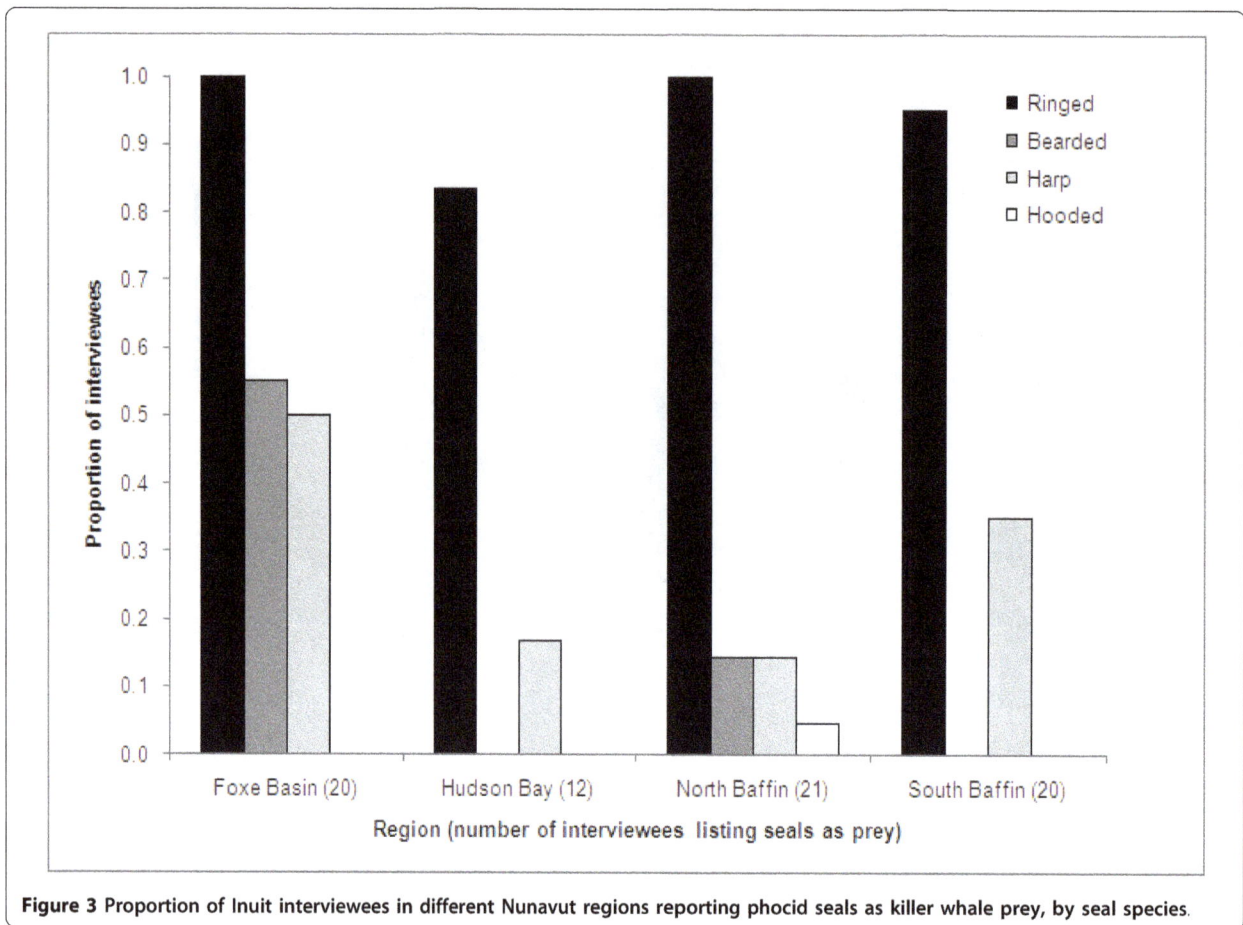

Figure 3 Proportion of Inuit interviewees in different Nunavut regions reporting phocid seals as killer whale prey, by seal species.

bowhead whales in a single summer, with higher numbers in some years. A total of 32 observations of dead bowheads were reported for the region (including one in Lyon Inlet), and after correcting for overlapping

reports, a minimum of 22 different dead bowheads were documented (with 8 found in 1999; data on file). In many cases, these whales were not eaten, or very little has been consumed (see below). In 1999, one was fresh

Table 2 Summary of Inuit interviewee information and responses regarding killer whale predation on fish in Nunavut waters.

Region	Community	Summary of interviewee responses about killer whales eating fish				
		No mention	Doesn't know, not sure	Maybe, probably	No, do not eat fish	Yes, do eat fish
Hudson Bay	Repulse Bay (17)	17				
	Rankin Inlet (10)	9		1		
	Arviat (5)	5				
Foxe Basin	Igloolik (16)	5	1	4	2	4
	Hall Beach (7)	1	2	1	2	1
North Baffin Island	Arctic Bay (11)	9	2			
	Pond Inlet (8)	8				
	Qikiqtarjuaq (8)	8				
South Baffin Island	Kimmirut (5)	5				
	Iqaluit (7)	4		2		1
	Pangnirtung (11)	9		1		1
Total		79	5	9	4	7

Table 3 Summary of Inuit interviewee information and observations of killer whale attacks on bowhead whales, and observations of dead bowhead whales that were killed by killer whales.

Observations and information on killer whale attacks on bowhead whales	Region				
	Foxe Basin	Hudson Bay	North Baffin	South Baffin	Total
Interviewees reporting first-hand observations	10	2	5	0	17
Interviewees reporting second-hand observations and stories	6	4	2	6	18
Interviewees providing descriptions of attacks and attack methods	10	4	3	4	21
Killer whales:					
...circle the bowhead, to keep it from escaping while others attack	1		1		2
...hold the whale underwater and/or cover the blowhole, to drown it	8	2	2		12
...bite and hold on to the whale by the front flippers and/or tail flukes	5		1	2	8
...ram the whale in the side "to break ribs", and tear chunks out of the belly	5	2	2	4	13
Interviewees reporting dead bowhead whales that were killed by killer whales	16	1	5		22
Interviewees providing descriptions of dead bowheads	9		2		11
Hunters find dead bowhead whales:					
...with killer whale teeth marks on the flippers, tail, belly, and baleen	3		2		5
...with large chunks torn out and their bellies ripped open	5		2		7
...that have broken bones, busted ribs	1				1
...that are fresh kills, still warm	3				3
...that are fresh enough for meat and muktuk to be collected from	2				2
...that were killed for fun and not eaten, wasted by the killer whales	4				4
...that are being scavenged by polar bears	1				1

enough for local Inuit to use as indicated by an interviewee in each of Igloolik and Hall Beach (Table 3).

Narwhal

Attacks on narwhal were observed by 24 interviewees, and four others reported stories they had heard from others. Five Foxe Basin interviewees had observed attacks, although most occurred in other areas (Table 4). Four Hudson Bay interviewees (all Repulse Bay) had observed attacks in their general area. Most observations (13) were reported by interviewees in north Baffin communities. Fourteen interviewees provided descriptive information on the methods used to kill narwhal. Prior to starting the attack, killer whales often herded the narwhal to a suitable location with deep enough water (see below), and also circled the group of whales to keep them stationary. Two interviewees noted that killer whales appear to tire the narwhal out prior to commencing an attack.

Interviewees reported killer whales ramming narwhal to "break their ribs", and killer whales will "play with" the narwhal, or pieces of them, and throw them around. One interviewee provided a second-hand description of killer whales "playing soccer" with the narwhal, and another observed killer whales killing narwhal by slapping them with their tail. Interviewees had observed killer whales biting narwhal, and sometimes carrying a narwhal in their mouth after biting them in the middle of the body. One Pond Inlet interviewee described two

killer whales biting a narwhal and pulling it apart, leaving the head and tail behind and taking the "meat in the middle". Two interviewees reported that killer whales will also drown narwhal. One Igloolik interviewee described an attack in Admiralty Inlet, where the killer whales herded the narwhal close to shore, and then the smaller killer whales would come close to the shore to grab a narwhal and then head back out into deeper water, where the larger killer whales stayed. Participants in Repulse Bay noted that when they are stalking prey, killer whales will slow down, move very deliberately, and remain as quiet as possible in order to reduce the wake and sound produced by their dorsal fin moving through the water. When they are close to their prey they pick up speed.

Dead narwhals were often found with crushed abdomens and broken ribs, bite marks, and with pieces missing, and this was reported by 17 interviewees (Table 4). Two North Baffin hunters reported collecting *maqtaq* (skin and blubber) from narwhal that were killed by killer whales. An interviewee from Repulse Bay indicated that killer whales normally do not go after tusked narwhal, and another indicated that narwhal with tusks are never found floating dead. A Pond Inlet interviewee had observed killer whales going after female narwhal, but not males. He recounted a story that tusked narwhal had killed killer whales in the past; the narwhal pierced the killer whale with its tusk, and become stuck, with

Table 4 Summary of Inuit interviewee information and observations of killer whale attacks on narwhals, and observations of dead narwhals that were killed by killer whales.

Observations and information on killer whale attacks on narwhals	Region				
	Foxe Basin*	Hudson Bay	North Baffin	South Baffin	Total
Interviewees reporting first-hand observations	5	4	13	2	24
Interviewees reporting second-hand observations and stories		1	3		4
Interviewees providing descriptions of attacks and attack methods	3		9	2	14
Killer whales:					
...herd narwhal to a suitable location with deep enough water, and circle to keep them stationary and prevent their escape	2		1		3
...tire out the narwhal prior to commencing an attack		1	1		2
...drown narwhal			2		2
...ram narwhal "to break their ribs"	1		5		6
...bite narwhal in the middle of the body, carry them in their mouth	2		3	1	6
...throw narwhal into the air, hit them with their tail and play with them, tear them apart and throw the pieces around	1		4	1	6
...leave lots of oil and scraps of blubber on the water	1	1	2		4
Interviewees reporting dead narwhals that were killed by killer whales	3	7	7		17
Interviewees providing descriptions of dead narwhals	2	6	6		14
Hunters find dead narwhals:					
...that are all busted up, with broken ribs	2	1	3		6
...that are covered with bite marks, with chunks and pieces missing		5	3		8
...that are killed but not eaten, killed for fun		2	1		3
...that are being scavenged by birds			1		1
...that meat and maqtaq can be collected from			3		3

* five Foxe Basin interviewees observed attacks on narwhal, but most occurred in other areas (Admiralty Inlet, 1; Pond Inlet, 1; Lyon Inlet, 2)

both animals dying as a result. Tusked narwhal may therefore represent a danger to killer whales; however, a Pangnirtung interviewee stated that they will attack narwhal with and without tusks, and two interviewees had observed killer whales kill tusked narwhal. An Arctic Bay hunter found a dead tusked narwhal covered with killer whale bite marks.

Beluga

Nineteen interviewees discussed killer whale attacks on beluga whales, with most from south Baffin (10, mostly (8) from Pangnirtung) (Table 5). Five interviewees told stories they had heard from others, the others recounted direct observations. Similar to attacks on narwhal, killer whales were reported to circle beluga before attacking, and killer whales have been observed ramming belugas with the presumed intent to cause internal damage that included broken ribs. Participants also reported that killer whales bit belugas in the mid-section, hold them in their mouth, and lift them out of the water, and will also toss them around. Predation events are often accompanied by oil slicks and scraps of blubber, and one interviewee noted gulls swarming around an attack site. A Pangnirtung interviewee told of a story he heard, where in the 1950s a person watched a killer whale eating a beluga that was still alive and moving.

Seven interviewees noted that hunters often see beluga scraps (skin and blubber) floating in the water after killer whale attacks, and carcasses were sometimes found. One Pangnirtung interviewee saw a dead beluga after it had been hit and killed, the blubber was missing and the ribs were broken on both sides. Another noted that killer whales would sometimes kill "hundreds" of belugas and not eat them all. When the killer whales had left the kill site, Inuit would collect the *maqtaq* from the numerous dead belugas.

Phocid seals

Twelve interviewees reported observations of attempted or successful predation on seals (ringed, harp and bearded seals), dominated by observations from south Baffin and Foxe Basin (Table 6). Four interviewees also noted second-hand observations and stories of killer whales attacking ringed and harp seals. Two north Baffin interviewees had observed killer whales slapping live ringed seals in the air with their tails, and four interviewees discussed the methods killer whales use to wash seals off small ice floes (reported for ringed and harp seals). Only two interviewees reported finding dead seals: a ringed seal in Foxe Basin (no additional details provided), and a harp seal (Iqaluit interviewee) that was

Table 5 Summary of Inuit interviewee information and observations of killer whale attacks on beluga whales, and observations of dead beluga whales that were killed by killer whales.

Observations and information on killer whale attacks on beluga whales	Region				
	Foxe Basin	Hudson Bay	North Baffin	South Baffin	Total
Interviewees reporting first-hand observations	4	3	1	7	15
Interviewees reporting second-hand observations and stories		3		3	6
Interviewees providing descriptions of attacks and attack methods	2	5	1	10	18
Killer whales:					
...circle the belugas before commencing an attack		1			1
...splash a lot when hunting belugas				1	1
...ram beluga whales in the side, "to break their ribs"			1	1	2
...toss belugas around, throw them in the air				3	3
...bite belugas in the midsection, lift them out of the water, carry them around	2			4	6
...leave lots of blubber scraps with bite marks, and oil slicks on the water		4		4	8
...kill beluga whales for fun, play with them, and sometimes do not eat them				2	2
...prefer the meat, and strip the blubber off the belugas				2	2
...will start to eat belugas when they are still alive				1	1
...sometimes eat belugas underwater, big pieces of blubber float to the surface		1			1
...leave blubber and dead whales behind that Inuit gather for food				1	1
Interviewees reporting dead belugas (or parts) that were killed by killer whales	2	1		4	7
Interviewees providing descriptions of dead belugas	2	1		4	7
Hunters find:					
...dead belugas with their ribs busted				1	1
...dead belugas with the blubber missing				1	1
...dead belugas covered in bite marks				1	1
...belugas that the killer whales have killed and left, wasting them				1	1
...scraps of blubber and maktak, all chewed up, floating in the water	2	1		4	7

found half eaten, by either a killer whale or a Greenland shark (*Somniosus microcephalus*).

All four observations of washing seals from ice were from south Baffin communities (two each from Kimmirut and Pangnirtung). Killer whales will circle a piece of ice and use their tails to create waves, washing seals off into the water. Two provided reports based on stories from others, but one hunter in each community

Table 6 Summary of Inuit interviewee information and observations of killer whale attacks on phocid seals, and observations of dead seals that were killed by killer whales.

Observations and information on killer whale attacks on phocid seals	Region				
	Foxe Basin	Hudson Bay	North Baffin	South Baffin	Total
Interviewees reporting first-hand observations	4	1	2	5	12
Interviewees reporting second-hand observations and stories		1		3	4
Interviewees providing descriptions of attacks and attack methods			2	4	6
Killer whales:					
...throw live ringed seals in the air with their tails			2		2
...drive harp seals in all directions when they are pursued				1	1
...wash seals (ringed and harp) off of ice floes				4	4
Interviewees reporting dead seals that were killed by killer whales	1		1		2
Interviewees providing descriptions of dead seals	1		1		2
Hunters find:					
...dead ringed seals with bite marks	1				1
...dead harp seals that are partially eaten			1		1

provided a first-hand observation. One Kimmirut inter-viewee noted that killer whales could create enough tur-bulence to make waves and try to knock seals off, no matter the thickness of the ice floe, and the other described this behaviour as an example of the intelli-gence killer whales exhibit when hunting.

Prey behaviour in the presence of killer whales, predator avoidance behaviour

When killer whales were present, all the different mar-ine mammal prey species sought refuge in shallow waters close to shore as a predator avoidance technique. This behaviour is well-known among Inuit hunters and is known as 'aarlirijuk' ("fear of killer whales") in the south Baffin dialect of Inuktitut ([39]; alternate spellings given by [40,41] are 'ardlingayuq', 'ardlungaijuq' or 'aar-lungajut'). The majority of interviewees (68%) reported that prey species will head to shallow waters and to the shoreline to avoid killer whales (with multiple species identified in some cases: 41 for seals, 34 for narwhal, 30 for beluga, and six for bowhead) (Figure 4). The propor-tion of interviewees identifying this behaviour was simi-lar for the four regions, ranging from 61 to 70% of the

total number interviewed. The prey species reported to exhibit this behaviour varied by region and community however, with narwhal observations predominating in North Baffin and Hudson Bay communities (primarily Arctic Bay and Repulse Bay, also Pond Inlet). Observa-tions of seal avoidance behaviour were reported in all communities, and beluga observations were mainly reported by South Baffin interviewees (in all three communities).

Eleven interviewees noted the behaviour bowhead whales use to avoid killer whale predation (three each from Foxe Basin and Hudson Bay, five from north Baf-fin). Three reported that bowhead whales will "run away", and five indicated that they will run to the ice, where killer whales will not go, or into shallow waters, inlets and fjords. One interviewee (Repulse Bay) reported that bowhead whales will dive deep to avoid killer whales, and that they can stay down longer due to their larger lungs. A Rankin Inlet interviewee described a bowhead whale being chased by killer whales, and the bowhead was jumping out of the water. In one Igloolik observation (from August 2007), a killer whale was observed chasing after a bowhead and her calf. At one

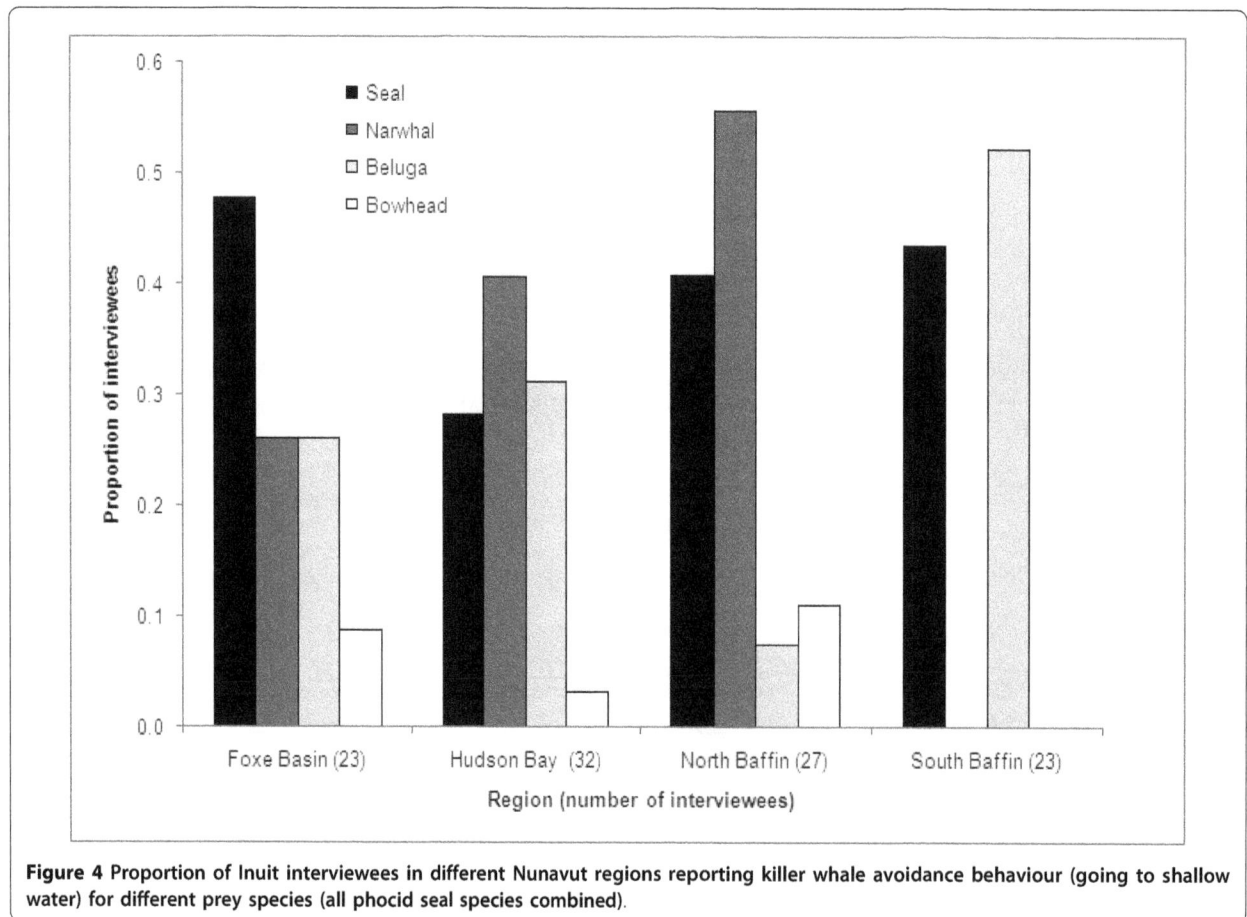

Figure 4 Proportion of Inuit interviewees in different Nunavut regions reporting killer whale avoidance behaviour (going to shallow water) for different prey species (all phocid seal species combined).

point during the chase, the entire bowhead calf was up out of the water. The interviewee thought that the cow was underneath holding the calf out of the water to protect it from the killer whale (the final outcome was not known).

Many interviewees had observed narwhal fleeing to shallow waters to avoid killer whales. One interviewee in Pond Inlet described seeing narwhal half-beached (the observation was also described by [42]), and one in Arctic Bay said there were once so many narwhal on shore that people could even touch them. Another Arctic Bay interviewee described a 2008 observation where so many narwhal fled into shallow water that it "looked liked waves on shore". A Pond Inlet interviewee reported that in 2008, he heard on the CB radio that killer whales were coming, and people headed to the area to hunt the narwhal that they knew would be gathering. Within 30 minutes the narwhal had gathered in the shallow water close to the shoreline. Killer whales are reported to prefer deep water, and heading to the shoreline is an effective anti-predator technique: an Arctic Bay hunter observed narwhal stay in the shallows long enough for the killer whales to give up and leave, and a Pond Inlet interviewee also observed this in the 1970s in Milne Inlet. Many killer whales came near, and were there for four days, but could not get the narwhal that were hiding in shallow waters and then left.

Three interviewees from Repulse Bay described differences in narwhal surfacing behaviour when killer whales were near. The narwhal will be more quiet than usual, and they stay lower in the water, just barely surfacing to breathe. One noted that narwhal will become aware of killer whales two to three days before they were seen by people. Another observed that narwhal swim faster when followed by killer whales than when hunted by boats. Narwhal will also use alternate travel routes when killer whales were near. One Repulse Bay hunter reported that narwhal would go to Wager Bay to avoid killer whales, although he had not seen this personally. Another hunter had observed narwhal in Wager Bay two years previously (2005), and noted that it was the first time he ever saw narwhal there, despite many trips to the area. Another Repulse Bay hunter indicated that narwhal have become less concentrated in certain predictable areas, and are more scattered in their distribution. Interviewees in Arctic Bay and Pond Inlet also described alternate routes and different staging areas that narwhal will use to avoid killer whales. As noted above, tusked narwhal were also reported to be able to defend themselves by stabbing or piercing an attacking killer whale, although this is likely to result in the death of both animals. An Igloolik interviewee suggested that belugas were arriving later in the season due to the presence of killer whales.

Prey consumption by killer whales

Some participants noted that killer whales will sometimes kill and eat only a little of their prey, leaving the rest, and sometimes even not eat anything at all. This observation varied with prey type, with less eaten of larger prey (i.e., bowhead whales). However, even for the smaller cetaceans, only some parts were eaten as indicated by square *maqtaq* pieces being observed by Inuit after kills. Many interviewees noted what they termed as wastage by killer whales choosing the choicest parts. Eleven interviewees in four communities (Igloolik, 4; Pond Inlet, 1; Rankin Inlet, 3; Repulse Bay, 3) discussed killer whale wastage of prey (beluga, 4; bowhead, 3; narwhal, 1). Eight interviewees noted that killer whales sometimes "kill for fun", kill without eating, or "play" with wildlife. Three interviewees described wastage in a positive manner due to Inuit getting *maqtaq* from killed whales, while five comments were negative towards killer whales because they wasted food. Three interviewees also noted that killer whales appeared to prefer meat to blubber, as floating chunks of skin and blubber were often seen in large square pieces (and sometimes collected for human use).

Discussion

Killer whales dominate marine ecosystems as top predators controlling top-down trophic interactions [6,43,44]. Thus, knowledge of food habits and predatory behaviour is key to conservation initiatives. Many Inuit interviewees noted that they did not have much knowledge about killer whales, in comparison to other marine mammals, because they are not hunted (or rarely hunted [27]). Despite these assertions, local Inuit have broad knowledge of killer whales as predators, and their observations represent a substantial addition to our knowledge base on this species in the Canadian Arctic, in terms of their distribution, prey selection and hunting behavior. This is the first dedicated survey of TEK on killer whales in Nunavut waters and both corroborates the results of previous investigators and provides new insights.

Killer whale distribution and prey selection

Direct historical comparisons are difficult given differences in interview techniques and interviewees, but the results suggest an increase in observations of killer whale predation in recent years. Stewart et al. [36] collected TEK in the early 1990s on narwhal and beluga in four communities and solicited information on killer whale predation. Three communities overlap with our study (Arctic Bay, Igloolik and Hall Beach), allowing some comparison of results. Seven hunters from two Foxe Basin communities did not see any attacks on beluga or narwhal, and only one observed scars caused by an unsuccessful attack on beluga. In contrast, five of our 23 Foxe Basin interviewees observed attacks on narwhal (although most occurred in other areas such as Admiralty Inlet, see Table 4) and four saw attacks on beluga. In Arctic Bay Stewart et al. [36] interviewed six hunters for their knowledge of beluga and narwhal. One had seen a killer whale successfully attack a beluga, and another hunter observed scars on a beluga from an unsuccessful attack. One also saw six killer whales attacking a narwhal, and another observed a narwhal killed by a killer whale. Scars from unsuccessful attacks were also reported. In contrast, five of our 11 Arctic Bay interviewees had observed attacks on narwhal [36]), and three of these had also seen dead narwhals that were killed by killer whales. None reported observations of attacks on beluga whales.

Differences in prey species taken by killer whales observed in this study were likely due to non-random distribution of the prey species and/or regional selectivity by particular killer whale groups familiar with different regions. Such regional disparities have been noted by other researchers. Thomsen [45] summarized hunter observations of attacks on belugas and narwhals in West Greenland. In Disko Bay, five of 59 hunters had observed beluga being attacked, compared to four of 14 hunters from the Uummannaq District, one of 34 in the Upernavik District, and none of the 22 hunters interviewed in the Avanersuaq District. For narwhal, none of the 40 Disko Bay hunters or 24 Uummannaq District hunters had seen attacks, and they had been observed by only two of 33 Upernavik District hunters. Attacks were most frequently observed in northwest Greenland, where 15 of 31 Avanersuaq District hunters had seen them. These results, combined with ours, indicate that the north Baffin region of Nunavut (Arctic Bay, but also Pond Inlet) and northeast Baffin region of west Greenland are key locations with concentrated interactions between narwhal and killer whales. In addition to the north Baffin region, the Repulse Bay area was also a key killer whale-narwhal interaction area where four of 17 hunters observed attacks and seven had found dead whales.

Our TEK results support the expectation that killer whales eat multiple prey species, but appear to restrict their hunting to marine mammals in the selected study region. TEK results also suggests that killer whales in eastern Canadian Arctic waters do not feed on walrus, which contrasts with what is known about killer whale feeding habits in Arctic waters north of the Pacific Ocean [46,47].

Killer whales are apex predators with high energy requirements that can result in conflicts with human interests due to competition with commercial fisheries and interactions with endangered species. Therefore, there is a need to understand killer whale ecology, distribution, abundance, and evolutionary history to inform management and conservation. Interviewees in Foxe Basin (also see [39,48]) noted a link between a growing bowhead population and increasing killer whale presence. In the eastern Canadian Arctic most cetacean and phocid populations have increased since large-scale commercial whaling and sealing ended in the early to mid-1900s, and improved foraging conditions therefore exist for killer whales. Ferguson et al. [11] tested alternative predator-prey relationships within the Hudson Bay geographic region, where evidence suggested killer whales seasonally concentrate feeding activities on the large-bodied bowhead whale. Results indicated that killer whales feed on narwhal and beluga whales early and late in the ice-free season whereas feeding was focused on bowhead whales during summer. Using TEK estimates of prey mortality, Ferguson et al. [11] estimated that on average 57 bowhead (range 28-90), 112 narwhal (range 10-234), 174 beluga (range 12-326), and 117 seals (range 12-322) are likely removed annually from the region by killer whales. Thus, Inuit TEK can provide information necessary for modeling predation impact and thereby inform management and conservation decisions.

Killer whale ecotypes are morphologically recognizable forms that prefer different prey and as a result they have different foraging strategies, acoustic behaviour, and habitat preferences. Divergence likely originates through natural selection of an intelligent social animal that transmits cultural heritages within stable family groups [49]. A likely explanation for these differences in killer whale prey preferences would consider the profitability and risk of attack choices. Smaller marine mammal prey, such as seals, is not as profitable as baleen whales but presumably have lower risk of injury or death to the predator during an attack. At lower latitudes in less productive tropical waters, killer whales may have to be generalists [50,51] but at higher latitudes in more productive marine ecosystems killer whales typically specialize and the number of ecotypes appears to increase with increasing prey density. As many as three sympatric ecotypes have been reported in different communities (data on file) and this is a reasonable working hypothesis for the northwest Atlantic where fish-eaters have been observed in southern Greenland and Newfoundland and Labrador waters, shark-eaters may occur offshore, and marine mammal specialists enter the region seasonally. For the eastern Canadian Arctic, during the open-water season, the mammal-eating ecotype predominates. And it is this type of killer whale that Inuit are familiar with and it is the foraging behaviour of this type that we have documented using Inuit TEK.

Baleen whales, including bowheads, and Arctic odontocetes show strong seasonal migration patterns that include intensive feeding during spring through fall building up blubber stores for a winter season where feeding is rare. The eastern Canadian Arctic marine-mammal-eating killer whales also cycle seasonally in both summer-winter ranges and energy intake. In particular, arrival into the Arctic during the open-water season coincides with timing of peak pupping and calving that would presumably provide more easily accessible prey for killer whales. It has been estimated that higher latitude killer whale food intake rates during summer are approximately double the rest of the year [50]. Another study estimated that killer whales at high latitudes input 50% of their annual intake during a 90 day summer period [11]. As with many high-latitude large-bodied mammals, killer whales undergo an intense feeding period during summer followed by a relatively inactive negative energy period over winter. Inuit knowledge and observations have provided substantial information necessary to estimate seasonal variation in killer whale feeding needed to better assess the regulatory role of killer whale predation on their prey.

Hunting behavior

A significant body of literature exists on the hunting behavior of killer whales and the responses of their prey. For example, numerous previous studies have described prey species fleeing to shallow waters to escape from killer whales [26,39,48,52-54]. Many of these sources also note the influence of killer whales on the movements of bowhead, narwhal and beluga whales [26,39,53-58]. The methods used by killer whales to kill marine mammal prey have also been described by interviewees in previous studies, as has the fact that killer whales do not always consume everything and leave pieces of skin and blubber behind which are sometimes collected by Inuit [26,36,39,45,59].

The results of our TEK study compliment the results of other studies and provides some new insights on foraging behavior in Nunavut waters. For example, Inuit observed killer whales attacking and killing large whales and then only consuming parts. Observations of killer whale eating baleen whales include ripping out only tongues and lips, the favored parts [17].

Most recorded attacks of killer whales on large whales are by relatively small groups of killer whales (1-5 [60]), as observed by Inuit [24]. Transient killer whales encountered around Southern California in April and May hunting gray whale calves involve groups of 2-4 individuals conducting the attacks [61]. Inuit have observed that killer whales are like canids (e.g., wolves *Canis lupus*) that hunt cooperatively and attack in a coordinated manner through communication when

preying on ungulates [62,63]. Inuit observations also indicate that killer whales are cooperative pack hunters, like wolves, when pursuing fast or dangerous prey. Inuit observed that killer whales use prey capture techniques that utilize prior knowledge of landscape such as attacking prey before they can seek shallow water refuge [64-66]. Inuit observed that killer whales wash seals off of sea ice as observed in Antarctica [16,67].

Inuit ecological knowledge also provides new information that indicates that cetacean prey use sea ice as a refuge from killer whale predation, which has only been suggested in scientific literature [68-70]. Although noted for other regions [71,72], Inuit described Arctic killer whale foraging behaviour that involves a strategy of searching, finding prey, giving chase, and if unsuccessful, then waiting to see if prey leave refugia and following the landscape shoreline topography for opportunities to attack. An Igloolik interviewee observed a killer whale chasing a bowhead whale cow and calf, with the calf being held out of the water. The interviewee thought that the cow was underneath the calf holding it out of the water to protect it. Grey whales (*Eschrichtius robustus*) have been observed to respond to killer whale attacks by rolling at the surface to keep their ventral area from being exposed to killer whales below, and mothers have been observed to roll over and hold their calf above the water on their ventral surface to keep it away from killer whales [73,74]. To our knowledge, this report from Igloolik is the first time this behaviour has been noted for bowhead whales.

Marine mammal scientists have observed and described sophisticated techniques used by Antarctic Type B killer whales to hunt seals on ice floes [16,75]. Eastern Arctic killer whales are occasionally observed near light to moderate ice, and south Baffin interviewees described similar hunting behaviour with killer whales washing seals off ice floes (also see [24]). Although it has not been observed or described by scientists, Inuit have long recognized different hunting techniques killer whales use to catch seals on ice floes. In 1790, Fabricius noted reports from West Greenland hunters where killer whales worked in unison to lift an ice floe from underneath and cause harp seals to fall off into the water [76]. On northern Baffin Island, hunters have similarly seen killer whales surround an ice floe while one pushes up from underneath to break the ice and force seals into the water [55]. These observations add to a growing body of knowledge on the use of advanced and sophisticated hunting behaviour by killer whales.

In most cases the information provided in different communities and regions was complementary (e.g., prey items, prey avoidance behaviour, attack techniques). However, there is disagreement among opinions and/or observations of different behavior. Several Repulse Bay interviewees suggested that killer whale do not normally attack tusked narwhal and that only females are found dead. The same information has been reported previously [26], possibly from the same local experts. These observations are contrasted by those of other interviewees who indicate that tusked narwhal are killed on occasion. In northwest Greenland (Avanersuaq District), hunters have also observed killer whales attacking and killing both male and female narwhal [45]. A Pond Inlet interviewee reported a story of a narwhal piercing a killer whale with its tusk, suggesting that male narwhal are dangerous to attack. Rosing [77] described an observation from Greenland in December 1924 where killer whales were observed killing narwhal and one was seen jumping out of the water with a narwhal stuck to its side, with its tusk penetrating to the root straight through the killer whale. During interviews for the Igloolik Oral History project [78], an elder provided a story [79] of a dead killer whale that was found in the waters of *qaqqalik* (possibly near Kimmirut) that had a narwhal tusk pierced through its mouth.

Conclusion

Inuit observations have provided a sound background of ecological information detailing Arctic killer whale feeding habits including hunting techniques that are specific to prey species and prey behavior associated with reducing the risk of killer whale attacks being successful. Such knowledge is critical to management and conservation efforts to ensure persistence of sustainable Inuit cultural hunting. Given the importance of long-term relationships between scientists and hunters for successful knowledge integration, this study provides an example of the potential for meaningful integration in short-term projects such as incorporating Inuit understanding of the ecology of killer whale predation in an assessment of the forms of active participation by TEK-holders in science development. We expect considerable improvement in knowledge of killer whale ecology relevant to management as scientists and Inuit combine forces to tackle a major conservation issue - the rapid shifting of the Arctic marine ecosystem structure associated with warming and loss of sea ice.

Acknowledgements

We thank the Inuit participants (interviewees and interpreters) that shared information on Nunavut killer whales during our tours of 11 communities (see Appendix 2). Funding was provided by Nunavut Wildlife Management Board Research Trust Fund, Fisheries & Oceans Canada, International Polar Year (Global Warming and Marine Mammals), the University of Manitoba, and Natural Sciences and Engineering Research Council. Special thanks to Tara Bortoluzzi for helping to organize logistics and to the many Hunters and Trappers Organizations that provided their knowledge of local hunters and assisted with administrating interviews. Considerable improvements were made to the manuscript following the advice of the editorial team.

Appendix 1

Inuit knowledge on killer whales
Part 1 - General information (Optional)
Name:
Age:
Community:
Years living in community:
How often do you go hunting or fishing (on the ocean, not land)?:
(Example: twice a week, on weekends, once every 2 weeks)
Is this more or less time than five year ago?: Ten years ago?
In a year, how many days do you spend on the water?:
How far do you travel during hunting trips?:
Part 2 - Killer whales
Have you seen killer whales in the past?:
Sightings:
Date (Year, Location Number of killer Behaviour** Comments month, etc)
(Use map) whales
* use the provided map to mark the location of the sighting, and give each map sighting a unique number which will be listed on the table
** What were the whales doing? swimming direction, preying on other species, etc
How often are killer whales observed in this area (e.g., every year, only some years)?:
What time of year are killer whales around?:
Do killer whales stay for a while or do they show up briefly and then leave again?:
Do killer whales use the same areas each year? Which areas?
What travel routes do killer whales use?
Are killer whale numbers increasing?:
Decreasing?
Not changing?
How many killer whales do you think use this area?
What animals do killer whales eat?:
Have you seen killer whales attacking other species? If yes, which species?
Do you think killer whales are having a negative impact on other species? If yes, which other species?
Have you ever taken photos of killer whales? Do you know anyone else who has?:

Appendix 2

We thank the following people for sharing their knowledge with us and facilitating the research, with communities and interviewees in alphabetical order:
Arctic Bay: M. Akumalik, A. Huges, N. Iqalukjuak, J. Kalluk, L. Kalluk, I. Kigutijuk, L. Koonark, I. Koonoo, K. Oyukuluk, S. Qaunaq, and I. Shooyook (interviewees); D. Koyukuluk (interpreter); Ikajutit HTO
Arviat: L. Angalik Sr., J. Kaludjak, P. Kaludjak, J. Karetak, and Rev. J. Muckpah (interviewees); Frank Nutarasungnik (interpreter), Arviat HTO
Hall Beach: A. Allianaq, S. Arnardjuak, D. Irqittuq, D. Issigaitok, S. Kaernerk, P. Pikuyak, and R. Siakuluk (interviewees); L. Ningmalik (interpreter); Hall Beach HTA
Iqaluit: I. Adamie, J. Adamie, C. Erkidguk, I. Inookee, A. Inookie, J. Kownirk, and A. Sata (interviewees); Adamie Inookie (interpreter); Amarok HTO
Igloolik: S. Allurut, S. Ammaq, A. Arnatsiaq, M. Arnatsiaq, E. Ipkarnak, D. Irngaut, H. Ittuksarjuat, A. Ivalu, J. Kopak, E. Kunuk, L. Makkik, C. Piugattuk, S. Qammanirq, A. Qrunnut, A. Ulayuruluk, and L. Uttak (interviewees); J. Kopak (interpreter); Igloolik HTA
Kimmirut: S. Akavak, T. Akavak, S. Aqpik, E. Padluq, and Q. Pudlat (interviewees); S. Aqpik (interpreter); Mayukalik HTA
Pangnirtung: M. Battye, J. Keakee, S. Keenainak, O. Kilabuk, P. Kilabuk, M. Kisa, E. Nashalik, P. Nauyuk, M. Noah, P. Quappik, and D. Veevee (interviewees); L. Kanayuk (interpreter); Pangnirtung HTA
Pond Inlet: P. Enooagak, L. Kadloo, P. Komangapik, T. Maktar, J. Muckpa, C. Nutorak, E. Panipakoocho, and J. Simonee (interviewees); T. Arnakallak and R. Soucie (interpreters); Mittimatalik HTA
Qikiqtarjuaq: L. Alikatuktuk, J. Alookie, J. Keyooktak, I. Kokseak, J. Newkingngak, T. Newkingnak, A. Kooneeliusie, and L. Nutaralak (interviewees); H. Olookie (interpreter); Nattivak HTA
Rankin Inlet: M. Innukshuk, P. Ipkorneak, H. Ittinuar, O. Ittinuar, J. Kabvitok, F. Kaput, N. Makayak, M. Tarparti, J. Tattuinee, and R. Tatty (interviewees); Norman Ford (interpreter); Kangiqliniq HTA

Repulse Bay: H. Aglukka, M. Akkuardjuk, L. Angogingoar, J. Ignerdjuk, J. Ivalutanar, S. Kidlapik, L. Kringayark, L. Malliki, Q. Malliki, S. Malliki, D. Milortok, L. Putulik, P. Sanertanut, A. Siatsiak, M. Tagornak, C. Tinashlu, and D. Tuktudjuk (interviewees); M. Tungalik (interpreter); Arviq HTO

Author details
[1]Fisheries and Oceans Canada, Central and Arctic Region, 501 University Crescent, Winnipeg, Manitoba R3T 2N6 Canada. [2]Department of Environment and Geography, University of Manitoba, 501 University Crescent, Winnipeg, Manitoba R3T 2N6 Canada. [3]Higdon Wildlife Consulting, 45 Pilgrim Avenue, Winnipeg, Manitoba R2M 0L3 Canada. [4]Oceans North Canada, 515-70 Arthur Street, Winnipeg, Manitoba R3B 1G7 Canada.

Authors' contributions
SHF was the project manager responsible for conceiving, coordinating, and obtaining the funding to carry out the survey of Inuit knowledge and wrote the final draft of the manuscript. JWH carried out surveys in two communities, performed the statistical analysis, and wrote the first draft of the manuscript. KHW conducted 9 of the community surveys and helped write the manuscript. All authors participated in planning, coordinating, and conceiving the study design and have read and approved the final manuscript.

Competing interests
The authors declare that they have no competing interests.

References
1. Barrett-Lennard LG, Heise KA: **The natural history and ecology of killer whales**. In *Whales, whaling, and ocean ecosystems*. Edited by: Estes JA, Demaster DP, Doak DF, Williams TM, Brownell RI. Berkeley, California: University of California Press; 2006:163-173, ISBN-13: 978-0-520-24884-7.
2. Leatherwood JS, Dahlheim ME: *Worldwide Distribution of Pilot Whales and Killer Whales* Naval Ocean Systems Center, Technical Note 39; 1978.
3. Heyning JE, Dahlheim ME: *Orcinus orca*. *Mammalian Species Account, American Society of Mammalogists* 1988, **304**:1-9.
4. Forney KA, Wade PR: **Worldwide distribution and abundance of killer whales**. In *Whales, whaling and ocean ecosystems*. Edited by: Estes JA et al. Berkeley: University of California Press; 2006:145-162.
5. Reeves RR, Berger J, Clapham PJ: **Killer whales as predators of large baleen whales and sperm whales**. In *Whales, whaling and ocean ecosystems*. Edited by: Estes JA, DeMaster DP, Doak DF, Williams TM, Brownell Jr. Berkeley: University of California Press; 2006:174-187.
6. Estes JA, Tinker MT, Williams TM, Doak DF: **Killer whale predation on sea otters linking oceanic and nearshore ecosystems**. *Science* 1998, **282**:473-476.
7. Springer AM, Estes JA, van Vliet GB, Williams TM, Doak DF, Danner EM, Forney KA, Pfister B: **Sequential megafaunal collapse in the North Pacific Ocean: An ongoing legacy of industrial whaling?** *Proc Nat Acad Sci* 2003, **100**:12223-12228.
8. Trites AW, Deecke VB, Gregr EJ, Ford JKB, Olesiuk PF: **Killer whales, whaling, and sequential megafaunal collapse in the North Pacific: a comparative analysis of the dynamics of marine mammals in Alaska and British Columbia following commercial whaling**. *Marine Mammal Science* 2007, **23**:751-765.
9. Roemer GW, Gompper ME, Van Valkenburgh B: **The ecological role of the mammalian mesocarnivore**. *BioScience* 2009, **59**:165-173.
10. Carter BTG, Nielsen EA: **Exploring ecological changes in Cook Inlet beluga whale habitat though traditional and local ecological knowledge of contributing factors for population decline**. *Marine Policy* 2011, **35**:299-308.
11. Ferguson SH, Kingsley MCS, Higdon JW: **Killer whale (Orcinus orca) predation in a multi-prey system**. *Population Ecology* 2011, Online early.
12. Ferguson SH, Higdon JW, Chmelnitsky EG: **The rise of killer whales as a major Arctic predator**. In *A little less Arctic: top predators in the world's largest northern inland sea, Hudson Bay*. Edited by: Ferguson SH, Loseto LL, Mallory ML. London: Springer; 2010:117-136.
13. Ford JKB, Ellis GM: *Transients: Mammal-hunting killer whales of British Columbia, Washington, and Southeastern Alaska* Vancouver: UBC Press; 1999.
14. Baird RW, Whitehead H: **Social organization of mammal-eating killer whales: group stability and dispersal patterns**. *Can J Zool* 2000, **78**:2096-2105.
15. Pitman RL, Durban JW, Greenfelder M, Guinet C, Jorgensen M, Olson PA, Plana J, Tixier P, Towers JR: **Observations of a distinctive morphotype of killer whale (Orcinus orca), type D, from subantarctic waters**. *Polar Biology* 2010, **34**:303-306.
16. Pitman RL, Durban JW: **Cooperative hunting behavior, prey selectivity and prey handling by pack ice killer whales (Orcinus orca), type B, in Antarctic Peninsula waters**. *Marine Mammal Science* 2011, doi: 10.1111/j.1748-7692.2010.00453.x.
17. Pitman RL: **Antarctic killer whales: top of the food chain at the bottom of the world**. *J Am Cetacean Soc* 2011, **40**:39-45.
18. Foote AD, Newton J, Piertney SB, Willerslev E, Gilbert MTP: **Ecological, morphological and genetic divergence of sympatric North Atlantic killer whale populations**. *Mol Ecol* 2009, **18**:5207-5217.
19. Poncelet E, Barbraud C, Guinet C: **Population dynamics of killer whales in Crozet Archipelago, southern Indian Ocean: exploiting opportunistic and protocol-based photographs in a mark-recapture study**. *Journal of Cetacean Research and Management* 2011, **11**:41-48.
20. Matthews CJD, Luque SP, Petersen SD, Andrews RD, Ferguson SH: **Satellite tracking of a killer whale (Orcinus orca) in the eastern Canadian Arctic documents ice avoidance and rapid, long-distance movement into the North Atlantic**. *Polar Biol* 2011, **7**:1091-1096.
21. Young BG, Higdon JW, Ferguson SH: **Killer whale (Orcinus orca) photo-identification in the eastern Canadian Arctic**. *Polar Research* 2011, **30**, Online 1-11.
22. Reeves RR, Mitchell E: **Distribution and seasonality of killer whales in the eastern Canadian Arctic**. *Rit Fiskideildar* 1988, **11**:136-160.
23. Higdon JW, Ferguson SH: **Loss of Arctic sea ice causing punctuated change in sightings of killer whales (Orcinus orca) over the past century**. *Ecol Appl* 2009, **19**:1365-1375.
24. Higdon JW, Hauser DDW, Ferguson SH: **Killer whales in the Canadian Arctic: distribution, prey items, group sizes, and seasonality**. *Mar Mammal Sci* 2011, Online DOI: 10.1111/j.1748-7692.2011.00489.x.
25. Campbell RR, Yurick DB, Snow NB: **Predation on narwhals, Monodon monoceros, by killer whales, Orcinus orca, in the eastern Canadian Arctic**. *Can Field Nat* 1988, **102**:689-696.
26. Gonzalez N: **Inuit traditional ecological knowledge of the Hudson Bay narwhal (Tuugaalik) population**. *Ottawa: Report prepared for Department of Fisheries and Oceans, Iqaluit, Nunavut* 2001.
27. Higdon JW: *Status of knowledge on killer whales Orcinus orca in the Canadian Arctic* Ottawa Canada: Canadian Science Advisory Secretariat Research Document 2007/048; 2007.
28. Stephenson SA, Hartwig L: **The Arctic Marine Workshop: Freshwater Institute Winnipeg, Manitoba, February 16-17, 2010**. *Can Manuscript Rep Fish Aquat Sci* 2010, **2934**:vi+67.
29. Gilchrist G, Mallory M, Merkel F: **Can local ecological knowledge contribute to wildlife management? Case studies of migratory birds**. *Ecology and Society* 2005, **10**:20, [online] URL: http://www.ecologyandsociety.org/vol10/iss1/art20/.
30. Clark DA, Lee DS, Freeman MMR, Clark SG: **Polar Bear Conservation in Canada: Defining the Policy Problems**. *Arctic* 2008, **61**:347-360.
31. Dowsley M: **Community clusters in wildlife and environmental management: using TEK and community involvement to improve co-management in an era of rapid environmental change**. *Polar Research* 2009, **28**:43-59.
32. Gearheard S, Shirley J: **Challenges in community-research relationships: Learning from natural science in Nunavut**. *Arctic* 2007, **60**:62-74.
33. Huntington HP: **Observations on the utility of the semi-directive interview for documenting traditional ecological knowledge**. *Arctic* 1998, **51**:237-242.
34. Huntington HP: **Using traditional ecological knowledge in science: methods and applications**. *Ecological Applications* 2000, **10**:1270-1274.
35. Freeman MMR: In *Inuit Land Use and Occupancy Project Report. Volume 3*. Ottawa: Supply and Services Canada; 1976.
36. Stewart DB, Akeeagok A, Amarualik R, Panipakutsuk S, Taqtu A: **Local knowledge of beluga and narwhal from four communities in Arctic Canada**. *Can Tech Rep Fish Aquat Sci* 1995, **2065**, viii + 48 p. + Appendices.

37. Goodman LA: **Snowball sampling.** *Annals of Mathematical Statistics* 1961, **32**:148-170.
38. Kitchin R, Tate NJ: *Conducting research into human geography: Theory, methodology and practice* Harlow: Pearson Education; 2000.
39. NWMB: **Final report of the Inuit Bowhead Knowledge Study, Nunavut, Canada. Iqaluit, Nunavut.** Ottawa: Nunavut Wildlife Management Board; 2000.
40. Finley KJ: **Isabella Bay, Baffin Island: an important historical and present-day concentration area for the endangered bowhead whale (*Balaena mysticetus*) of the eastern Canadian Arctic.** *Arctic* 1990, **43**:137-152.
41. Finley KJ: **Natural history and conservation of the Greenland Whale, or bowhead, in the northwest Atlantic.** *Arctic* 2001, **54**:55-76.
42. Ford JKB, Nichol LM, Canvanagh DM: *Preliminary assessment of the value of underwater vocalization in population studies of narwhals in the Canadian Arctic. Whales Beneath the Ice Program* Vancouver, Canada: World Wildlife Fund Canada; 1986, 44.
43. Pace ML, Cole JJ, Carpenter SR, Kitchell JF: **Trophic cascades revealed in diverse ecosystems.** *TREE* 1999, **14**:483-488.
44. Frank KT, Petrie B, Choi JS, Leggett WC: **Trophic cascades in a formerly cod-dominated.** *Ecosystem Science* 2005, **308**:1621-1623.
45. Thomsen ML: **Local knowledge of the distribution, biology and hunting of beluga and narwhal: a survey among Inuit hunters in West and North Greenland.** *Inuit Circumpolar Conference: Report prepared for Greenland Hunters' and Fishermen's Association, Greenland Home Rule Authorities, and Inuit Circumpolar Conference* 1993, 98 pp + appendices.
46. Zenkovich BA: **O kosatke ili kite-ubiitse.** *Priroda* 1938, 109-112.
47. Mizroch SA, Rice DW: **Have North Pacific killer whales switched prey species in response to depletion of the great whale populations?** *Mar Ecol Prog Ser* 2006, **310**:235-246.
48. Piugaattuk N: **IE-303.** Igloolik: Igloolik Nunavut: Archives of the Inullariit Society, Igloolik Research Centre; 1994.
49. Gowans S, Würsig B, Karczmarski L: **The Social Structure and Strategies of Delphinids: Predictions Based on an Ecological Framework.** *Adv Mar Biol* 2007, **53**:195-294.
50. Baird RW: **Predators, prey, and play: killer whales and other marine mammals.** *J Am Cetacean Soc* 2011, **40**:54-57.
51. Guinet C, Tixier P: **Crozet: Killer whales in a remote but changing environment.** *J Am Cetacean Soc* 2011, **40**:33-38.
52. Qipanniq P: **IE-206.** Igloolik: Archives of the Inullariit Society, Igloolik Research Centre; 1991.
53. Westdal K: **Movement and diving of northern Hudson Bay narwhals (*Monodon monoceros*): relevance to stock assessment and hunt co-management.** University of Manitoba: M.Env. Thesis. Department of Environment and Geography; 2008, 103.
54. Westdal KH, Richard PR, Orr JR: **Migration route and seasonal home range of the northern Hudson Bay narwhal (*Monodon monoceros*).** In *A little less Arctic: changes to top predators in the world's largest nordic inland sea, Hudson Bay*. Edited by: Ferguson SH, Losetto LL, Mallory ML. Amsterdam: Springer; 2010:71-92, doi: 10.1007/978-90-481-9121-5_6.
55. Brody H: **Land occupancy: Inuit perceptions.** In *Inuit land use and occupancy project. Volume One. Land use and occupancy*. Edited by: Freeman M. Ottawa: Ministry of Supply and Services; 1976:185-242.
56. DFO: **Proceedings of the workshop on traditional and contemporary knowledge of Nunavik belugas.** Maurice Lamontagne Institute, Mt. Joli, PC. Fisheries and Oceans Canada, Mt. Joli, PC; 1994.
57. Kilabuck P: **A study of Inuit knowledge of the southeast Baffin beluga.** *Iqaluit Nunavut: Nunavut Wildlife Management Board* 1998, 1-74.
58. Remnant RA, Thomas ML: **Inuit traditional knowledge of the distribution and biology of High Arctic narwhal and beluga.** Winnipeg: Unpubl rep. prep. by North/South Consultants Inc.; 1992, 96.
59. Aqatsiaq V: **IE-376.** Igloolik: Archives of the Inullariit Society, Igloolik Research Centre, Igloolik, Nunavut; 1996.
60. Jefferson TA, Stacey PF, Baird RW: **A review of killer whale interactions with other marine mammals: predation to co-existence.** *Mammal Review* 1991, **21**:151-180.
61. Schulman-Janiger A, Black N, Ternullo R: **Killer whales of California.** *J Am Cetacean Soc* 2011, **40**:46-47.
62. Silber GK, Newcomer MW, Pérez-Cortés HM: **Killer whales (Orcinus orca) attack and kill a Bryde's whale (Balaenoptera edeni).** *Can J Zool* 1990, **68**:1603-1606.
63. Beck S, Kuningas S, Esteban R, Foote AD: **The influence of ecology on sociality in the killer whale (Orcinus orca).** *Behav Ecol* 2011, Online first.
64. Baldridge A: **Killer whales attack and eat a gray whale.** *J Mammal* 1972, **53**:898-900.
65. Würsig B: **Cetaceans.** *Science* 1989, **244**:1550-1557.
66. Barrett-Lennard LG, Ford JKB, Heise KA: **The mixed blessing of echolocation: differences in sonar use by fish-eating and mammal-eating killer whales.** *Anim Behav* 1996, **51**:553-565.
67. Pitman RL, Ensor P: **Three forms of killer whales (Orcinus orca) in Antarctic waters.** *J Cetacean Res Manag* 2003, **5**:131-139.
68. Kovacs KM, Lydersen C: **Climate change impacts on seals and whales in the North Atlantic Arctic and adjacent shelf seas.** *Science Progress* 2008, **91**:117-150.
69. Laidre KL, Stirling I, Lowry LF, Wiig Ø, Heide-Jørgensen MP, Ferguson SH: **Quantifying the sensitivity of Arctic marine mammals to climate-induced habitat change.** *Ecological Applications* 2008, **18**:S97-S125.
70. Ferguson SH, Dueck L, Loseto LL, Luque SP: **Bowhead whale (Balaena mysticetus) seasonal selection of sea ice.** *Marine Ecology Progress Series* 2010, **411**:285-297.
71. Srinivasana M, Granta WE, Swannacka TM, Rajanb J: **Behavioral games involving a clever prey avoiding a clever predator: An individual-based model of dusky dolphins and killer whales.** *Ecol Model* 2010, **221**:2687-2698.
72. Deecke VB, Nykänen M, Foote AD, Janik VM: **Vocal behaviour and feeding ecology of killer whales Orcinus orca around Shetland, UK.** *Aquatic Biol* 2011, **13**:79-88.
73. Walker LW: **Nursery of the gray whales.** *Natural History* 1949, **58**:248-256.
74. Ford JKB, Reeves RR: **Fight or flight: antipredator strategies of baleen whales.** *Mammal Rev* 2008, **38**:50-86.
75. Visser IN, Smith TG, Bullock ID, Green GD, Carlsson OGL, Imberti S: **Antarctic peninsula killer whales (Orcinus orca) hunt seals and a penguin on floating ice.** *Marine Mammal Science* 2008, **24**:225-234.
76. Kapel FO: **Otto Fabricius and the Seals of Greenland.** *Danish Polar Center, Copenhagen: Meddelelser om Grønland, Bioscience* 2005, **55**:150.
77. Rosing J: **The Unicorn of the Arctic Sea: The Narwhal and Its Habitat.** Manotick, Canada: Penumbra Press; 1999.
78. MacDonald J: **The Igloolik Oral History Project.** Ottawa: Canadian Polar Commission. Meridian Fall/Winter; 2001, 1-3, Online: http://www.polarcom.gc.ca/.
79. Kappianaq G: **Interview IE-456.** Igloolik Research Centre, Igloolik, Nunavut; Igloolik: Archives of the Inullariit Society; 2000.

Evolutionary patterns of carbohydrate transport and metabolism in *Halomonas boliviensis* as derived from its genome sequence: influences on polyester production

Daniel Guzmán[1,2], Andrea Balderrama-Subieta[1], Carla Cardona-Ortuño[1], Mónica Guevara-Martínez[1], Nataly Callisaya-Quispe[1] and Jorge Quillaguamán[1]*

Abstract

Background: *Halomonas boliviensis* is a halophilic bacterium that is included in the γ-Proteobacteria sub-group, and is able to assimilate different types of carbohydrates. *H. boliviensis* is also able to produce poly(3-hydroxybutyrate) (PHB) in high yields using glucose as the carbon precursor. Accumulation of PHB by microorganisms is induced by excess of intracellular NADH.

The genome sequences and organization in microorganisms should be the result of evolution and adaptation influenced by mutation, gene duplication, horizontal gen transfer (HGT) and recombination. Furthermore, the nearly neutral theory of evolution sustains that genetic modification of DNA could be neutral or selected, albeit most mutations should be at the border between neutrality and selection, i.e. slightly deleterious base substitutions in DNA are followed by a slightly advantageous substitutions.

Results: This article reports the genome sequence of *H. boliviensis*. The chromosome size of *H. boliviensis* was 4 119 979 bp, and contained 3 863 genes. A total of 160 genes of *H. boliviensis* were related to carbohydrate transport and metabolism, and were organized as: 70 genes for metabolism of carbohydrates; 47 genes for ABC transport systems and 43 genes for TRAP-type C4-dicarboxylate transport systems. Protein sequences of *H. boliviensis* related to carbohydrate transport and metabolism were selected from clusters of orthologous proteins (COGs). Similar proteins derived from the genome sequences of other 41 archaea and 59 bacteria were used as reference. We found that most of the 160 genes in *H. boliviensis*, c.a. 44%, were obtained from other bacteria by horizontal gene transfer, while 13% of the genes were acquired from haloarchaea and thermophilic archaea, only 34% of the genes evolved among Proteobacteria and the remaining genes encoded proteins that did not cluster with any of the proteins obtained from the reference strains. Furthermore, the diversity of the enzymes derived from these genes led to polymorphism in glycolysis and gluconeogenesis. We found further that an optimum ratio of glucose and sucrose in the culture medium of *H. boliviensis* favored cell growth and PHB production.

Conclusions: Results obtained in this article depict that most genetic modifications and enzyme polymorphism in the genome of *H. boliviensis* were mainly influenced by HGT rather than nearly neutral mutations. Molecular adaptation and evolution experienced by *H. boliviensis* were also a response to environmental conditions such as the type and amount of carbohydrates in its ecological niche. Consequently, the genome evolution of *H. boliviensis* showed to be strongly influenced by the type of microorganisms, genetic interaction among microbial species and its environment. Such trend should also be experienced by other prokaryotes. A system for PHB production by *H. boliviensis* that takes into account the evolutionary adaptation of this bacterium to the assimilation of

* Correspondence: jorgeqs@supernet.com.bo
[1]Centro de Biotecnología, Facultad de Ciencias y Tecnología, Universidad Mayor de San Simón, Cochabamba, Bolivia
Full list of author information is available at the end of the article

combinations of carbohydrates suggests the feasibility of a bioprocess economically viable and environmentally friendly.

Keywords: *Halomonas boliviensis*, Halophilic bacterium, *Halomonas*, *Halomonadaceae*, Biopolyesters, Polyhydroxyalkanoates, Genome evolution, Population genetics

Background

Cellular evolution and adaptation have imprinted patterns in microbial genomes through mutation, gene duplication, horizontal gen transfer (HGT) and recombination [1,2]. The genomes of microorganisms of the three domains of life have experienced such genetic modifications to succeed on their permanence in a particular habitat, where environmental conditions and the size of the microbial populations might influence the organization and number of genes in a particular species throughout the time [1,3]. Furthermore, the nearly neutral theory of evolution points out that genetic modification of DNA could be neutral or selected, albeit most mutations should be at the border between neutrality and selection, i.e. slightly deleterious base substitutions in DNA are followed by a slightly advantageous substitutions [1].

The increasing number of genome sequences of different organisms is helping to discern how microbial species diverged. Recent reports on the evolutionary traits followed by different bacteria and archaea have demonstrated that the transfer of genes among these organisms, also referred as horizontal gene transfer, has led to net-like relationships among their genomes [2,4,5]. Nevertheless, the phylogenetic association among prokaryotes derived from the sequences of proteins encoded by 102 different genes was consistent to the taxonomic differentiation observed when 16 rRNA sequences of microorganisms are analyzed [4]. The 102 proteins were mainly related to translation and transcription, although proteins involved in the transport and metabolism of amino acids, metal ions and carbohydrates revealed such taxonomic information as well [4].

The aforementioned studies included the genome sequences of extremely halophilic archaea such as *Haloarcula marismortui*, *Haloquadratum walsbyi* and a *Halobacterium* sp. [4]. These studies on the genome sequences did not include halophilic bacteria. However, a report on the genes of poly(3-hydroxybutyrate) (PHB) polymerases, PHB depolymerases and ectoine synthesis by *Halomonas* sp. TD01, a halophilic bacterium, suggested that HGT has a role to play on the genome organization of the microorganism [6]. Halophilic microorganisms require salt (NaCl) to grow; a halophile should grow optimally at NaCl concentrations of 5% (w/v) or higher, and tolerate at least 10% (w/v) salt [7]. There are five genome sequences of halophilic bacteria available in public data bases. The sequences of *Chromohalobacter salexigens* and

Halorhodospira halophila were first published followed by the sequence of *Halomonas elongata* [8,9], *Halomonas* sp. TD01 [6] and *Halomonas* sp. HAL1 [10]. *Chromohalobacter* and *Halomonas* species are included in the family *Halomonadaceae* within the γ-Proteobacteria subgroup. The family *Halomonadaceae* contains only halophilic and halotolerant aerobic heterotrophs; some of them are able to grow in media with up to 30% (w/v) NaCl [7]. Halophilic bacteria maintain low concentrations of salt intracellularly by accumulating organic compounds of low molecular weight, also known as osmolytes or "compatible solutes" such as ectoine [11].

Understanding the evolution and levels of polymorphism among genes is attracting much attention in evolutionary biology and biotechnology. Evolution of energy-producing pathways, particularly glycolysis and gluconeogenesis, posses relevance since they determine the type of carbon sources that a species is able to assimilate, and link to metabolic routs that may generate compounds of biotechnological interest [12]. Theories on the evolution of the metabolisms of organisms consider that enzyme polymorphism—alleles for the different enzymes or allozymes—in metabolic pathways was related to genetic mutations [12-14]. A proposal states that the fitness of the pathways associated with an increasing flux is influenced by selected mutations of genes that enhance enzyme activities, albeit enzyme improvements do not continue indefinitely [12,14]. Mutations will reach a point at which the incremental gains of fitness for a new mutation will be equaled by the noise caused by the random genetic variation [12,14]. At this stage, the genes or enzymes might evolve under a nearly neutral trend [12,14]. Moreover, metabolic control in the organisms is also to regulate molecular evolution as well [12,14]. The proposal assumes no contextual changes such as a change in the functional conditions of an enzyme originated by either epistasis or the environment; or a change in the effective population size of the species [12].

Halomonas boliviensis is a halophilic bacterium that can develop under a wide range of NaCl concentrations (i.e. 0-25% (w/v)), pH (5-11) and temperatures (0-45°C) [15]. It can also assimilate several carbohydrates as carbon source for growth [15]. Bioprocesses have been designed to attain high productivities of a polyester and osmolytes by *H. boliviensis* using glucose as the carbon precursor [16,17]. The polyester accumulated by the bacterium is poly(3-hydroxybutyrate) (PHB), which is used as carbon and energy reservoir [18]. PHB is synthesized by several bacteria from acetyl-CoA when an excess of NADH is present in the bacterial cytoplasm [19]. Such excess can be generated when a high concentration of a carbon source is added to a culture medium and cell growth is limited by the depletion of an essential nutrient, e.g. nitrogen, oxygen, trace elements among others [19]. PHB is attracting much attention in biotechnology because it is a biodegradable plastic-like material, and possesses potential in biomedical applications such as tissue engineering, organ transplants and drug delivery systems [20]. Moreover, the efficiency and economics of the manufacturing process of PHB are determined by the carbon source, fermentation process, and downstream processing of the polymer. The development of cultivation conditions for microorganisms that allow high PHB content and productivity from cheap and renewable carbon sources is therefore important [21,22].

The present research work reports the genome sequence of *H. boliviensis*. It also depicts the evolutionary trends that proteins of *H. boliviensis* have experienced to allow the transport of carbohydrates and their assimilation to achieve acetyl-CoA. The conclusions drawn from these studies were used to create an alternative production system of PHB by *H. boliviensis* using a combination of carbohydrates. This system should lead to a more economically and environmentally beneficial bioprocess.

Methods
Genome sequencing
The fine high coverage genome sequence, gene prediction, repetitive sequence, COGs and KEGG annotation of *Halomonas boliviensis* LC1^T (= DSM 15516^T) were obtained at BGI-Hongkong Co., Hong Kong. For this, Illumina HiSeq 2000 technology was used to conduct paired-end sequencing for DNA samples, and constructed a 1,000 bp library with extended data of 500 Mb. Genome coverage based on k-mer was 95.4%, and genome coverage based on reads mapping was 99.9%. Glimmer 3.0 software package was used to conduct *de novo* gene prediction [23]. The functional annotation was accomplished by analysis of protein sequences. Genes of *H. boliviensis* were aligned to others in databases to attain its corresponding functional annotation. To ensure the biological meaning, only one high-quality information as annotation to the genes from many results was chosen. BLAST was used to accomplish functional annotation combined with different databases. BLAST version: blastall 2.2.21 software (provided by the National Center for Biotechnology Information, NCBI) was used for these studies. Alignment results were obtained using the following databases: KEGG, COG, SwissProt, TrEMBL, NR. This whole genome shotgun project was deposited at DDBJ, EMBL and GenBank under the accession number

AGQZ00000000. The version described in this paper is the first version, AGQZ01000000.

Evolutionary analysis
A total of 6,901 alignments of clusters of orthologous proteins (COGs) of 59 bacteria and 41 archaea, as classified in COGs [24] and EggNOG [25] data bases, were gently provided by Puigbò, Wolf and Koonin (2009). The protein sequences of these 100 microorganisms were used as reference for the evolutionary analysis. Protein sequences of *H. boliviensis* related to carbohydrate transport and metabolism were selected and aligned along with the references for each corresponding COG (Additional file 1: Table S1, supplementary data) using the Muscle program [26] included in the MEGA 5 software package [27] with default parameters. Unrooted maximum likelihood phylogenetic trees were constructed using MEGA 5 under a WAG with frequencies (+F) model, with uniform mutation rates among amino acid sites and complete deletion of gaps and missing data.

Analysis and assembly of supernetworks
Supernetworks were constructed by combining the phylogenetic trees of proteins of the glycolysis and gluconeogenesis metabolisms in *H. boliviensis* and reference strains using the SplitsTree4 program [28,29] with default parameters. Three analyses were performed for these studies: 1) A supernetwork obtained from three COGs (0126, 0149 and 0837). Both COG0126 and COG0149 are considered among the 102 genes that contain taxonomic information that discriminate well bacteria and archaea in already known families and genera [4]; 2) A supernetwork obtained after combining six COGs (0126, 0149, 0837, 0469, 0696 and 837); and 3) A supernetwork obtained after combining twenty two COGs (0057, 0126, 0148, 0149, 0166, 0191, 0205, 0235, 0365, 0469, 0508, 0696, 0837, 1012, 1063, 1109, 1249, 1454, 1866, 2017, 2609 and 4993). Supernetworks were analyzed according to method described by Huson *et al.* in 2006 [28].

Culture media composition
Seed culture and PHB production media were formulated as described previously [16]. Seed culture contained% (w/v): NaCl, 2.5; MgSO$_4$•7H$_2$O, 0.25; K$_2$HPO$_4$, 0.05; NH$_4$Cl, 0.23; FeSO$_4$•7H$_2$O, 0.005; sucrose 1; monosodium glutamate (MSG), 0.3 and TRIS, 1.5. The PHB production medium included % (w/v): NaCl, 2.5; MgSO$_4$•7H$_2$O, 0.5; K$_2$HPO$_4$, 0.22; NH$_4$Cl, 0.4; FeSO$_4$•7H$_2$O, 0.005; MSG, 0.2; and the following concentration of carbohydrates % (w/v): 1) 2.5 sucrose, 2) 2.0 sucrose and 0.5 glucose, 3) 1.5 sucrose and 1 glucose, 4) 1.0 sucrose and 1.5 glucose, 5) 0.3 sucrose, 0.7 glucose and 1.5 dried molasses and 6) 2.5 dried molasses for 6 different assays, respectively. The

composition of the molasses used was 78.1% sucrose, 15.3% glucose and 6.6% of other uncharacterized solids. A low amount of MSG is added to the production medium to induce its depletion by *H. boliviensis* during the cultivation.

H. boliviensis growth and PHB production in flasks

H. boliviensis was grown in 100 ml of seed culture medium in 1,000-ml flasks with rotary shaking at 220 rpm, 30°C for 13 h. The pH of the medium was adjusted to 7.5 using concentrated HCl. Subsequently, 5 ml of the seed culture were inoculated in 1,000-ml Erlenmeyer flasks containing 95 ml of PHB production medium. The pH of the PHB production medium was initially adjusted to 7.5 using 5 M NaOH. The cultures were incubated at 35°C with shaking at 220 rpm, and samples were withdrawn at different time intervals during the cultivation.

Quantitative analyses

Cell dry weight (CDW) and PHB content in *H. boliviensis* were determined as reported previously [18]. Residual cell mass (RCM) concentration was calculated as the difference between the CDW and PHB concentration, while PHB content (wt%) was obtained as the percentage of the ratio of PHB concentration to the CDW as defined by Lee *et al.* in 2000 [30]. All analyses were performed in triplicate.

Glutamate concentration was determined by high performance liquid chromatography (HPLC) analysis, as described previously [31], using a Perkin-Elmer HPLC system with an Aminex HPX-87 C column (Biorad) and a UV detector at 65°C. Calcium chloride solution (5 mM) was used as mobile phase at a flow rate of 0.5 ml/min. Glutamate was monitored at 210 nm. Glucose and sucrose were determined using the same HPLC system with a Polypore CA column (Perkin-Elmer), a RI detector at 80°C and water as mobile phase at a flow rate of 0.3 ml/min.

Results and discussion
Genome of *H. boliviensis*

Table 1 provides a description of the genome composition of *H. boliviensis*. The chromosome size of *H. boliviensis* (4 119 979 bp) was slightly longer than those determined for *H. elongata* (4 061 296 bp) [9], *Halomonas* sp. TD01 (4 092 837 bp) [6] and *Chromohalobacter salexigens* (3 696 649 bp) (Accession number: CP000285.1). The % of G+C content showed in Table 1 is similar to that determined experimentally for *H. boliviensis*, i.e. 52.6% [15], and is lower than that found for the genome of *H. elongata* (63.6%) and that evaluated for the description of *C. salexigens* (64.2%) [32]. Such wide difference between the G+C content of different *Halomonas* and *Chromohalobacter* species is a feature of the family *Halomonadaceae* [33]. Moreover, the genes constitute most part of the

Table 1 Genome of *H.boliviensis*

Chromosome	1
DNA, total number of bases	4 119 979
% G+C content	54.69
Number of genes	3 863
Length occupied by genes (bp)	3 673 824
% G+C content in the gene region	55.64
% Gene/Chromosome	89.17
Length occupied by the intergenic region (bp)	446 146
% G+C content in the intergenic region	46.86
% Intergenic length/Chromosome	10.82

chromosome of *H. boliviensis* and the %G+C content for the region containing the genes was similar to that found in its chromosome (Table 1). On the other hand, the number of genes in the genome of *H. boliviensis* (3 863) is slightly higher than that reported for *H. elongata* (3 555) [9].

Inferring the evolution of proteins involved in the uptake and metabolism of carbohydrates

Protein sequences of *H. boliviensis* related to carbohydrate transport and metabolism were obtained from clusters of orthologous proteins (COGs), as classified in COGs and EggNOG data bases [24,25]. A total of 160 genes of *H. boliviensis* encoded proteins for these clusters: 70 genes were related to the metabolism of carbohydrates; 47 genes were related to ABC transport systems and encoded 14 permease proteins, 23 ATPase proteins and 10 periplasmic proteins; and 43 genes were related to TRAP-type C4-dicarboxylate transport systems and encoded 15 large permease proteins, 11 small permease proteins and 17 periplasmic proteins (Additional file 1: Table S1, supplementary data). Similar proteins were selected from COGs derived from the genome sequences of other 41 archaea and 59 bacteria. To perform evolutionary analyses, unrooted phylogenetic trees were constructed based on a maximum likelihood approach using the sequences of the proteins of *H. boliviensis* and proteins of other 100 microorganisms for each corresponding COG.

Figure 1 presents three phylogenetic trees that were selected to exemplify the genetic modifications experienced by the genome of *H. boliviensis*. Figure 1 shows a phylogenetic tree for a COG related to the ABC type transport system for ribose, xylose, arabinose and galactoside. *H. boliviensis* has three alternative forms of genes, i.e. alleles, for this tree. The first allele (*H. boliviensis* A1) was clustered with thermophilic archaea (Figure 1), hence implying a long distance HGT [5]. After comparing the closest identities of this allele to other sequences in pubic data bases, we found that the sequence corresponded to a periplasmic binding protein. We also found that the

Evolutionary patterns of carbohydrate transport and metabolism in Halomonas boliviensis...

21

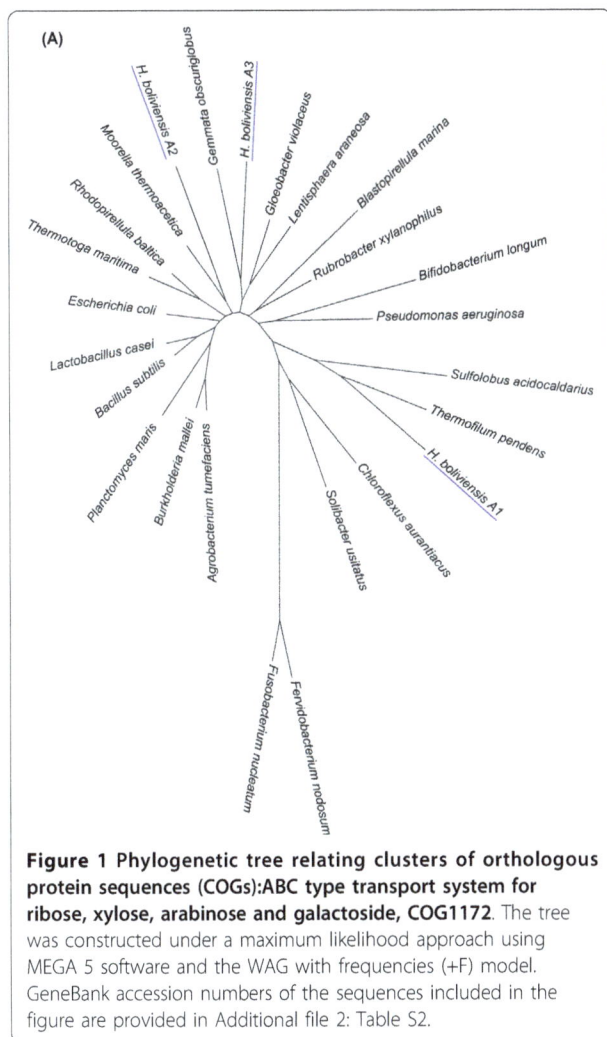

Figure 1 Phylogenetic tree relating clusters of orthologous protein sequences (COGs):ABC type transport system for ribose, xylose, arabinose and galactoside, COG1172. The tree was constructed under a maximum likelihood approach using MEGA 5 software and the WAG with frequencies (+F) model. GeneBank accession numbers of the sequences included in the figure are provided in Additional file 2: Table S2.

temperatures and the second active at high temperatures. The third allele (*H. boliviensis* A3) was clustered along with trans-membrane proteins of Planctomycetes, Cyanobateria and Lentisphaerae (Figure 1). Both allele 2 and 3 suggest HGT among bacteria [5].

On the other hand, Figure 2 depicts a phylogenetic tree for a COG corresponding to keto-3-deoxy-6-phosphogluconate aldolases. The first allele of *H. boliviensis* in this phylogenetic tree was closely related to proteins of halophilic archaea, while the second diverged among enzymes of Proteobacteria with a close relationship to γ-Proteobacteria (Figure 2). *H. boliviensis* has also adapted to a wide range of NaCl concentrations. At different NaCl concentrations, *H. boliviensis* should thrive in environments with distinct type of microorganisms varying from non-halophilic to extreme halophilic archaea [7]. *H. boliviensis* showed to be able to acquire genes from other organisms that share its habitat (Figures 1, 2). Moreover, Figure 3 depicts two alleles that are closely related to proteins of γ-Proteobacteria, as might be expected. The same evolutionary analyses were followed with all COGs related to carbohydrate transport and metabolism. The results obtained are summarized in Additional file 1: Table S1 and in Figure 4. Interestingly, most of the proteins (44%) of *H. boliviensis* involved in carbohydrate transport and metabolism were

amount of acidic amino acids of *H. boliviensis* A1 (10.1% of the total residues in the protein) was significantly higher than the basic amino acids (3.5%), resulting in a ratio of 2.9 of acidic to basic amino acids. Various scientific articles reported that some extracellular and periplasmic enzymes of halophilic bacteria tolerate high temperatures, and possess a relatively high content of acidic amino acids [8,34]. The result obtained in Figure 1 denotes that *H. boliviensis* had originally obtained the binding protein from a thermophile and it diverged later. The second allele (*H. boliviensis* A2) was also identified as a periplasmic binding protein with similar functions to those of the first allele. However, the second allele shared a closer affiliation to proteins of bacteria other than Proteobacteria (Figure 1). The ratio of acidic (9.4% of the total residues in the protein) to basic (5.5%) amino acids for allele 2 was 1.7. *H. boliviensis* is a microorganism that is able to grow at low (0°C) and high (45°C) temperatures, whereby it should be useful for this organism to hold two different proteins that can accomplish a similar task; one active at low

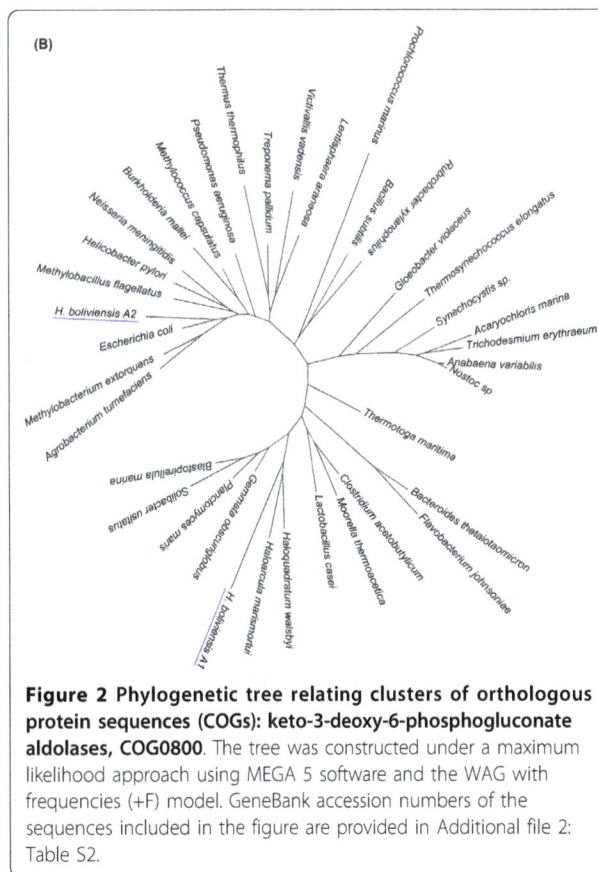

Figure 2 Phylogenetic tree relating clusters of orthologous protein sequences (COGs): keto-3-deoxy-6-phosphogluconate aldolases, COG0800. The tree was constructed under a maximum likelihood approach using MEGA 5 software and the WAG with frequencies (+F) model. GeneBank accession numbers of the sequences included in the figure are provided in Additional file 2: Table S2.

Figure 3 Phylogenetic tree relating clusters of orthologous protein sequences (COGs): Pyruvate kinases, COG0469. The tree was constructed under a maximum likelihood approach using MEGA 5 software and the WAG with frequencies (+F) model. GeneBank accession numbers of the sequences included in the figure are provided in Additional file 2: Table S2.

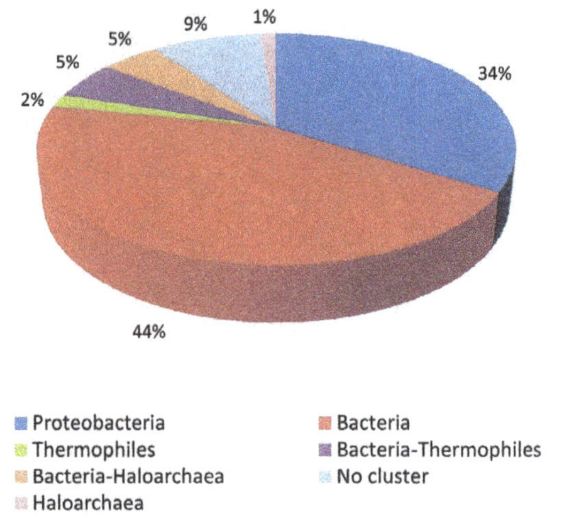

Figure 4 Percentage of proteins that *H. boliviensis* obtained from other bacteria, halophilic archaea and thermophilic archaea by horizontal gene transfer (HGT), and proteins that evolved among proteins of other Proteobacteria. Some proteins of *H. boliviensis* did not form a cluster with any of the proteins of the microorganisms used as reference. The proteins analyzed were related to COGs of carbon transport and metabolism as listed in Additional file 1: Table S1.

obtained from other bacteria; only a 34% of the proteins evolved among Proteobacteria (Figure 4). Moreover, transfer of genes from thermophilic organisms and halophilic archaea has had an effect on the genome evolution of *H. boliviensis* as well (Figure 4). Yet, 9% of the proteins of *H. boliviensis* did not cluster with any of the proteins belonging to the reference microorganisms (Figure 4). They may form phylogenetic groups with proteins of microorganisms that were not included in these studies.

Under the neutral mutation-random drift theory, it is assumed that a certain fraction of new mutation are free of constraint or are selectively neutral, while the rest have deleterious effects and are selectively eliminated [35]. Nevertheless, Figures 1 and 4 imply that most mutations found in proteins related to carbohydrate transport and metabolism were a result of HGT, which agree on some criteria that point out that genetic drift is not sufficient for claiming neutrality [36], and on a resent observation that estimated that about 60% of the genome evolution of prokaryotes is dominated by HGT [2]. Furthermore, HGT can be related to adaptation of *H. boliviensis* to its environment (Figures 1, 2) and might, therefore, be selected to attain an optimum physiological response of the species to its habitat [3]. Yet, nearly neutral mutations could be inferred from Figure 3 and Additional file 1: Table S1, suggesting a continuous evolution of the proteins [1].

Metabolic assimilation of carbohydrates by *H. boliviensis*
The metabolic routes in *H. boliviensis* for the assimilation of carbohydrates were obtained by matching the

highest identities of enzymes derived from its genome with the KEGG pathway database [37]. The studies began searching for enzymes of *H. boliviensis* that form part of the starch and sucrose metabolism pathway. Although *H. boliviensis* is unable to hydrolyze starch [15], it assimilates maltose, glucose and other oligosaccharides obtained after the hydrolysis of starch [38]. Furthermore, *Halomonas boliviensis* can grow on sucrose [15]. We deduced in this study that both maltose and sucrose are metabolized via α-D-glucose-mono-phosphorylated before entering the glycolysis and gluconeogenesis pathway (Figure 5). Figure 5 shows further a high level of polymorphism for most of the enzymes in this metabolic route.

Similar polymorphism was observed in glycolysis and gluconeogenesis pathways of *H. boliviensis* (Figures 6, 7). For the first part of the metabolism, *H. boliviensis* obtained most of its gens by HGT among bacteria (Figure 6), whereas the enzymes at the bottom of the pathway were mainly related to enzymes of the Proteobacteria (Figure 7). Glycolysis in *H. boliviensis* concluded with the 2-oxoglutarate dehydrogenase complex (PDHC) (Figure 7), that is part of the pyruvate dehydrogenases family. This route is characterized by generation of NADH, and is commonly found in Gram-negative bacteria [39]. The enzymes that form part of PDHC in *H. boliviensis* were obtained in part from thermophilic archaea and mesophilic bacteria (Figure 7). Archaea do not utilize a pyruvate

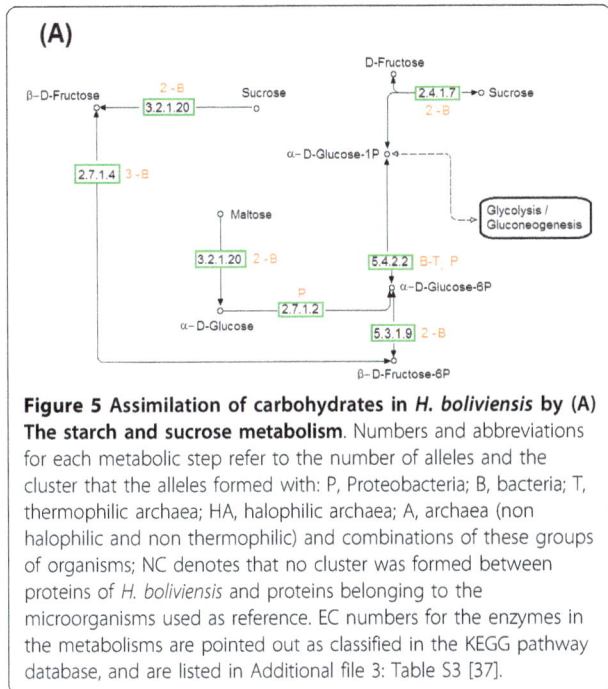

Figure 5 Assimilation of carbohydrates in *H. boliviensis* by **(A) The starch and sucrose metabolism**. Numbers and abbreviations for each metabolic step refer to the number of alleles and the cluster that the alleles formed with: P, Proteobacteria; B, bacteria; T, thermophilic archaea; HA, halophilic archaea; A, archaea (non halophilic and non thermophilic) and combinations of these groups of organisms; NC denotes that no cluster was formed between proteins of *H. boliviensis* and proteins belonging to the microorganisms used as reference. EC numbers for the enzymes in the metabolisms are pointed out as classified in the KEGG pathway database, and are listed in Additional file 3: Table S3 [37].

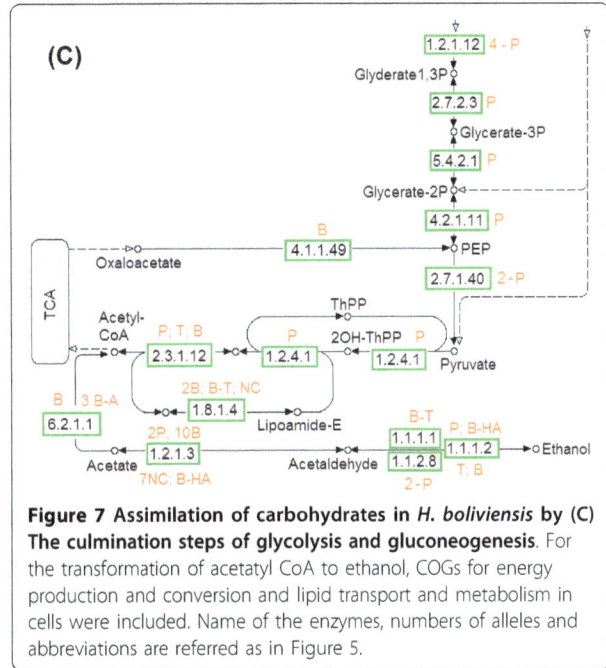

Figure 7 Assimilation of carbohydrates in *H. boliviensis* by **(C) The culmination steps of glycolysis and gluconeogenesis**. For the transformation of acetatyl CoA to ethanol, COGs for energy production and conversion and lipid transport and metabolism in cells were included. Name of the enzymes, numbers of alleles and abbreviations are referred as in Figure 5.

dehydrogenase complex to transform pyruvate to acetyl-CoA rather they accomplish the transformation using 2-oxoacid oxidoreductases [40]. However, putative enzyme sequences that form part of the PDHC can be found in the genome sequences of thermophilic [24,25] and halophilic archaea [24,25,41]. The protein divergence and gene duplication (Figure 3 and 7) may provide evidence of adaptive evolution of the metabolism [13]. The diversity of enzymes that can accomplish the same function in *H. boliviensis* explains its versatility in the assimilation of carbohydrates and other carbon sources (e.g. acetate and short chain fatty acids) [18,38]. The polymorphic metabolism of *H. boliviensis* might also lead to an efficient generation of energy (ATP) and the reducing agent NADH to correlate with its fast cell growth and its capability to metabolize different carbon sources to PHB [18,38]. Besides the synthesis of PHB, excess of NADH could potentially be oxidized by *H. boliviensis* via a fermentative route to allow the formation of ethanol (Figure 7).

Relationship of the enzymes involved in glycolysis and gluconeogenesis among Prokaryotes

Considering the degree of polymorphism in the metabolic routes of *H. boliviensis*, we wonder whether this trend could be followed by other microorganisms in its environment. To address this question, supernetworks were constructed by combining the phylogenetic trees related to glycolysis and gluconeogenesis. Figure 8 shows a supernetwork obtained after combining three phylogenetic trees; two of them derived from COGs that are considered among the 102 genes that contain taxonomic information that discriminate well bacteria and archaea in already known families and genera [4]. The internetwork relationship among microorganisms shown in Figure 8 denotes that HGT occurred among bacteria, archaea and between bacteria and archaea. A supernetwork obtained from six trees reflected a higher effect of HGT among microorganisms (Figure 9). In Figure 9, taxonomic differentiation between proteins of bacteria and archaea was barely

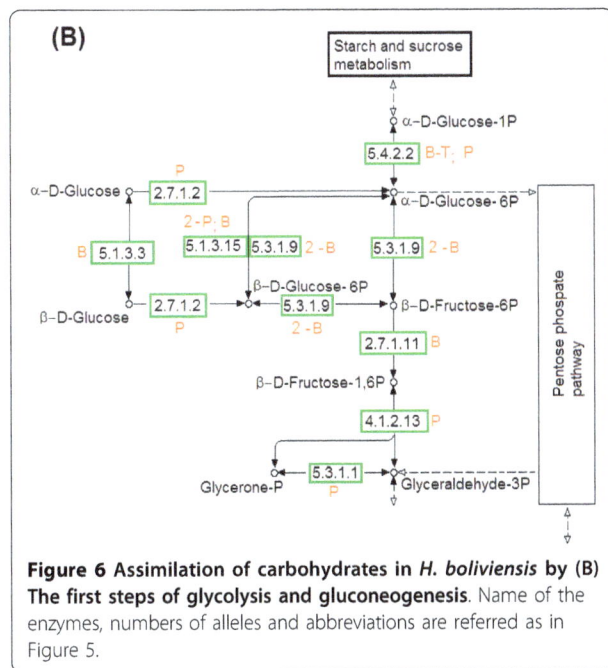

Figure 6 Assimilation of carbohydrates in *H. boliviensis* by **(B) The first steps of glycolysis and gluconeogenesis**. Name of the enzymes, numbers of alleles and abbreviations are referred as in Figure 5.

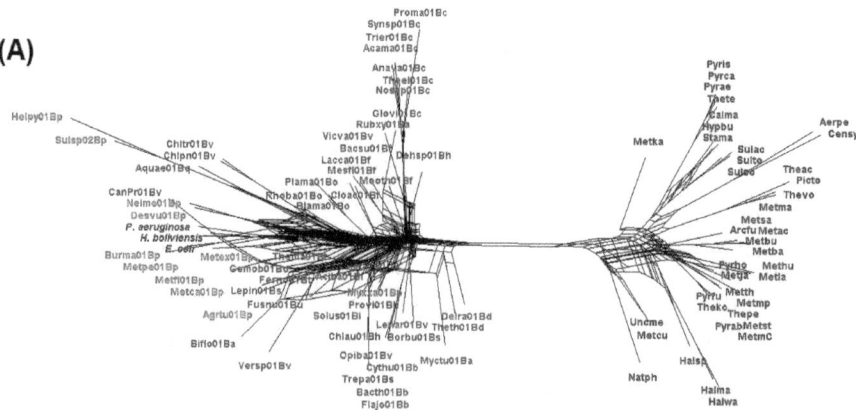

Figure 8 Supernetworks constructed using the Split4tree program after combining phylogenetic trees of proteins of the glycolysis and gluconeogenesis metabolisms in *H. boliviensis* and reference strains. The figures included combination of trees corresponding to COGs **(A)** 0126, 0149 and 0837. Names of the archaeal and bacterial species corresponding to each abbreviation used in the figure are listed in Additional file 4: Table S4.

observed, although *H. boliviensis* was still clustered with other γ-Proteobacteria, i.e. *Escherichia coli* and *Pseudomonas aeruginosa*. Finally, 22 phylogenetic trees related to glycolysis and gluconeogenesis were used to attain a supernetwork (Figure 10). Taxonomic differentiation among the proteins of the microorganisms was no longer observed (Figure 10), hence suggesting that flow of genes involved in glycolysis and gluconeogenesis among Prokaryotes was significant. Experimental analysis demonstrated polymorphism for the enzymes that form part of glycolysis in *E. coli* [42], whereas various different *Halomonas* and *Chromohalobacter* species grow on several common carbon sources and are able to produce PHB [43]. On the whole, our studies imply that the availability of the selection of a particular carbohydrate by a microbial species should be related to the rate of evolution of the enzymes, and might be linked not only to the size of the microbial

population, as stated by the nearly neutral theory of evolution [1], but also to the groups of microorganisms able to thrive in a particular niche.

Use of combination of carbohydrates for the production of PHB by *H. boliviensis*

The aforementioned results revealed that the rate of evolution, mutations and the molecular interaction between *H. boliviensis* and other microorganisms in its environment influenced significantly the evolution of the carbohydrate transport and metabolism in this bacterium–a similar evolutionary pattern might be expected in other prokaryotes. However, phenotypic traits concerning microbial growth on different carbon sources are stamps of different phylogenetic groups and species. We hypothesized that the amount and type of carbon sources in a particular environment should also influence the

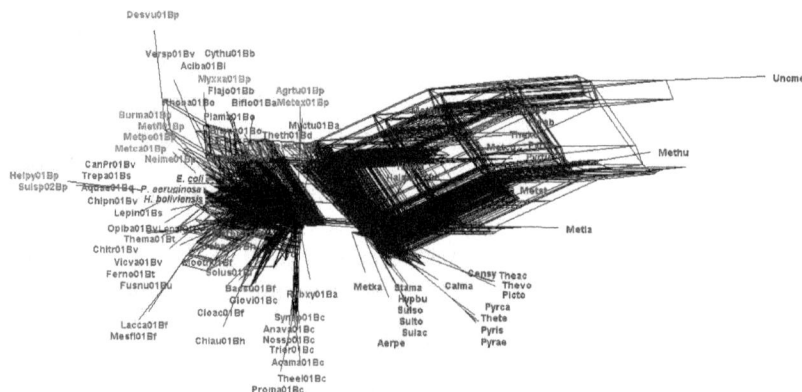

Figure 9 Supernetworks constructed using the Split4tree program after combining phylogenetic trees of proteins of the glycolysis and gluconeogenesis metabolisms in *H. boliviensis* and reference strains. The figures included combination of trees corresponding to COGs **(B)** 0126, 0149, 0837, 0469, 0696 and 0837. Names of the archaeal and bacterial species corresponding to each abbreviation used in the figure are listed in Additional file 4: Table S4.

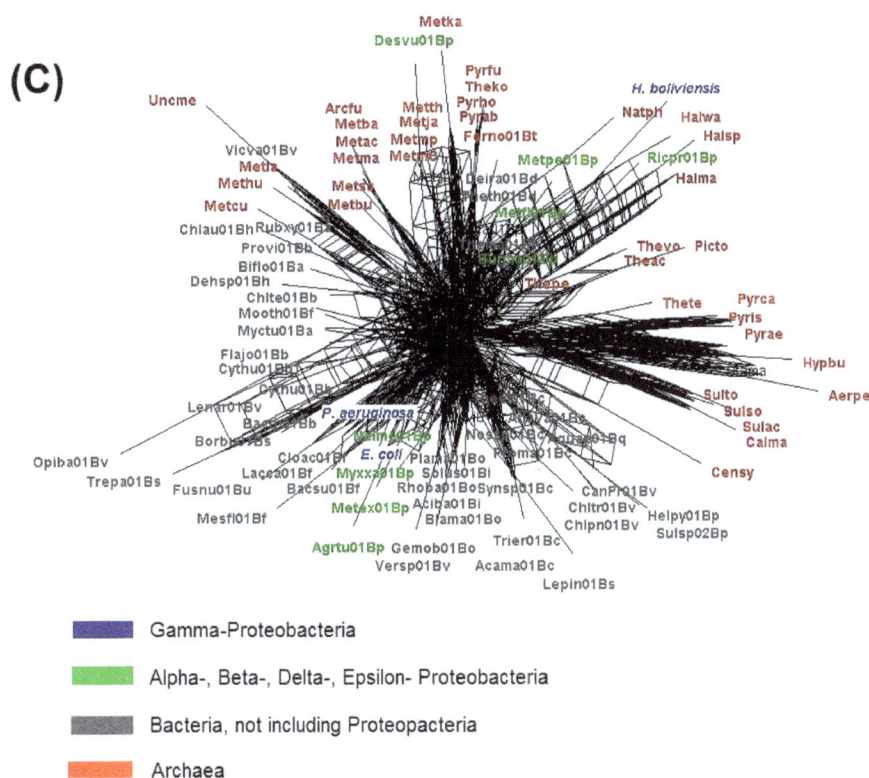

(C)

■ Gamma-Proteobacteria

■ Alpha-, Beta-, Delta-, Epsilon- Proteobacteria

■ Bacteria, not including Proteopacteria

■ Archaea

Figure 10 Supernetworks constructed using the Split4tree program after combining phylogenetic trees of proteins of the glycolysis and gluconeogenesis metabolisms in *H. boliviensis* and reference strains. The figures included combination of trees corresponding to COGs **(C)** 0057, 0126, 0148, 0149, 0166, 0191, 0205, 0235, 0365, 0469, 0508, 0696, 0837, 1012, 1063, 1109, 1249, 1454, 1866, 2017, 2609 and 4993. Names of the archaeal and bacterial species corresponding to each abbreviation used in the figure are listed in Additional file 4: Table S4.

fitness of glycolysis and gluconeogenesis fluxes in *H. boliviensis*. The influence of the environment on the functional features of enzymes is a context not commonly evaluated in evolutionary theories of metabolic pathways [12,14]. For this reason, we decide to use various combinations of glucose and sucrose concentrations as precursors for PHB synthesis in *H. boliviensis* (Figures 11, 12).

Three alleles of the PHB synthases were found in the genome of *H. boliviensis* (Additional file 5: Figure S1). The three alleles are closely related to PHB synthases of Proteobacteria. Moreover, *H. boliviensis* A2 was clustered with two alleles of PHB polymerases of *Halomonas* sp. TD01 (Figure S1); one of these alleles (phaC1) was previously reported [6]. However, a third allele of *Halomonas* sp. TD01 (named phaC2) showed a distant phylogenetic relationship to the PHB synthases of Proteobacteria (Figure S1); phaC2 might have been acquired by HGT [6]. Research on the PHB polymerization and depolymerization pathways in *H. boliviensis* is in progress. PHB production by *H. boliviensis* was performed in shake flask experiments under nitrogen limitation conditions (i.e. a low concentration sodium glutamate was added to the culture medium to limit the

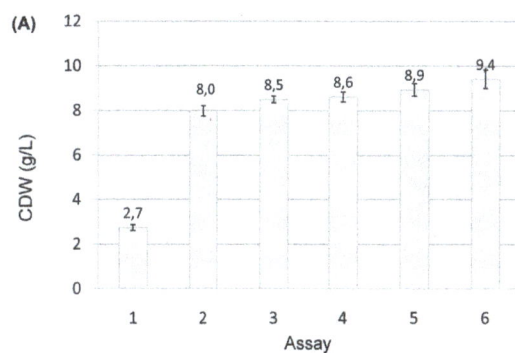

Figure 11 Cell growth and PHB production by *H. boliviensis* using different combinations of carbohydrates and molasses. **(A)** Cell dry weight (CDW) in batch culture using 1.5% (w/v) sucrose and 1% (w/v) glucose. Assay numbers refer to the concentration of carbohydrates added to the medium, %(w/v): 1) 2.5 sucrose, 2) 2.0 sucrose and 0.5 glucose, 3) 1.5 sucrose and 1 glucose, 4) 1.0 sucrose and 1.5 glucose, 5) 0.3 sucrose, 0.7 glucose and 1.5 dried molasses and 6) 2.5 dried molasses. All experiments were performed in shake flasks at 35°C and 220 rpm of agitation. All experiments were performed in triplicate. The error bars refer to the SD of the average values.

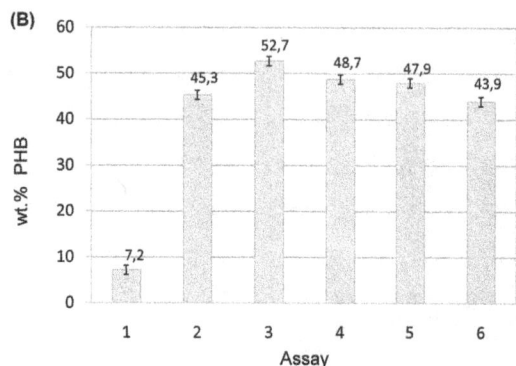

Figure 12 Cell growth and PHB production by *H. boliviensis* using different combinations of carbohydrates and molasses. **(B)** PHB accumulation in batch culture using 1.5% (w/v) sucrose and 1% (w/v) glucose. Assay numbers refer to the concentration of carbohydrates added to the medium, %(w/v): 1) 2.5 sucrose, 2) 2.0 sucrose and 0.5 glucose, 3) 1.5 sucrose and 1 glucose, 4) 1.0 sucrose and 1.5 glucose, 5) 0.3 sucrose, 0.7 glucose and 1.5 dried molasses and 6) 2.5 dried molasses. All experiments were performed in shake flasks at 35°C and 220 rpm of agitation. All experiments were performed in triplicate. The error bars refer to the SD of the average values.

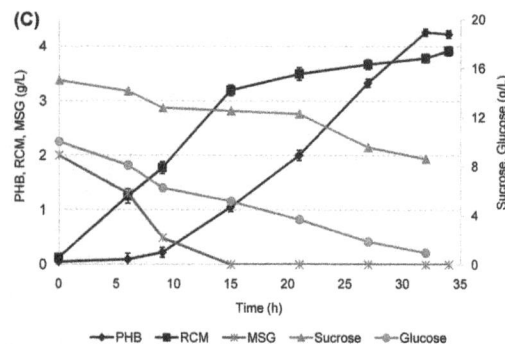

Figure 13 Cell growth and PHB production by *H. boliviensis* using different combinations of carbohydrates and molasses. **(C)** Carbohydrate assimilation and PHB production in batch culture using 1.5% (w/v) sucrose and 1% (w/v) glucose. All experiments were performed in shake flasks at 35°C and 220 rpm of agitation. The abbreviations RCM and MSG in the figure refer to residual cell mass and mono sodium glutamate, respectively. All experiments were performed in triplicate. The error bars refer to the SD of the average values.

cell growth). When sucrose was used as the sole carbon source, the accumulation of PHB in *H. boliviensis* (7.2 wt%) and cell growth (2.7 g/L) were low compared to those obtained with combinations of sucrose and glucose (Figures 11,12). Cell growth increased as the amount of glucose was higher in the medium to reach 8.6 g/L (Figure 11), while the maximum PHB content in *H. boliviensis* was 52.7 wt% when 1.5% (w/v) sucrose and 1.0% (w/v) glucose were included in the medium composition (Figure 12). The use of molasses enhanced to some extent the cell growth, c.a. 9.4 g/L, but the PHB accumulated in the cells was lower, 43.9 wt% (Figures 11, 12). Both the cell density and the maximum PHB yield attained by *H. boliviensis* are higher to those reported using glucose as carbon source, i.e. 5.3 g/L and 45 wt% respectively, under similar culture conditions [16].

Glucose and sucrose uptake and assimilation were analyzed further using the optimum sucrose and glucose ratio, i.e. 1.5:1, for PHB production (Figure 13). Residual cell mass (RCM), which is the cell biomass without the polymer inclusions, was used to analyze the active cell growth. Glucose and sucrose were assimilated parallelly during the exponential phase of growth of *H. boliviensis* (Figure 13). However, glucose consumption rate by *H. boliviensis* shows a linear decrease after 9 hours of cultivation when sodium glutamate was almost depleted from the medium and PHB synthesis was triggered (Figure 13). Sucrose concentration in the medium was constant from

9 to 21 h of cultivation and was only reduced when the concentration of glucose in the medium was low (Figure 13). Glucose generates a higher amount of energy and NADH in *H. boliviensis* than sucrose because the CDW and PHB reached by *H. boliviensis* using glucose were much higher than those achieved when sucrose was used as the sole carbon source (Figures 11, 12) [16]. Nevertheless, an overflow of NADH is known to obstruct the tricarboxylic acid cycle because of inhibition of citrate synthase [44]. An adequate ratio of glucose and sucrose in the medium promoted an appropriate balance of energy for an active cell growth (Figures 11, 13), albeit Figure 13 reflects that glucose is preferred during the PHB accumulation in *H. boliviensis* due to an excess of NADH in the cytoplasm of cells improves the polymer synthesis [19,44].

The maximum PHB concentration and volumetric productivity reached by *H. boliviensis* were 4.3 g/L and 0.13 g/L/h, respectively; they are comparable to those reached by *Cupriavidus necator*, i.e. 5.1 g/L and 0.11 g/L/h [45], and to those reported for a recombinant *E. coli* strain, c. a. 7.2 g/L and 0.15 g/L/h [46]. The medium for *C. necator* and *E. coli* contained glucose as the carbon source for experiments performed in shake flasks. Under similar culture conditions, *Azotobacter vinelandii* led to a PHB concentration of 7.5 g/L and a productivity of 0.30 g/L/h [47]. These bacteria attained among the highest productions of PHB, and are recognized for their potential utilization at industrial scales [19,21].

The viability of the commercialization of PHB is dependent upon the reduction of the total production costs [22]. The price of the carbon source supplied in the culture medium may account up to 40% of the total production costs [22]. Sucrose is at least two times cheaper than glucose while molasses are cheaper than sucrose. The results obtained for the production of PHB by *H. boliviensis* (Figures 11, 13) suggest that an agricultural surplus such as molasses could be used during the bioprocess scale up to stimulate the cell growth; furthermore an optimum ratio of sucrose and glucose should be added in the culture medium of the largest bioreactor used in a process to induce a high polymer production. Replacing partially glucose by sucrose and molasses should surly reduce the production costs of the polymer and lead also to an environmentally friendly bioprocess. Nevertheless, fed-batch cultivations systems are yet to be performed with *H. boliviensis* using combinations of carbohydrates to reveal their potential in a large scale process.

Conclusions

The genome size and number of genes found in *H. boliviensis* were similar to those determined for other halophilic bacteria of the family *Halomonadaceae*. The ability of *H. boliviensis* to grow on different carbon sources is explained by the high number of genes related to the carbohydrate uptake and metabolism. Interestingly, most of these genes were obtained from other bacteria by HGT, only 34% of the genes evolved as proteins belonging to Proteobacteria, while 13% of the genes were transferred from haloarchaea and thermophilic archaea. Furthermore, the diversity of enzymes that have the same physiological function led to polymorphism in the metabolic routs. Results obtained in this article depict that most genetic modifications and enzyme polymorphism in the genome of *H. boliviensis* were mainly influenced by HGT rather than nearly neutral mutations. Molecular adaptation and evolution experienced by *H. boliviensis* were also a response to environmental conditions such as the type and amount of carbohydrates in its ecological niche. Consequently, the genome evolution of *H. boliviensis* showed to be strongly influenced by the type of microorganisms, genetic interaction among microbial species and its environment. Such trend should also be experienced by other prokaryotes. A system for PHB production by *H. boliviensis* that takes into account the evolutionary adaptation of this bacterium to the assimilation of combinations of carbohydrates suggests the feasibility of a bioprocess economically viable and environmentally friendly.

Halomonas boliviensis genome sequence

This whole genome shotgun project was deposited at DDBJ, EMBL and GenBank under the accession number AGQZ00000000. The version described in this paper is the first version, AGQZ01000000.

Additional material

Additional file 1: Table S1. Clusters of orthologous genes (COGs) of *H. boliviensis* related to carbon transport and metabolism.

Additional file 2: Table S2. List of GenBank accession numbers for the microorganisms shown in the phylogenetic trees in Figure 1, 2, 3.

Additional file 3: Table S3. EC numbers and COG classification of the enzymes involved in starch metabolism, glycolysis and gluconeogenesis in *H. boliviensis*.

Additional file 4: Table S4. List of species and abbreviations of the 100 microorganisms (59 Bacteria and 41 Archaea) used as reference. Abbreviations are named as described by Puigbò, et al. 2009.

Additional file 5: Figure S1.

Acknowledgements
The authors would like to thank the Swedish International Development Cooperation Agency (Sida) for supporting our research work.

Author details
[1]Centro de Biotecnología, Facultad de Ciencias y Tecnología, Universidad Mayor de San Simón, Cochabamba, Bolivia. [2]Department of Biotechnology, Lund University, P.O. Box 124, SE-221 00 Lund, Sweden.

Authors' contributions
DG, AB-S, CC-O, MG-M and JQ performed the evolutionary analyses on the COGs corresponding to the carbohydrate transport and metabolism of *H. boliviensis*. JQ constructed the supernetworks. JQ, DG and NC-Q studied the PHB production by *H. boliviensis*. All authors wrote the manuscript and approved the final version.

Competing interests
The authors declare that they have no competing interests.

References
1. Ohta T: **The nearly neutral theory of molecular evolution.** *Annu Rev Ecol Syst* 1992, **23**:263-286.
2. Puigbò P, Wolf YI, Koonin EV: **The tree and net components of prokaryote evolution.** *Genome Biol Evol* 2010, **2**:745-756.
3. Orr H: **The genetic theory of adaptation: a brief history.** *Nature Reviews* 2005, **6**:119-127.
4. Puigbò P, Wolf YI, Koonin EV: **Search for a 'Tree of Life' in the thicket of the phylogenetic forest.** *J Biology* 2009, **8**:1-17.
5. Schliep K, Lopez P, Lapointe FJ, Bapteste E: **Harvesting evolutionary signals in a forest of prokaryotic gene trees.** *Mol Biol Evol* 2010, **28**:1393-1405.
6. Cai L, Tan D, Aibaidula G, Dong X, Chen J, Tian W, Chen G: **Comparative genomics study of polyhydroxyalkanoates (PHA) and ectoine relevant genes from *Halomonas* sp. TD01 revealed extensive horizontal gene transfer events and co-evolutionary relationships.** *Microb Cell Fact* 2011, **10**:88.
7. Oren A: **Microbial life at high salt concentrations: phylogenetic and metabolic diversity.** *Saline Systems* 2008, **4**:2.
8. Oren A, Larimer F, Richardson P, Lapidus A, Csonka LN: **How to be moderately halophilic with broad salt tolerance: clues from the genome of *Chromohalobacter salexigens*.** *Extremophiles* 2005, **9**:275-279.
9. Schwibbert K, *et al*: **A blueprint of ectoine metabolism from the genome of the industrial producer *Halomonas elongata* DSM 2581[T].** *Environ Microbiol* 2010, **13**:1973-1994.
10. Lin Y, *et al*: **Draft genome sequence of *Halomonas* sp. strain HAL1, a moderately halophilic arsenite-oxidizing bacterium isolated from gold-mine soil.** *J Bacteriol* 2011, **194**:199-200.
11. Roberts MF: **Organic compatible solutes of halotolerant and halophilic microorganism.** *Saline Systems* 2005, **1**:5.
12. Eanes WF: **Molecular population genetics and selection in the glycolytic pathway.** *J Exp Biol* 2011, **214**:165-171.

13. Eanes WF: Analysis of selection on enzyme polymorphism. *Annu Rev Ecol Syst* 1999, **30**:301-326.

14. Hartl D, Dykhuizen D, Dean A: Limits of adaptation: the evolution of selective neutrality. *Genetics* 1985, **111**:655-674.

15. Quillaguamán J, Hatti-Kaul R, Mattiasson B, Alvarez MT, Delgado O: *Halomonas boliviensis* sp. nov., an alkalitolerant, moderate halophile bacterium isolated from soil around a Bolivian hypersaline lake. *Int J Syst Evol Microbiol* 2004, **54**:721-725.

16. Quillaguamán J, Van-Thuoc D, Guzmán H, Guzmán D, Martín J, Akaraonye E, Hatti-Kaul R: Poly(3-hydroxybutyrate) production by *Halomonas boliviensis* in fed-batch culture. *Appl Microbiol Biotechnol* 2008, **78**:227-232.

17. Van-Thuoc D, Guzmán H, Quillaguamán J, Hatti-Kaul R: High productivity of ectoines by *Halomonas boliviensis* using a combined two-step fed-batch culture and milking process. *J Biotechnol* 2010, **147**:46-51.

18. Quillaguamán J, Delgado O, Mattiasson B, Hatti-Kaul R: Poly(β-hydroxybutyrate) production by a moderate halophile, *Halomonas boliviensis* LC1. *Enzyme Microb Technol* 2006, **38**:148-154.

19. Steinbüchel A, Füchtenbush B: Bacterial and other biological systems for polyester production. *Trends Biotechnol* 1998, **16**:419-427.

20. Philip S, Keshavarz T, Roy I: Polyhydroxyalkanoates: biodegradable polymers with a range of applications. *J Chem Technol Biotechnol* 2007, **82**:233-247.

21. Lee SY: Plastic bacteria? Progress and prospects for polyhydroxyalkanoate production in bacteria. *Trends Biotechnol* 1996, **14**:431-438.

22. Choi J, Lee S: Factors affecting the economics of polyhydroxyalkanoate production by bacterial fermentation. *Appl Microbiol Biotechnol* 1998, **51**:13-21.

23. Delcher A, Bratke K, Powers E, Salzberg S: Identifying bacterial genes and endosymbiont DNA with Glimmer. *Bioinformatics* 2007, **23**:673-679.

24. Tatusov R, *et al*: The COG database: an updated version includes eukaryotes. *BMC Bioinforma* 2003, **4**:41.

25. Jensen L, Julien P, Kuhn M, von Mering C, Muller J, Doerks T, Bork P: eggNOG: automated construction and annotation of orthologous groups of genes. *Nucleic Acids Res* 2008, **36**:D250-D254.

26. Edgar R: MUSCLE: multiple sequence alignment with high accuracy and high throughput. *Nucleic Acids Res* 2004, **32**:1792-1797.

27. Tamura K, Peterson D, Peterson N, Stecher G, Nei M, Kumar S: MEGA5: molecular evolutionary genetics analysis using maximum likelihood, evolutionary distance, and maximum parsimony methods. *Mol Biol Evol* 2011, **10**:2731-2739.

28. Huson D, Bryant D: Application of phylogenetic networks in evolutionary studies. *Mol Biol Evol* 2006, **23**:254-267.

29. Huson D, Dezulian T, Klopper T, Steel M: Phylogenetic supernetworks from partial trees. *IEEE/ACM Trans Comput Biol Bioinform* 2004, **1**:151-158.

30. Lee SY, Wong HH, Choi J, Lee SH, Lee SC, Han CS: Production of medium-chain-length polyhydroxyalkanoates by high-cell-density cultivation of *Pseudomonas putida* under phosphorus limitation. *Biotechnol Bioeng* 2000, **68**:466-470.

31. Onraedt A, Wlcarius B, Soetaert W, Vandamme E: Optimization of ectoine synthesis through fed-batch fermentation of *Brevibacterium epidermis*. *Biotechnol Prog* 2005, **21**:1206-1212.

32. Arahal DR, García MT, Vargas C, Cánovas D, Nieto JJ, Ventosa A: *Chromohalobacter salexigens* sp. nov., a moderately halophilic species that includes *Halomonas elongata* DSM 3043 and ATCC 33174. *Int J Syst Evol Microbiol* 2001, **51**:1457-1462.

33. Arahal DR, Ventosa A: The family *Halomonadaceae*. In *The Prokaryotes. A handbook on the biology of bacteria*. Edited by: Dworkin M, et al. New York: Springer; 2006:811-835.

34. Gandbhir M, Rashed I, Marlière P, Mutzel R: Convergent evolution of amino acid usage in archaebacterial and eubacterial linages adapted to high salt. *Res Microbiol* 1995, **146**:113-120.

35. Kimura M: Evolutionary rate at the molecular level. *Nature* 1968, **217**:624-626.

36. Kreitman M: The neutral theory is dead. Long live the neutral theory. *Bioessays* 1996, **18**:678-683.

37. Kanehisa M, *et al*: From genomics to chemical genomics: new developments in KEGG. *Nucleic Acids Res* 2006, **34**:D354-D357.

38. Quillaguamán J, Hashim S, Bento F, Mattiasson B, Hatti-Kaul R: Poly(β-hydroxybutyrate) production by a moderate halophile, *Halomonas*

39. de Kok A, Hengeveld AF, Martin A, Westphal AH: The pyruvate dehydrogenase multi-enzyme comples from Gram-negative bacteria. *Biochim Biophys Acta* 1998, **1385**:353-366.

40. Danson MJ: Central metabolism of the Archaea. *New Compr Biochem* 1993, **26**:1-24.

41. Jolley KA, *et al*: 2-oxoacid dehydrogenase mutienzyme complexes in the halophilic Archaea? Gene sequences and protein structural predictions. *Microbiology* 2000, **146**:1061-1069.

42. Sawyer S, Dykhuizen D, Hartl D: Confidence interval for the number of selectively neutral amino acid polymorphism. *Proc Natl Acad Sci USA* 1987, **84**:6225-6228.

43. Quillaguamán J, Guzmán H, Van-Thuoc D, Hatti-Kaul R: Synthesis and production of polyhydroxyalkanoates by halophiles: current potential and future prospects. *Appl Microbiol Biotechnol* 2010, **85**:1687-1696.

44. Babel W, Ackermann JU, Breuer U: Physiology, regulation, and limits of the synthesis of poly (3HB). In *Advances in Biochemical Engineering/ Biotechnology: Biopolyesters*. Edited by: Scheper T, Babel W, Steinbüchel A. Berlin: Springer; 2001:125-157.

45. Doi Y, Tamaki A, Kunioka M, Soga K: Production of copolyesters of 3-hydroxybutyrate and 3-hydroxyvalerate by *Alcaligenes eutrophus* from butyric and pentanoic acids. *Appl Microbiol Biotechnol* 1988, **28**:330-334.

46. Lee SY, Lee KM, Chang HN, Steinbüchel A: Comparison of recombinant *Escherichia coli* strains for synthesis and accumulation of poly-(3-hydroxybutyric acid) and morphological changes. *Biotechnol Bioeng* 1994, **44**:1337-1347.

47. Page WJ: Production of poly-β-hydroxybutyrate by *Azotobacter vinelandii* UWD in media containing sugars and complex nitrogen. *Appl Microbiol Biotechnol* 1992, **38**:117-121.

boliviensis LC1 using starch hydrolysate as substrate. *J Appl Microbiol* 2005, **99**:151-157.

Culturable halophilic archaea at the initial and crystallization stages of salt production in a natural solar saltern of Goa, India

Kabilan Mani, Bhakti B Salgaonkar and Judith M Braganca[*]

Abstract

Background: Goa is a coastal state in India and salt making is being practiced for many years. This investigation aimed in determining the culturable haloarchaeal diversity during two different phases of salt production in a natural solar saltern of Ribandar, Goa. Water and sediment samples were collected from the saltern during pre-salt harvesting phase and salt harvesting phase. Salinity and pH of the sampling site was determined. Isolates were obtained by plating of the samples on complex and synthetic haloarchaeal media. Morphology of the isolates was determined using Gram staining and electron microscopy. Response of cells to distilled water was studied spectrophotometrically at 600nm. Molecular identification of the isolates was performed by sequencing the 16S rRNA.

Results: Salinity of salt pans varied from 3-4% (non-salt production phase) to 30% (salt production phase) and pH varied from 7.0-8.0. Seven haloarchaeal strains were isolated from water and sediment samples during non-salt production phase and seventeen haloarchaeal strains were isolated during the salt production phase. All the strains stained uniformly Gram negative. The orange-red acetone extract of the pigments showed similar spectrophotometric profile with absorption maxima at 393, 474, 501 and 535 nm. All isolates obtained from the salt dilute phase were grouped within the genus *Halococcus*. This was validated using both total lipid profiling and 16S rRNA data sequencing. The isolates obtained from pre-salt harvesting phase were resistant to lysis. 16S rRNA data showed that organisms belonging to *Halorubrum*, *Haloarcula*, *Haloferax* and *Halococcus* genera were obtained during the salt concentrated phase. The isolates obtained from salt harvesting phase showed varied lysis on suspension in distilled water and /or 3.5% NaCl.

Conclusion: Salterns in Goa are transiently operated during post monsoon season from January to May. During the pre-salt harvesting phase, all the isolates obtained belonged to *Halococcus sp*. During the salt harvesting phase, isolates belonging to *Halorubrum*, *Haloarcula*, *Haloferax* and *Halococcus* genera were obtained. This study clearly indicates that *Halococcus sp*. dominates during the low salinity conditions.

Keywords: Archaea, Haloarchaea, Hypersaline, Solar saltern

Findings

Marine solar salterns are thalassohaline hypersaline environments located in tropical and subtropical areas worldwide, consisting of shallow ponds for the production of common salt from seawater during summer. The method of making salt through natural evaporation dates back to pre-historic times. This traditional approach of salt production involves construction of series of rectangular ponds, each connected to the other through a common opening [1-4].

Goa (15°34′60N, 74°0′0E) is a coastal state in India and salt making is being practiced for many years. Saltpans are found in Pernem, Bardez, Tiswadi and Salcete talukas of Goa. Saltpans are inundated by sea water from estuaries during high tides. Sea water is retained in every pond for certain time to facilitate evaporation. As concentration of NaCl gradually increases, first component to precipitate is

* Correspondence: judith@bits-goa.ac.in
Department of Biological Sciences, BITS PILANI, K K Birla Goa Campus, NH 17 B, Zuarinagar, Sancoale, Goa 403 726, India

calcium ion (Ca^{2+}) in the form of gypsum. Then the concentrated sea water is allowed into final crystallizer pond, where NaCl crystals precipitate out [1]. The whole process of concentrating sea water begins usually in January - February and salt crystals are harvested during March - May. During the remaining months of June - December, ponds are inundated by sea/rain water. The crude salt produced from these saltpans is being used domestically for cooking, ice plants, as fertilizers, as termite repellent and for curing dry fish.

Microbial life is found in various extreme environments and salt pans are no exception. Depending on salt concentration, salt pans are inhabited by different groups of microbes thriving symbiotically [5]. Various microbes that inhabit the salt pans range from prokaryotes like Bacteria (*Salinibacter* spp.) and Archaea (*Halobacterium* spp.) to eukaryotes like Fungi (*Hortaea* spp.) and Algae (*Dunaliella* spp.) [6,7].

Haloarchaea are a group of extreme halophiles which require at least 2.5 M NaCl for their growth and are placed in the order *Halobacteriales* under family *Halobacteriaceae* [8]. At the time of writing, family *Halobacteriaceae* accommodated 36 recognized genera, members of which inhabiting both thalassohaline and athalassohaline environments [9,10]. Many studies have shown that there is a great variation in the diversity and dominance of haloarchaeal genera within various geographical locations [11-16]. Novel haloarchaeal microorganisms have been isolated from various econiches such as acidic and alkaline regions, animal hides, salted fishes and also from commercial salt [17-23].

The aim of the present investigation was to evaluate the diversity of culturable haloarchaeal members in saltpans of Goa during two stages of salt production. For this investigation, the diversity of haloarchaea found within a single solar saltern at Ribandar (15°30′N, 73°51′E), Goa was examined.

Ribandar salterns are located on the banks of the river Mandovi, surrounded by mangrove vegetation in the adjoining marshy area. These salt pans cover an area about 12 dm^2 and lie between the cities of Panaji and Old Goa (Figure 1). Ribandar experiences tropical monsoon climate with maximum temperature around 30-36°C in summer and minimum around 20-28°C in winter. This region receives a heavy monsoon rainfall averaging around 300 cm. The salt pans are located at 9 ft above sea level and experience strong coastal winds during summer facilitating the evaporation of water in salterns. The salt pans are surrounded by raised mud borders called as *bunds* (dykes). These *bunds* help in containing the sea water within the pan. The sluice gate at the inlet regulates the inflow of water from the Mandovi estuary (Additional file 1: Figure S1, supplementary data). The salterns are seeded with crude salt to speed up the

crystallization process. During April and May the salt is harvested daily from these salterns.

Water samples and sediment samples were collected from 0-10 cm distance from the surface Sampling was carried out twice during the two phases of salt production. First sampling was carried out in February 2010, when the saltpans are full of sea water. The second sampling was carried out in April 2010, when salt harvesting is at its peak. Water and sediment samples were collected by scooping from the surface as well as at a depth of 10 cm. Salinity and pH of sampling sites were measured using conductivity meter (EQUIP-TRONICS MODEL EQ-682) and pH meter (EQUIP-TRONICS MODEL EQ-632). Conductivity was correlated to salinity using the equation given by Williams, 1986 [24].

$$S = 0.4665 \times^{1.0878} (r^2 = 0.98799)$$

The samples were collected in sterile tubes, sterile 1 L bottles and stored at 4°C and processed within 48 h.

Direct plating and enrichment techniques were employed for isolation of haloarchaeal organisms from water and sediment samples. Two media were used in the study, NaCl Tryptone Yeast Extract (NTYE) medium and NaCl Tri-Na-citrate (NT) medium [25-27], both containing 25% NaCl. The main difference with both the media is the presence of trisodium citrate in NT medium, which can support the growth of fastidious organisms as compared to NTYE medium. In direct plating method, one hundred μl of water sample or a loopful of sediment sample was directly spread plated on media. In the enrichment technique method, one ml of water sample or one gram of sediment sample was aseptically transferred to 50ml media and incubated at 37°C for up to 5 days. Then ten μl aliquots were plated on media. Plates were incubated at room temperature (30°C) for 30 to 45 days until red-orange pigmented colonies appeared. Colonies were selected based on their morphology and/or pigmentation and purified through repeated sub-culturing.

Morphology of the isolates was determined using Gram staining and electron microscopy. Cell suspensions were prepared on glass slides in a drop of 15% (w/v) NaCl solution and air-dried. The cells were desalted with 2% acetic acid followed by Gram staining and observed using phase-contrast microscope (Olympus BX41). For electron microscopy, the cell pellets were dispensed in NaCl Synthetic Media (NSM) to an absorbance of 0.8 at 600nm [28]. One hundred μl of suspension was mounted onto circular glass cover slips, fixed with 2.0% glutaraldehyde fixative (prepared in NSM) at room temperature (30°C) overnight. The coverslips were then exposed, to a series of increasing gradient of acetone-water, corresponding to 30%, 50%, 70%, 90% for 10 min respectively and finally in 100%

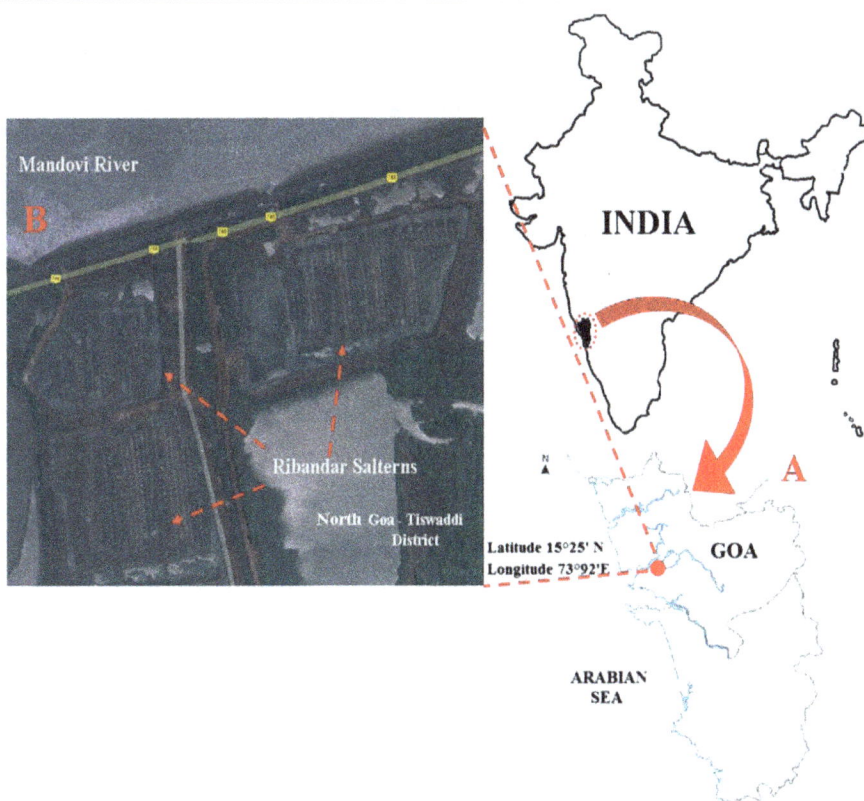

Figure 1 Solar Salterns of Ribandar, Goa. A) Location of Ribandar Salterns in Goa. B) Aerial image of the Ribandar Salterns. [source: Google Earth, 2010].

acetone, for 30 min, air dried and then viewed under scanning electron microscope (JEOL-5800 LV SEM).

Pigments were extracted by sonicating the cells for 30 min at a pulse rate of 0.5 s in acetone or chloroform: methanol (2:1). Identification of pigment was done through spectrophotometric scans in both UV and visible range (190-800nm). Total polar lipids were analysed as described by Litchfield et al in 2000 and Oren et al in 1996 [29,30].

Response of cells to distilled water was studied spectrophotometrically at 600nm (Shimadzu, Japan). Cell viability assay was performed by resuspending the cells in distilled water, 3.5% NaCl and 30% NaCl and then incubating them for time periods of 24 h to 10 days respectively [31]. The cell suspensions were then plated and observed for viable colony formation.

Molecular characterization of the isolates was performed by extracting genomic DNA using phenol-chloroform method and amplifying gene for 16S rRNA with primers A109 (F) AC(G/T)GCTCAGTAACACGT and 1510(R) GGTTACCTTGTTACGACTT [32]. Each PCR reaction contained 2 U Taq Polymerase, 10X Taq buffer, 2 mM $MgCl_2$, 10 mM of dNTPs (Sigma), 10 μM of each primer (IDT technologies) and 1 μl of template DNA. Final reaction was made up to 50 μl with ultra-pure distilled water. The amplification was performed under the following conditions:

Initial denaturation for 5 min at 94°C, denaturation for 30 s at 94°C, annealing for 40 s at 53.5°C, elongation for 60 s at 68°C (35 cycles) and final elongation at 68°C for 5 min. Amplified products were purified and then sequenced using an automated DNA sequencer (Applied Biosystems). The sequencing results of the amplified 16S rRNA fragments were subjected to BLAST analysis. Multiple sequence alignment was done out using MUSCLE and phylogenetic tree was constructed with MEGA 5.0 by neighbor-joining method with bootstrap analysis using 1000 replicates [33,34].

Specific conductance (conductivity) is a measure of the electrical current of a solution. The greater the salinity, greater is the conductivity. Salinity of salt pans varied from 3-4% (non-salt production phase) to 30% (salt production phase) and pH varied from 7.0-8.0. To determine the nature of elements contributing to the salinity, total chemical analysis of the brine water revealed that Na^+ and Cl^- are the dominant ions. This indicates the thalassohaline nature of the brine. The main cations were Na^+ (747 g/l), Ca^{2+} (121 g/l), K^+ (80 g/l) and Mg^{2+} (171 g/l) and the main anion was Cl^- (611 g/l).

Plating of sediments and/or water samples on NTYE agar plates, resulted in 30 cream/yellow/white and 14 orange/red colored colonies during the salt dilute phase and about 60-920 cream/yellow/white and 70 - 110 orange/red colonies during salt harvesting phase, on incubation of the plates for 20-30 days at room temperature (30°C). On NT medium during pre-salt harvesting phase matt growth of cream/yellow/white and pale pink colonies were obtained. During salt harvesting phase about 20 cream colonies appeared within 24 h of incubation. On further incubation of NT media plates for 8-20 days, light orange, orange red to brick red pigmented colonies appeared. Some of the orange-red colonies growing on NTYE and NT plates were accompanied by white/cream colonies, which persisted even after several subcultures. However these white/cream colonies were eliminated when streaked on NTYE and NT plates containing 50 μg/ml of ampicillin, indicating their bacterial origin (Additional file 1: Figure S2, supplementary data).

Since carotenoid or bacteriruberin pigment is one of the characteristic features of haloarchaea, orange-red colored colonies were selected on basis of their difference in colony morphology and pigmentation.

During the pre-salt harvesting phase, orange-red colonies appeared after 20 -30 days of incubation. These colonies appeared as a uniform pure culture in all the water/sediment samples plated. However based on visual differentiation seven different types of orange-red isolates were picked from NTYE medium and were designated as BK3, BK6, BK7, BK11, BK18, BK19 and BK20. The isolates were maintained on NTYE agar slopes.

During the salt harvesting phase, fifteen visually different pigmented isolates were obtained on NT agar medium and were designated as BS1, BS2, BS3, BS4, BS5, BS6, BS7, BS8, BS11, BS13, BS15, BS16, BS17, BS19 and BS20. Two strains BBK1and BBK2, orange/red in color, were isolated from NTYE agar medium (Table 1).

All the strains stained uniformly Gram negative. All of the BK and BBK series cultures appeared as cocci either as singles, pairs, chains or groups. Cultures in BS series (BS1, BS2, BS3, BS5, BS6, BS7 BS13, BS15 and BS20) appeared as tiny cocci whereas cultures BS4, BS11, BS17 and BS19 appeared as short rods. The coccoid morphology of the cultures was further confirmed by scanning electron microscopy. Most of the coccoid isolates appeared as single cells or diplococci with exception of BK19 which exhibited classical *Sarcina* like packets. The isolates BBK2 and BS16 exhibited unique morphology which appeared as flattened involuted discs (Figure 2).

The orange-red acetone or chloroform: methanol extract of the pigment showed similar spectrophotometric profile with absorption maxima at 393, 474, 501 and 535 nm. These peaks correspond to bacteriruberin pigments which are typical pigments of haloarchaea [35,36].

The BLAST analysis of the 16S rRNA gene fragments of the six isolates BK3, BK6, BK7, BK11, BK18 and BK19 obtained from the pre-salt harvesting phase showed 98-99% similarity to *Halococcus salifodinae* and *Halococcus saccharolyticus* (Figure 3). Among isolates from the salt

harvesting phase, only BBK1 (AB588755) was very close to *Halococcus sp.* with 98-99% similarity whereas cultures BS1, BS2, BS3, BS5, BS6 and BS7 were assigned to the genera *Haloarcula* with similarity of 98-99%. BBK2 and BS16 showed highest similarity to *Haloferax alexandrinus* of about 97-98% and the 16S rRNA sequence of BS17 and BS19 was closely related to the genera *Halorubrum* with 98% similarity.

As noted earlier, orange – red colonies from pre-salt harvesting phase were almost uniform and appeared to be in pure form. All these isolates showed 99% similarity to the genera *Halococcus*. It was of interest to see if there were any differences in these strains. Hence the total polar lipids were studied. The lipid profile obtained showed 2 spots of phosphotidylglycerol (PG), methyl ester of phosphatidyl glycerophosphate (PGP-Me) corresponding to R_f values 0.828 and 0.517 respectively which are the signature polar lipids of haloarchaea. The glycolipid, sulfated diglycosyl diether lipids (S-DGD) having an R_f of 0.368 was seen in all the strains of BK and BBK series which is typical of genus *Halococcus* and *Haloferax* [37]. No differentiation at the strain level could be done based on the lipid profile. Most likely it was the same strain that had been isolated multiple times, due to its abundance in the environment.

The isolates obtained from pre-salt harvesting phase (BK series) were resistant to lysis in distilled water and 3.5% NaCl. However isolates obtained from salt harvesting phase showed varied lysis on suspension in distilled water with the exception of BBK1 which was resistant to lysis. The isolate BBK2 lysed immediately, where as BS4 and BS5 were observed to have delayed lysis. Viability assay was performed for BK6 obtained during pre-salt harvesting phase and BBK2 obtained during salt harvesting phase. On suspension of cells in distilled water, 3.5% NaCl and 30% NaCl, followed by plating revealed that isolate BK6 was viable even after 10 days on suspension in distilled water. Cells of BK6 also retained their coccoid shape, when observed microscopically. The isolate BBK2 survived for up to 24 h in 3.5% NaCl, but lysed immediately in distilled water showing no growth on plating.

Interestingly the isolates of the BK series failed to lyse in acetone for extraction of pigment even on sonication but lysed on suspension in Chloroform : methanol (2:1 v/v).

It is interesting to see that Halococci were recovered almost as a pure culture in culturable form during the salt dilute phase (whereas other culturable genera could be recovered during the salt concentrated phase). Recent studies have shown that haloarchaea are being isolated from less hypersaline environments. Salt-marsh sediments, sulfur-rich spring and deteriorated ancient wall paintings has been investigated and found to have haloarchaeal members, predominantly belonging to the genera *Halococcus*, *Haloferax* and *Halogeometricum* [38-41]. A

Table 1 Halophilic archaeal isolates obtained from Ribandar solar salterns of Goa, India

Saltern phase and econiche	Isolates	Pigmentation	Gram character and morphology	Lysis in		Identification	Accession No.
				Distilled water	3.5% salt solution		
Pre salt harvesting phase/Initial Stage/Salt Dilute Stage							
Water samples	BK3	Bright Orange-red	Gram negative cocci	-	-	*Halococcus salifodinae*	HQ455793
	BK6	Bright Orange-red	Gram negative cocci	-	-	*Halococcus salifodinae*	AB588757
	BK7	Bright Orange-red	Gram negative cocci	-	-	*Halococcus salifodinae*	HQ455794
	BK11	Bright Orange-red	Gram negative cocci	-	-	*Halococcus salifodinae*	HQ455795
Sediment samples	BK18	Orange	Gram negative cocci	-	-	*Halococcus salifodinae*	HQ455796
	BK19	Bright Orange-red	Gram negative cocci	-	-	*Halococcus salifodinae*	AB588758
	BK20	Light Orange	Gram negative cocci	-	-	Not sequenced	Not sequenced
Salt harvesting phase/Crystallization Stage							
Brine samples	BBK1	Orange	Gram negative cocci	-	-	*Halococcus salifodinae*	AB588755
	BBK2	Orange	Gram negative cocci	+	+/-	*Haloferax volcanii*	AB588756
	BS1	Bright red	Gram negative cocci	+	+/-	*Haloarcula argentinensis*	HQ455797
	BS2	Bright red	Gram negative cocci	+	+/-	*Haloarcula japonica*	HQ455798
	BS3	Bright red	Gram negative cocci	+	+/-	*Haloarcula sp.*	HQ455799
	BS5	Bright red	Gram negative pleomorphic	+	+/-	*Haloarcula argentinensis*	AB588759
	BS6	Bright red	Gram negative short rods	+	+/-	*Haloarcula hispanica*	HQ455801
	BS7	Bright red	Gram negative pleomorphic	+	+/-	*Haloarcula japonica*	HQ455802
Sediment samples	BS17	Bright red	Gram negative cocci	+	+/-	*Halorubrum sp.*	ND
	BS16	Light Orange	Gram negative cocci	+	+/-	*Haloferax alexandrinus*	HQ455803

"-" No Lysis; "+" Lysis; "ND" Not Deposited.

study by Fukushima et al. (2007), showed that the cells of *Halococcus* survived in sea water (salinity of which is 3.5%) upto 9 days without losing its cell rigidity. It is also possible that haloarchaea are trapped in the salt crystals and get deposited in the sediments. The salinity of sediments obtained at about 10 cm was 6-10%. Therefore it is quite possible that even though the haloarchaeal members cannot flourish they can still retain their viability. Another possibility is presence of clay in these salterns. These clay particles have micropores on which the salt fluid gets filled along with the haloarchaeal members. These micropores could serve as a salt rich environment for the survival of haloarchaea [42-44]. Diversity studies of hypersaline areas around the world have indicated that *Halorubrum*

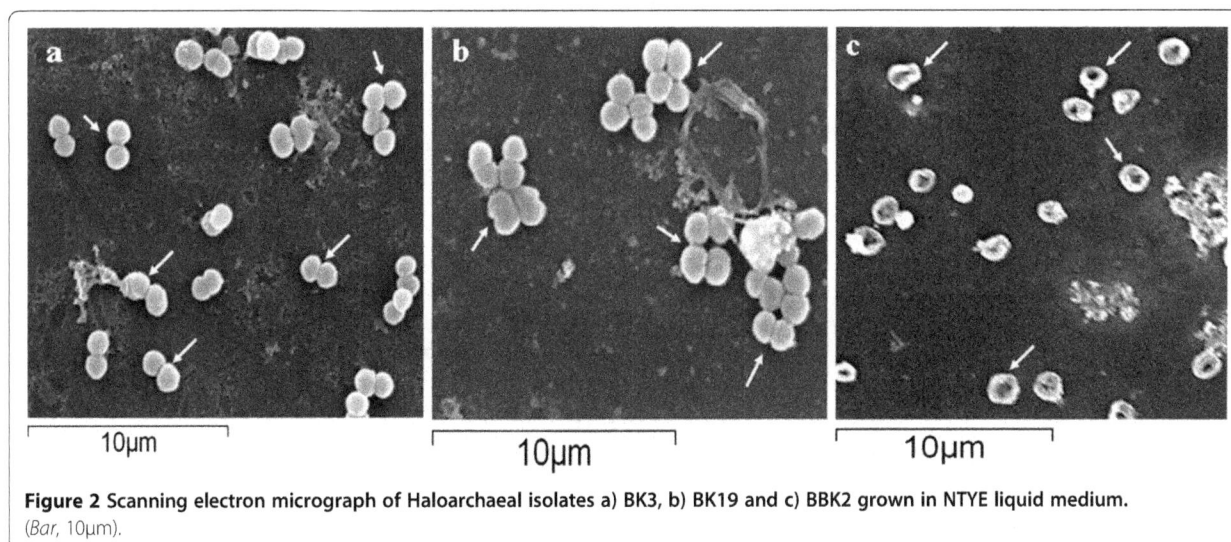

Figure 2 Scanning electron micrograph of Haloarchaeal isolates a) BK3, b) BK19 and c) BBK2 grown in NTYE liquid medium. (*Bar*, 10μm).

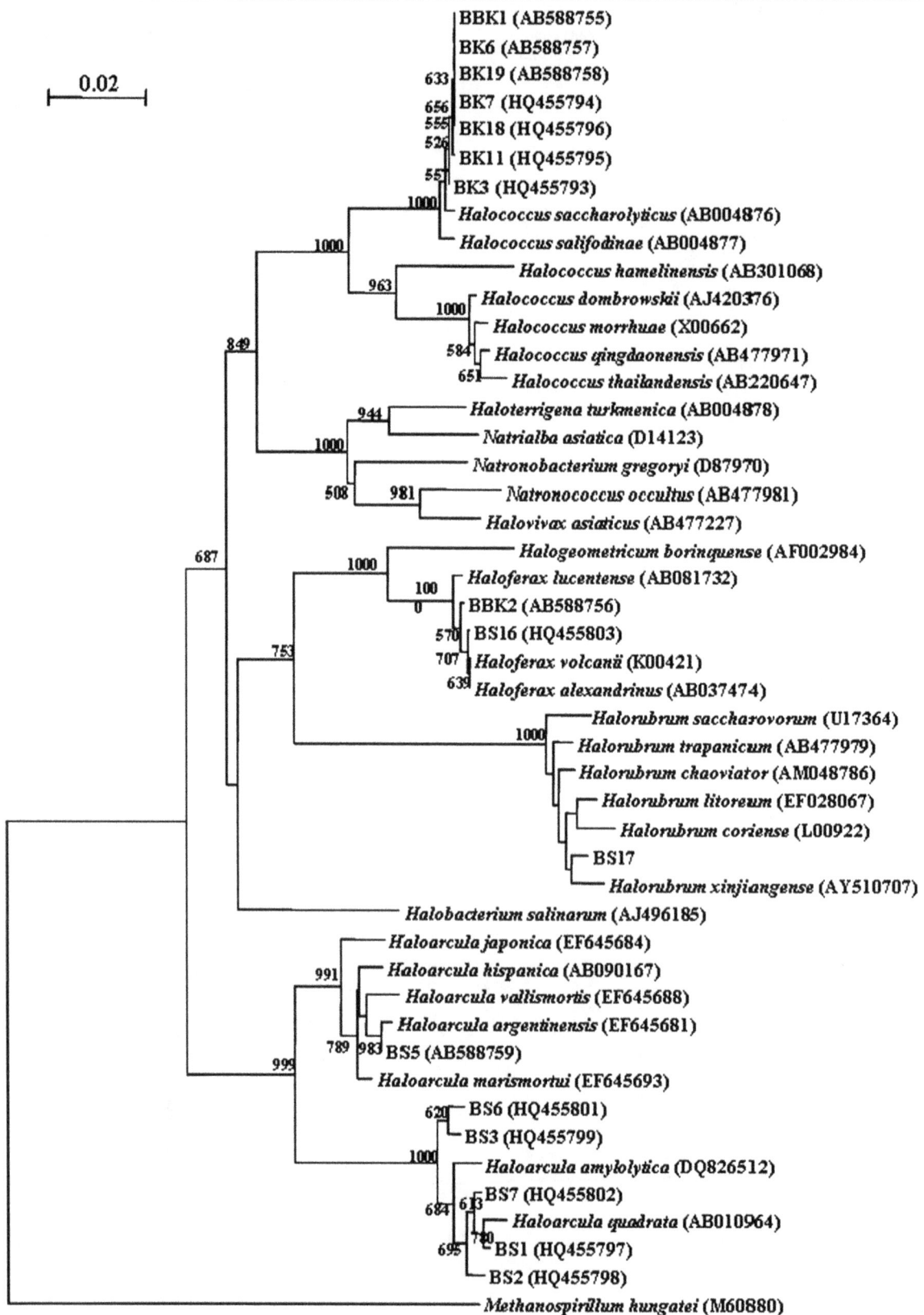

Figure 3 Phylogenetic tree showing the positions of isolated haloarchaeal strains constructed with MEGA 5.0 by using NJ method. *Methanospirillum hungatei* was used as an outgroup.

and *Haloarcula* are the dominant culturable members of haloarchaea [45-50]. This investigation revealed that haloarchaeal members belonging to the genus *Halococcus,* thrive in less saline environments and are the dominant culturable haloarchaea during the pre-salt harvesting phase or the salt dilute phase.

The salt pan under study is transient and operated only during the post monsoon season of January to May. During monsoons, the saltpans are inundated with sea water as well as rain water. Most of haloarchaeal strains are known to lyse in distilled water as they require at least 10% NaCl to maintain the integrity of their outer membrane. It is interesting to note that Halococcal microorganisms were most abundant, during the pre-salt harvesting phase as they are resistant to lysis in lowering salinities than their counterparts. However, as salinity gradually increases, other members of haloarchaeal community start to colonize the saltpans. Of the thirty six defined genera in the family *Halobacteriaceae,* four different genera namely, *Halococcus, Haloferax, Haloarcula* and *Halorubrum* were found to be represented in this study. This investigation provides valuable information about the change in culturable haloarchaeal diversity under variable salt conditions.

Nucleotide sequence data can be accessed from DDBJ and NCBI database under the accession numbers HQ455793 - HQ455803 and AB588755 - AB588759.

Additional file

Additional file 1: Figure S1. Solar Salt production at the Ribandar salterns a) bed preparative stage showing series of rectangular beds (January) b) Sluice gate / inlet point for entry of saline water c-d) rectangular beds inundated with saline water (February - March) e) tool used for extracting salt f) crude salt heaped up at the corners of the *bandhs* g) collection of brine sample h) crude salt collected and piled up on the sides i) collection of sediment sample. Figure S2. Media plates (NTYE and NT) showing diversity of halophilic microorganisms obtained during initial stage and crystallization stages of salt production. Interested bright orange culture obtained on media containing ampicillin.

Competing interests
The authors declare that they have no competing interests.

Authors' contributions
JB conceived the idea and designed the study. KM, BBS and JB conducted the field operations. KM carried out the laboratory work of physico-chemical analysis of the samples, molecular characterisation of the isolates and drafted the manuscript. BBS isolated the haloarchaeal strains and performed pigment characterisation and cell viability assays. KM and BBS carried out the SEM studies. JB edited and revised the manuscript. All authors have read and approved the final manuscript.

Acknowledgement
The authors thank Dr Hiroaki Minegishi, Bio-Nano Electronics Research Center, Toyo University, Japan for identifying the haloarchaeal isolates. This work was supported by University Grants Commission, India (UGC) Major Research Project No: 34-500/2008(SR).

References

1. Javor BJ: Industrial microbiology of solar salt production. *J Ind Microbiol Biotechnol* 2002, **28**:42–47.
2. Hoffman BA, Dawes CJ: Vegetational and Abiotic Analysis of the Salterns of Mangals and Salt Marshes of the West Coast of Florida. *J Coastal Res* 1997, **13**:147–154.
3. Rodrigues CM, Bio A, Amat F, Vieira F: Artisanal salt production in Aveiro/ Portugal - an ecofriendly process. *Saline Syst* 2011, **7**:3.
4. Rocha RM, Costa DFS, Lucena-Filho MA, Bezerra RM, Medeiros DHM, Azevedo-Silva AM, Araújo CN, Lauro Xavier-Filho L: Brazilian solar saltworks - ancient uses and future possibilities. *Aqua Biosyst* 2012, **8**:8.
5. Bardavid RE, Khristo P, Oren A: Interrelationships between *Dunaliella* and halophilic prokaryotes in saltern crystallizer ponds. *Extremophiles* 2008, **12**:5–14.
6. Oren A: The ecology of the extremely halophilic archaea. *FEMS Microbiol Rev* 1994, **13**:415–439.
7. Oren A: Microbial life at high salt concentrations: phylogenetic and metabolic diversity. *Saline Syst* 2008, **4**:2.
8. Grant WD, Kamekura M, McGenity TJ, Ventosa A: Order I. *Halobacteriales* Grant and Larsen 1989b, 495[VP]. In *Bergey's manual of systematic bacteriology.* 2nd edition. Edited by Boone DR, Castenholz RW, Garrity GM. New York: Springer; 2001:294–301.
9. Makhdoumi-Kakhki A, Amoozegar MA, Bagheri M, Ramezani M, Ventosa A: *Haloarchaeobius iranensis* gen. nov., sp. nov., an extremely halophilic archaeon isolated from the saline lake Aran-Bidgol, Iran. *Int J Syst Evol Microbiol* 2011, **62**:1021–1026.
10. Oren A: Taxonomy of the family *Halobacteriaceae*: a paradigm for changing concepts in prokaryote systematics. *Int J Syst Evol Microbiol* 2012, **62**:263–271.
11. Ahmad N, Sharma S, Khan FG, Kumar R, Johri S, Abdin MZ, Qazi GN: Phylogenetic analyses of Archaeal Ribosomal DNA sequences from salt pan sediment of Mumbai, India. *Curr Microbiol* 2008, **57**:145–152.
12. Manikandan M, Kannan V, Pasic L: Diversity of microorganisms in solar salterns of Tamil Nadu, India. *World J Microbiol Biotechnol* 2009, **25**:1001–1017.
13. Munson MA, Nedwell DB, Embley TM: Phylogenetic diversity of Archaea in sediment samples from a coastal salt marsh. *Appl Environ Microbiol* 1997, **63**:4729–4733.
14. Ochsenreiter T, Pfeiffer F, Schleper C: Diversity of Archaea in hypersaline environments characterized by molecular-phylogenetic and cultivation studies. *Extremophiles* 2002, **6**:267–274.
15. Oh D, Porter K, Russ B, Burns D, Dyall-Smith M: Diversity of *Haloquadratum* and other haloarchaea in three, geographically distant, Australian saltern crystallizer ponds. *Extremophiles* 2010, **14**:161–169.
16. Zafrilla B, Martinez-Espinosa RM, Alonso MA, Bonete MJ: Biodiversity of Archaea and floral of two inland saltern ecosystems in the Alto Vinalopó Valley, Spain. *Saline Syst* 2010, **6**:10.
17. Minegishi H, Mizuki T, Echigo A, Fukushima T, Kamekura M, Usami R: Acidophilic haloarchaeal strains are isolated from various solar salts. *Saline Syst* 2008, **4**:16.
18. Minegishi H, Echigo A, Nagaoka S, Kamekura M, Usami R: *Halarchaeum acidiphilum* gen. nov., sp. nov., a moderately acidophilic haloarchaeon isolated from commercial solar salt. *Int J Syst Evol Microbiol* 2010, **60**:2513–2516.
19. Roh SW, Bae JW: *Halorubrum cibi* sp. nov., an extremely halophilic archaeon from salt-fermented seafood. *J Microbiol* 2009, **47**:162–166.
20. DasSarma S, DasSarma P: *Halophiles, Encyclopedia of life Sciences.* London: Wiley; 2006.
21. DasSarma P, Coker JA, Huse V, DasSarma S: Halophiles, industrial applications. In *Encyclopedia of industrial biotechnology: bioprocess, bioseparation and cell technology.* Edited by Flickinger MC. Hoboken, NJ: John Wiley & Sons, Inc; 2010.
22. McGenity TJ, Gemmell RT, Grant WD, Stan-Lotter H: Origins of halophilic microorganisms in ancient salt deposits. *Environ Microbiol* 2000, **2**:243–250.
23. Mormile MR, Biesen MA, Gutierrez MC, Ventosa A, Pavlovich JB, Onstott TC, Fredrickson JK: Isolation of *Halobacterium salinarum* retrieved directly from halite brine inclusions. *Environ Microbiol* 2003, **5**:1094–1102.
24. Williams WD: Conductivity and salinity of Australian salt lakes. *Aust J Mar Fresh Res* 1986, **37**:177–182.
25. Braganca JM, Furtado I: Isolation and characterization of Haloarchaea from low-salinity coastal sediments and waters of Goa. *Curr Sci* 2009, **96**:1182–1184.

26. Elevi R, Assa P, Birbir M, Ogan A, Oren A: **Characterization of extremely halophilic Archaea isolated from the Ayvalik Saltern, Turkey.** *World J Microbiol Biotechnol* 2004, **20:**719–725.

27. Salgaonkar BB, Kabilan M, Braganca JM: **Sensitivity of Haloarchaea to eubacterial pigments produced by *Pseudomonas aeruginosa* SB1.** *World J Microbiol Biotechnol* 2011, **27:**799–804.

28. Raghavan TM, Furtado I: **Expression of carotenoid pigments of haloarchaeal cultures exposed to aniline.** *Environ Toxicol* 2005, **20:**165–169.

29. Litchfield CD, Irby A, Kis-Papo T, Oren A: **Comparisons of the polar lipid profiles of two solar salterns located in Newark, California, U.S.A., and Eilat, Israel.** *Extremophiles* 2000, **4:**259–265.

30. Oren A, Duker S, Ritter S: **The polar lipid composition of Walsby's square bacterium.** *FEMS Microbiol Lett* 1996, **138:**135–140.

31. Fukushima TT, Usami R, Kamekura M: **A traditional Japanese-style salt field is a niche for haloarchaeal strains that can survive in 0.5% salt solution.** *Saline Syst* 2007, **3:**2.

32. Birbir M, Calli B, Mertoglu B, Bardavid RE, Oren A, Ogmen MN, Ogen A: **Extremely halophilic Archaea from Tuz Lake, Turkey and the adjacent Kaldirim and Kayacik salterns.** *World J Microbiol Biotechnol* 2007, **23:**309–316.

33. Tamura K, Peterson D, Peterson N, Stecher G, Nei M, Kumar S: **MEGA5: Molecular Evolutionary Genetics Analysis using maximum likelihood, evolutionary distance, and maximum parsimony methods.** *Mol Biol Evol* 2011, **28:**2731–2739.

34. Wright AG: **Phylogenetic relationships within the order Halobacteriales inferred from 16S rRNA gene sequences.** *Int J Syst Evol Microbiol* 2006, **56:**1223–1227.

35. Stan-Lotter H, Pfaffenhuemer M, Legat A, Busse HJ, Radax C, Gruber C: *Halococcus dombrowskii* **sp. nov., an archaeal isolate from a Permian alpine salt deposit.** *Int J Syst Evol Microbiol* 2002, **52:**1807–1814.

36. Wang QF, Yang WLH, Liu YL, Cao HH, Pfaffenhuemer M, Stan-Lotter H, Guo GQ: *Halococcus qingdaonensis* **sp. nov., a halophilic archaeon isolated from a crude sea-salt sample.** *Int J Syst Evol Microbiol* 2007, **57:**600–604.

37. Oren A, Arahal DR, Ventosa A: **Emended descriptions of genera of the family *Halobacteriaceae*.** *Int J Syst Evol Microbiol* 2009, **59:**637–642.

38. Elshahed MS, Najar FZ, Roe BA, Oren A, Dewers TA, Krumholz LR: **Survey of archaeal diversity reveals an abundance of Halophilic Archaea in a low-salt, sulfide- and sulfur-rich spring.** *Appl Environ Microbiol* 2004, **70:**2230–2239.

39. Pinar G, Saiz-Jimenez C, Schabereiter-Gunter C, Blanco-Valera TM, Lubitz W, Rolleke S: **Archaeal communities in two disparate deteriorated ancient wall paintings: detection, identification, and temporal monitoring by denaturing gradient gel electrophoresis.** *FEMS Microbiol Ecol* 2001, **37:**45–54.

40. Purdy KJ, Cresswell-Maynard TD, Nedwell DB, McGenity TJ, Grant WD, Timmis KN, Embley TM: **Isolation of haloarchaea that grow at low salinities.** *Environ Microbiol* 2004, **6:**591–595.

41. Rölleke S, Witte A, Wanner G, Lubitz W: **Medieval wall painting– a habitat for archaea: Identification of archaea by denaturing gradient gel electrophoresis (DGGE) of PCR amplified gene fragments coding for 16S rRNA in a medieval wall painting.** *Int Biodeter Biodegr* 1998, **41:**85–92.

42. Benlloch S, López-López A, Casamayor EO, Øvreås L, Goddard V, Daae FL, Smerdon G, Massana R, Joint I, Thingstad F, Pedrós Alió C, Rodríguez-Valera F: **Prokaryotic genetic diversity throughout the salinity gradient of a coastal solar saltern.** *Environ Microb* 2002, **4:**349–360.

43. Jiang H, Dong H, Yu B, Liu X, Li Y, Ji S, Zhang CL: **Microbial response to salinity change in Lake Chaka, a hypersaline lake on Tibetan plateau.** *Environ Microbiol* 2007, **9:**2603–2621.

44. Norton CF, Grant WD: **Survival of halobacteria within fluid inclusions in salt crystals.** *J Gen Microbiol* 1988, **134:**1365–1373.

45. Özcan B, Özcengiz G, Colleri A, Cokmus C: **Diversity of halophilic Archaea from six hypersaline environments in Turkey.** *J Microbiol Biotech* 2007, **17:**985–992.

46. Pan HL, Zhou C, Wang HL, Xue YF, Ma YH: **Diversity of halophilic archaea in hypersaline lakes of Inner Mongolia, China.** *Acta Microbiol Sin* 2006, **46:**1–6.

47. Benlloch S, Acinas SG, Martínez-Murcia AJ, Rodríguez-Valera F: **Description of prokaryotic biodiversity along the salinity gradient of a multipond solar saltern by direct PCR amplification of 16S rDNA.** *Hydrobiologia* 1996, **329:**19–31.

48. Antón J, Roselló-Mora R, Rodríguez-Valera F, Amann RI: **Extremely halophilic Bacteria in crystallizer ponds from solar salterns.** *Appl Environ Microbiol* 2000, **66:**3052–3057.

49. Munson MA, Nedwell DB, Embley TM: **Phylogenetic diversity of archaea in sediment samples from a coastal salt marsh.** *Appl Envion Microbiol* 1997, **63:**4729–4733.

50. Burns DG, Camakaris HM, Janssen PH, Dyall-Smith ML: **Combined use of cultivation-dependent and cultivation-independent methods indicates that members of most haloarchaeal groups in an Australian Crystallizer pond are cultivable.** *Appl Environ Microbiol* 2004, **70:**5258–5265.

Brazilian solar saltworks - ancient uses and future possibilities

Renato De Medeiros Rocha[1], Diógenes FS Costa[1,2*], Milton A Lucena-Filho[1], Rodolfo M Bezerra[1], David HM Medeiros[1], Antonio M Azevedo-Silva[1], Cristian N Araújo[1] and Lauro Xavier-Filho[3]

Abstract

Coastal solar saltworks of Brazil are exploited for sea salt, which becomes progressively concentrated by evaporation. This study aimed to review the current and new potential uses of these systems, in order to provide more dynamic for this activity. The first evaporation ponds are also used for artisanal fisheries, ensuring the livelihood of many families. All the brine rich in secondary salts (bittern) can be widely used by the chemical industry, while the Brazil shows an incipient production of "flower of salt", a salt with distinct characteristics with higher market value than sodium chloride. On the other hand, the saltponds have a high potential for management and obtaining of large populations of *Artemia* spp., purifying the brine through the action as biological filter. This microcrustacean occurs naturally in intermediate salinity ponds, being commonly used in aquaculture. Species of microalgae and halobacteria found in the saltworks are employed for extraction of beta-carotene and glycerol, used in an extensive list of products with high commercial value. These ecosystems represent refuge zones for many species of migratory birds, becoming imperative to promote the conservation of these hypersaline wetlands.

Keywords: Wetland, Salt production, Management, Brazil, Conservation

Background

Coastal solar saltworks are anthropogenic supratidal habitats exploited for sea salt, which becomes progressively concentrated by evaporation [1]. This multi-pond ecosystem provides a range of environments with different salinity levels, from that of seawater up to sodium chloride saturation and sometimes even beyond [2]. As water evaporates and salinity increases, water is pumped or fed by gravity to the next pond, so that the salinity in each particular pond is kept within narrow limits, essentially constant [3].

Brazilian largest saltworks are located in coastlines, more specifically on the shores of estuaries in the northern coast of Rio Grande do Norte State. The establishment of the saltworks in the coastal zones of Rio Grande do Norte occurred with the use of salt plains of rivers. The saltwork ponds located in this stretch of the Brazilian coast account for 95% of sea salt produced in the country and exported, directly influencing the local and regional economies by creating jobs and payment of taxes [4].

Considering the form of exploration and harvesting of salt, we can classify the Brazilian solar saltwork ponds into two categories: craft (manual) and mechanized. The craft harvest is small, with an average area of 2-50 ha, divided into 10-20 tanks (evaporators and crystallizers), with manual harvesting of salt and a production of approximately 200-20,000 ton.year^{-1}. The short retention time of the brine in each evaporator prevents the full development of a stable ecosystem [1-3,5]. The result is a salt regarded as second quality by chemical, especially its content of organic and inorganic impurities, and lead tiny crystal and brittle [5,6]. On the other hand, a typical mechanized saltwork pond usually produces over 150,000 ton.year^{-1}, with a production area over 500 ha. The retention time of the brines is longer and the sea water abstracted by the saltwork arises as concentrated brine in the area of crystallization approximately three months later. In this way, this paper shows a brief review, whereby besides sodium chloride production, this system

* Correspondence: diogenesgeo@yahoo.com.br
[1]Departamento de Geografia, Universidade Federal do Rio Grande do Norte, Campus de Caicó, Joaquim Gregório, s/n, Penedo 59.300-000, Caicó-RN, Brasil
Full list of author information is available at the end of the article

(solar saltwork) has potential for many other economic activities, being possible to associate the salt production with other multiple potential uses of the salt ponds.

Artisanal fisheries

The consumption of fish and derivatives has been documented as having beneficial effects on human health due to the presence of omega-3 polyunsaturated fatty acid (PUFA) [7]. For the poor populations from coastal regions of the semiarid which suffer from nutritional deficiency, artisanal fisheries are one of the only sources of supply and obtaining income. In another sphere, associated with the entrance of fish in the saltworks by means of the pumping stations, the artisanal fishing in the evaporators of the saltworks is an activity that has been developed since the construction period. Several families who live near the saltworks depend directly on this activity for survival. For this, they also control the populations of fish inserted in the food chain of the saltworks ecosystems in Brazil [4].

This activity remarkably shows the direct relationship between the semi-artificial ecosystems of the saltwork [2,5] with the local fishermen. Fishing is also done in the estuary and tidal channels but, the catch of most species occurs faster and in greater quantity in the initial evaporators of the saltworks. This easiness concerns the fact that several species of fish, mollusk and crustacean are caught at juvenile stages in the pumping stations of the estuarine water, pulled into the saltworks by powerful pumps that abstract water into for these initial evaporators [1,8]. Most of these species does not survive to gradual increase of salinity along production circuit, perishing soon after the transference of water in the initial evaporators, when the salinity begins to reach 10-12° Bé. Fishermen act as important controlling elements of the ecosystem, once not caught, the dead biota are deposited in the margins of the evaporators, causing strong odor and increasing considerably the organic matter (and consequent eutrophication) inside the saltwork [4].

Extraction of secondary salts from "Mother Waters" ("Bitterns") used for chemical industry

The sequence of salts deposited by the evaporation of sea water is in accordance to the solubility of its several compounds. Thus, the precipitation of salts includes the less soluble compounds in the base to the more soluble at the top of the sequence, in the following order: limestone ($CaCO_3$), gypsum ($CaSO_4$), halite (NaCl), potassium salts sylvinite (NaCl-KCl system), and magnesium salts (bischofite - $MgCl_2 \cdot 6H_2O$); it is also considered the presence of other compounds, according to physical and chemical variations of the brine during the various stages of evaporation [9].

NaCl crystal (halite) is formed when the total salt concentration reaches value above 300 gL^{-1}. After most of the NaCl precipitates to the bottom of the crystallizer ponds, the remaining concentrated brine (the "bitterns") contains mainly Mg^{2+}, K^+, Cl^- and SO_4^{2-} (Oren, 2002). The bittern remaining after the crystallization of halite is nutrient-rich, but apparently devoid of life, as no organisms tolerate the extremely high Mg^{2+} concentration [10].

When all calcium carbonate, calcium sulfate, and 83% of the halite is crystallized from seawater by solar concentration, a bittern of a specific gravity of about 1.26 is obtained. This bittern with few exceptions is placed back to the sea. In some cases, as in Spain, is used for recovering some epsomite, bischofite, and bromine, but not for the production of potassium salts. On further evaporation, a complex mixture of halite, sylvite, and double salts of potassium, sodium, and magnesium start to crystallize; the recovery of marketable products becomes difficult and inefficient. However, in the absence or near absence of sulfate, the bittern may be readily processed to recover high-purity sylvite and bischofite with excellent efficiency [11].

The partially desulfated seawater bittern obtained from the epsomite plant is readily amenable for recovering sylvite and magnesium chloride hexahydrate by a combination of solar evaporation and fractional crystallization. Despite the very complex chemical phase of seawater bittern, a simple crystallization method may be employed for the efficient recovery of high-purity epsomite and sylvite [11].

Magnesium is profusely present in seawater evaporites as chloride (9.44%), sulfate (6.5%) and bromide (0.22%). The raw material of the magnesium industry is, however, magnesium hydroxide. This is then treated with hydrochloric acid to obtain magnesium chloride. The potential value of magnesium chloride as raw material is established, but involves separation of different salts to obtain magnesium chloride in a relatively pure form. Magnesium chloride occurs in the nature as bischofite ($MgCl2 \cdot 61120$) and as carnallite ($KCI \cdot MgC12 \cdot 6H20$), both from oceanic origin [12].

The production of crystalline magnesium chloride hexahydrate by solar evaporation of low-sulfate-containing inland bittern has yielded a product suitable for electrolytic production of magnesium metal. Using the sea bittern for the production of such crystalline magnesium chloride hexahydrate was not attempted, probably due to the high value of sulfate content of about 3.5 per cent at sp. gr. 1.350. In the arid and semiarid tropical regions, solar evaporation of sea bittern reaches the equilibrium density of sp. gr. 1.377, and at equilibrium relative humidity of 32% [13]. This clearly approaches the equilibrium of the pure system of magnesium chloride hexahydrate and water enabling to take advantage of solar evaporation in the process [12].

"Flower of salt" production

"Flower of salt" is a thin layer that forms on the surface of the salt tide, during the continuous evaporation. The salt does not suffer any transformation, besides the natural drying in the sun, which eliminates the rose tone. The flower of salt contains all 84 trace elements and micronutrients found in the sea, being a natural source of potassium, calcium, copper, zinc and magnesium [14]. An adequate level of this salt is very important for the body functioning, and is highly requested by the market of international gastronomy, replacing the refined salt [15].

This mineral product has extremely white color, with rigid crystalline structure and high moisture content. Although apparently made up by small crystals of salt, actually this form of halite has a structure organized into microcrystal clusters.

In relation to the production, this mineral is formed on the brine surface, only in a thin layer of salt crystals, which are harvested daily and dried in the sun. The harvesting is made daily on the hottest days in traditional saltworks [14]. The flower of salt is packed with no other processing, unlike what happens to with the sea salt for consumption that undergoes a process of washing, centrifugation and drying by the heat of combustion, ground and sieved.

The importance given to this product concerns the area required for production. While large saltwork companies need several kilometers for installations, operated by business groups, the flower of salt can be obtained in ponds with total area smaller than 01 hectare. Another fact concerns the production: since it is a handmade product, these small salterns can be operated by familiar groups, becoming a new income source for populations living in hypersaline areas of the country, or even to innovate small artisanal salterns still remaining.

Mass culture of artemia for aquaculture

With the development of fish and shellfish hatchery aquaculture, the use of the brine shrimp *Artemia* as a diet for larval culture of many species has become widespread due to convenience of use and high nutritional value [16]. Dormant cysts of *Artemia* are available year-round in large quantities along the shorelines of hypersaline lakes, coastal lagoons and solar saltworks spread over the five continents [7].

After harvesting and processing, cysts are made available in cans and stored 'on demand' live food. However, the expansion of aquaculture production made the demand for *Artemia* cysts now exceeds the supply. Prices have risen exponentially, turning *Artemia* into a bottleneck for the expansion of the hatchery aquaculture of marine fish and crustacean. In particular, many

developing countries can barely afford to import the very expensive cysts [7].

The use of a device such a Solar Pond (a green and renewable energy source) would save energy and time, in speeding up both *Artemia* cysts hatching time and *Artemia* nauplii development. Since the newly hatched *Artemia* nauplii are attracted to light, they easily concentrate in one area allowing the harvesting by shining a flashlight at the exit of the bioreactor.

Microalgae and halobacteria cultures for extraction of beta-carotenes and glycerol

The halophilic unicellular green algae *Dunaliella* is grown worldwide as a source of valuable chemicals. The most important product is β-carotene but other uses have been explored as well, including the production of glycerol and the pyrolysis of *Dunaliella* biomass for the production of oil [17].

The positive effect of the presence of dense communities of red halophilic Archaea in saltern crystallizer ponds has been recognized for a long time. The red coloration that develops in these ponds is mainly caused by Archaea, but strains of *Dunaliella*, and possibly even red halophilic bacteria of the genus *Salinibacter* contribute as well toward the absorption of light energy. By trapping solar radiation these microorganisms raise the temperature of the brine and the rate of evaporation, thereby increasing salt production [2,10].

In order to improve salt production in salterns without enough dense archaeal community, the fertilization with organic nutrient has been suggested [2]. C50 bacterioruberin derivatives are the main carotenoids of the Halobacteriaceae. However, additional carotenoids may be present that have proven economic value. An isolate from a seawater evaporation pond near Alexandria, Egypt, produces considerable amounts of the *ketocarotenoid canthaxanthin* [18]. Exploitation of this organism for commercial canthaxanthin production has already been suggested [18].

Dunaliella protein has a similar composition to soybean meal, but with higher lysine content [19]. It is therefore suitable for use as feedstock in mariculture (crab, shrimp, shellfish) and for livestock such as chickens. As a result of the absence of cell walls, the cells are digestible [17]. Catalytic pyrolysis of *Dunaliella* cell material at 200-240°C produces an oil-like substance. The overall process is exothermic, and thus most of the thermal energy required to initiate the reaction may be regained. A conversion of 22.3% of the algal protein was obtained at 350°C to a product that contains 69.9% carbon, 7.7% hydrogen, and 7.3% nitrogen. Addition of KCl, $MgCl_2$ and $MnCl_2$ increased the yield to 27% with 75.5% carbon, 8.5% hydrogen, and 6.8% nitrogen [20]. At an estimated price of about $ 40 per barrel [21] this process is not economically feasible at present.

Dunaliella is also being used as an additive in cosmetic anti-wrinkle skin creams in combination with Dead Sea minerals [22]. The algal cell preparation allegedly binds Ca^{2+} and Mg^{2+} ions. However, the authors stated that "the low biosorption of calcium and magnesium obtained from the algal biomass, and the tendency to a low release of minerals at the normal pH of human skin (5.5) led to the conclusion that the advantage of these algae as a mineral vehicle for Ca and Mg is limited [17]."

The role of biological system on maintenance of brine quality

The study on phototrophic communities inhabiting salterns is not only of purely scientific interest: the benthic cyanobacterial mats that develop in saltern ponds of intermediate salinity effectively seal the bottom of these ponds and prevent leakage of brine; on the other hand, unicellular Cyanobacteria in these mats and in the brine sometimes produce massive amounts of polysaccharide slime that unfavorably affects the salt production process [5,23].

The red pigmentation of the dense microbial communities in crystallizer ponds is caused by both the β-carotene accumulated by the green algae *Dunaliella salina*, which is the main or sole primary producer in these ponds, as well as by the carotenoid and retinal protein-based pigments of the heterotrophic community of prokaryotes that develop at the expense of photosynthetically fixed carbon derived from *Dunaliella* [24]. This red pigmentation increases light absorption by the brine and increases its temperature, thus enhancing the salt production process [10]. Even purely aesthetic considerations have been used as incentive to study the highly diverse communities of phototrophic microorganisms in salterns [5].

Recently assertions were made that the halophilic Archaea present in the crystallizer brines may be directly involved in the formation of halite crystals. It was suggested that halobacteria influence crystal growth rate and crystal habit, and that the cells and their envelope S-layers may serve as templates in the nucleation and halite crystal formation [25].

The indispensability of *Artemia* (brine shrimp) for the salt production lies on the ability of the animals to clear brine from particles up to 50 micrometers diameter, to metabolize large amounts of ingested organic matter to carbon dioxide, to deposit wastes in fecal pellets that become incorporated in the benthic community, and to provide highly suitable food for the *Halobacterium salinarium* populations in the downstream ponds [6].

Extracellular polysaccharide production by the cyanobacteria may be activated as a result of nutrient limitation as a way to dispose of the excessive fixed carbon photosynthetically produced [26]. A prominent feature of the microbial mats within the gypsum crusts, as well as in the evaporation ponds of lower salinity, is the often copious amounts of polysaccharide slime associated with the growth of the unicellular Euhalothece-Aphanothece Cyanobacteria. These organisms also spread into the overlying water in some saltern systems [5]. Massive slime formation can negatively affect the salt production process [6,26]. To control excessive blooms of these cyanobacteria, the introduction of grazing brine shrimp (*Artemia*) has been suggested as an effective management procedure [5,6].

Solar saltworks as refuge zones for migratory birds

In a scenario of intensive occupation of coastal zone, these large aquatic ecosystems represent important refuge zones for many species of migratory birds beyond being habitat for many endemic species of hypersaline environments.

Throughout the world, certain waterbirds use saltworks as places for rest, feeding and breeding [1,27]. This guild of species is one of the most frequently considered with regard to appraisal the natural value of these wetlands for conservation as protected areas.

Salt work ponds are considered to be high-quality feeding habitats for many non-breeding shorebird species, merely based on the high number of feeding birds that they support, but it is possible that birds could also be found at high densities in habitats of low quality. There are empirical confirmation that saltworks are indeed suitable feeding habitats for several migrating shorebird species that rely on intertidal habitats [27].

When saltworks are found near important wintering and/or staging areas, further saltworks loss could cause a movement and even an increase in the mortality of the displaced birds through density dependent forces. At Cadiz Bay, for example, the coincidence of saltworks loss, and the decline and redistribution of some shorebird species has led to the suggestion of a causal link between them [28].

From a functional point of view, the key factor for the Mediterranean saltworks is the gradient of salinity. The salt production process determines ecological partitioning within the system. This ecological segregation is very important for conservation of these environments because spatial heterogeneity can provide to species a high diversity of habitats, suitable for migratory waterbirds. Such habitats are nodes of ecological connectivity [1].

Conclusion

There is an urgent need to establish a strategy that promotes the inclusion of the Brazilian solar saltworks as conservation zones, in whose boundaries only can be developed activities that do not damage the ecological stability of these important and unique ecosystems. These environments have an ecological dynamics in space and time, where the existing knowledge about the diversity and potential use of natural resources found are still incipient. Therefore, the integrated management of solar saltwork ponds has implied the need for ongoing monitoring and conducting further studies on the feasibility of these other potential uses in Brazilian saltwork ponds.

Acknowledgements

The authors wish to thank to associates of SIESAL (Sindicato da Indústria de Extração de Sal Marinho do Rio Grande do Norte) that provided all the resources at our disposal to accomplish this study.

Author details

[1]Departamento de Geografia, Universidade Federal do Rio Grande do Norte, Campus de Caicó, Joaquim Gregório, s/n, Penedo 59.300-000, Caicó-RN, Brasil. [2]Departamento de Biologia, Universidade de Aveiro, Campus de Santiago, 3810-193 Aveiro, Portugal. [3]Instituto de Tecnologia e Pesquisa, Universidade Tiradentes, Av. Murilo Dantas, 300. Bairro Farolândia, 49032-490 Aracaju, Sergipe, Brasil.

Authors' contributions

Considering the continental area of Brazil, this analyze needed of a research group of diverse areas. All review it was write by RMR, LXF and DFSC, but they work specifically in themes about "Microalgae and halobacteria cultures for extraction of beta-carotenes and glycerol", "Mass culture of Artemia for aquaculture" and "The role of biological system on maintenance of brine quality". For other side, all the authors help us during bibliographical research. RMB are responsible for the topic of "Artisanal fisheries", DHMM for "Extraction of secondary salts from ≈Mother Waters≈ (Bitterns) used for chemical industry", AMAS for the theme "Flower of salt production", and MALF and CNA coleted many data and references about the "Solar saltworks as refuge zones for migratory birds". "All authors read and approved the final manuscript".

Competing interests

The authors declare that they have no competing interests.

References

1. López E, Aguilera PA, Schmitz MF, Castro H, Pineda FD: **Selection of ecological indicators for the conservation, management and monitoring of Mediterranean coastal salinas.** *Environ Monit Assess* 2010, **166(3)**:241-256.
2. Davis J: **Structure, function, and management of the biological system for seasonal solar saltworks.** *Global Nest* 2000, **2(3)**:217-226.
3. Pedrós-Alió C, Calderón-Paz JI, MacLean MH, Medina G, Marrasé C, Gasol JM, Guixa-Boixereu N: **The microbial food web along salinity gradients.** *FEMS Microbiol Ecol* 2000, **32**:143-155.
4. Rocha Renato de Medeiros, da Silva Costa Diógenes Félix, de Lucena Filho Milton Araújo, Bezerra Rodolfo Michel, de Medeiros David Hélio Miranda, da Silva Antonio Marcos Azevedo, de Araújo Cristian Nogueira, Xavier-Filho Lauro: **Tropical solar salt works - influence and challenges in the coexistence with traditional populations in the brazilian northeast region.** In *Proceedings of the 9th International Symposium on Salt, 4-7 September 2009*. Edited by: Zuoliang S. Beijing: Gold Wall Press; 2009:877-881.
5. Oren A: **Saltern evaporation ponds as model systems for the study of primary production processes under hypersaline conditions.** *Aquat Microb Ecol* 2009, **56**:193-204.
6. De Medeiros Rocha R, Câmara MR: **Prediction, monitoring and management of detrimental algal blooms on solar salt production in northeast Brazil.** In *Proceedings of the Seventh Symposium on Salt, 6-9 April 1992*. Edited by: Kakihana H, Hardy HR, Hoshi T, Toyokura K. Kyoto: Elsevier Science Publishers; 1993:657-660.
7. Gouveia L, Sousa J, Marques A, Tavares C, Giestas M: **Solar Pond devices: free energy or bioreactors for Artemi biomass production?** *J Ind Microbiol Biotechnol* 2009, **36**:1035-1045.
8. Vieira N, Bio A: **Spatial and temporal variability of water quality and zooplankton in an artisanal salina.** *J Sea Res* 2011, **65**:293-303.
9. Baseggio G: **The Composition of Sea Water and Its Concentrates.** In *Proceedings of the Fourth Symposium on Salt, 8-12 April 1973*. Edited by: Coogan Alan H. Houston: Open Library; 1973:351-358.
10. Javor BJ: *Hypersaline environments Microbiology and biogeochemistry* New York: Springer Verlag; 1989.
11. Fernândez-Lozano JK: **Recovery of epsomite and sylvite from seawater bittern by crystallization.** In *Proceedings of the Fourth Symposium on Salt, 8-12 April 1973*. Edited by: Coogan Alan H. Houston: Open Library; 1973:501-510.
12. Jadhav MH: **Recovery of Crystalline Magnesium Chloride-Hexahydrate by Solar Evaporation of Sea Bitterns.** In *Proceedings of the Sixth International Symposium on Salt, May 24-28 May 1983*. Edited by: Schreiber BC, Harner HL. Toronto: Salt Institute; 1983:417-449.
13. Derby IH, Victor I: **The dissociation tensions of certain hydrated chlorides and the vapour pressure of their saturated solutions.** *I Amer Chem Soc* 1916, **38**:1439.
14. Donadio C, Bialecki A, Valla A, Dufossé L: **Carotenoid-derived aroma compounds detected and identified in brines and speciality sea salts (fleur de sel), produced in solar salterns from Saint-Armel (France).** *Journal of Food Composition and Analysis* 2010, doi:10.1016/j.jfca.2011.03.005.
15. He FJ, MacGregor GA: **A Comprehensive review on salt and health and current experience of worldwide salt reduction programmes.** *Journal of Human Hypertension* 2009, **23**:363-384.
16. Zmora O, Avital E, Gordin H: **Results of an attempt for mass production of Artemi in extensive ponds.** *Aquaculture* 2002, **213**:395-400.
17. Oren A: **Biotechnological applications and potentials of halophilic microorganism.** In *Halophilic microorganisms and their environments Cellular origin, life in extreme habitats and astrobiology. Volume 5*. Edited by: Oren A. New York: Springer; 2002:357-388.
18. Asker D, Ohta Y: **Production of canthaxanthin by extremely halophilic bacteria.** *J Biosci Bioengin* 1999, **88**:617-621.
19. Galinski EA, Tindall BJ: **Biotechnological prospects for halophiles and halotolerant microorganisms.** In *Molecular biology and biotechnology of extremophiles*. Edited by: Herbert RA, Sharp RJ. New York: Blackie, Glasgow; 1992:76-114.
20. Goldman Y, Garti N, Sasson Y, Ginzburg BZ, Bloch MR: **Conversion of halophilic algae into extractable oil. 2. Pyrolysis of proteins.** *Fuel* 1981, **80**:90-92.
21. Ginzburg BZ: **Liquid fuel (oil) from halophilic algae: a renewable source of non-polluting energy.** In *General and applied aspects of halophilic microorganisms*. Edited by: Rodriguez-Valera F. New York: Plenum Press; 1991:389-395.
22. Ma'or Z, Simon-Meshulam G, Yehudah S, Gavrieli JA: **Antiwrinkle and skin-moisturizing effects of a mineral-algal-botanical complex.** *J Cosmet Sci* 2000, **51**:27-36.
23. Davis JS, Giordano M: **Biological and physical events involved in the origin, effects, and control of organic matter in solar saltworks.** *Int J Salt Lake Res* 1996, **4**:335-347.
24. Litchfield CD: **Red - the magic color for solar salt production.** In *Das Salz in der Rechts - und Handelsgeschichte*. Edited by: Hocquet JC, Palme R. Schwaz: Berenkamp; 1991:403-412.
25. López-Cortés A, Ochoa JL: **The biological significance of halobacteria on nucleation and sodium chloride crystal growth.** In *Adsorption and its applications in industry and environmental protection Studies in surface science and catalysis. Volume 120*. Edited by: Dubrowski A. Amsterdam: Elsevier; 1998:903-923.

26.   Roux JM: **Production of polysaccharide slime by microbial mats in the hypersaline environment of a Western Australian solar saltfield.** *Int J Salt Lake Res* 1996, **5**:103-130.

27.   Masero JA: **Assessing alternative anthropogenic habitats for conserving waterbirds: salinas as buffer areas against the impact of natural habitat loss for shorebirds.** *Biodiversity and Conservation* 2003, **12**:1157-1173.

28.   Perez-Hurtado A, Hortas F, Ruiz F, Solis F: **Importancia de la Bahıa de Cadiz para las poblaciones de limıcolas invernantes e influencia de las transformaciones humanas.** *Ardeola* 1993, **40**:133-142.

5

Use of vital wheat gluten in aquaculture feeds

Emmanuelle Apper-Bossard[*], Aurélien Feneuil, Anne Wagner and Frédérique Respondek

Summary

In aquaculture, when alternative protein sources of Fish Meal (FM) in diets are investigated, Plant Proteins (PP) can be used. Among them, Vital Wheat Gluten (VWG) is a proteinaceous material obtained from wheat after starch extraction. "It is mainly composed of two types of proteins, gliadins and glutenins, which confer specific visco-elasticity that's to say ability to form a network providing suitable binding. This will lead to specific technological properties that are notably relevant to extruded feeds". Besides these properties, VWG is a high-protein ingredient with an interesting amino-acid profile. Whereas it is rather low in lysine, it contains more sulfur amino acids than other PP sources and it is high in glutamine, which is known to improve gut health and modulate immunity. VWG is a protein source with one of the highest nitrogen digestibility due to a lack of protease inhibitor activity and to the lenient process used to make the product. By this way, addition of VWG in diet does not adversely affect growth performance in many fish species, even at a high level, and may secure high PP level diets that can induce health damages.

Keywords: Vital wheat gluten, Fish, Protein, Digestibility, Performance, Health

Introduction

Intensive production of farmed fish, fed with compound feeds, has been largely increased mainly due to the growth of aquaculture production, but also because it is the most efficient way of production [1]. In such feeds, Fish Meal (FM) used to be the major source of proteins, especially for marine fish and salmonids [2]. Nevertheless, because of the limited amount of available FM in the market, its lack of sustainability, and its increasing price, its inclusion in diets has been progressively reduced. In order to achieve a low FM incorporation (below 10%) without impairing growth performance, an active research was conducted on plant proteins (PP), which represent an interesting alternative to FM. In this context, many studies were undertaken to evaluate the effects of replacing FM by different types of PP on fish growth and health [3-5]. Nowadays, several studies are exploring the possibility to decrease FM in a large extent by replacing them with a mixture of several PP [6].

Among the tested PP being considered to replace FM, Vital Wheat Gluten (VWG) is a PP source that has been given very promising results. Indeed, VWG can act as a pellet binder in extruded feed. Furthermore, it is a high quality protein source, highly digestible, with an interesting profile of amino-acids, especially a high level of

glutamine. The action of antinutritional substances was not observed when

wheat gluten was used as fish meal replacement [7]. As a result, growth performance and feed efficiency are not modified when up to 50% of FM are replaced with VWG in the diet of salmon [8], trout [9], and sea bream [10]. Furthermore, when compared with soybean-meal (SBM), VWG does not damage gut structure in Atlantic salmon [8]. The use of VWG has to be emphasised in a global context tending to decrease incorporation of FM. Thus, this paper aims at reviewing the use of VWG, its functionalities and properties regarding farmed fish.

Vital wheat gluten: technological and nutritional properties

VWG can be defined as "a cohesive, visco-elastic, protein-aceous material prepared as a co-product of the isolation of starch from wheat flour" [11]. It is obtained from wheat flour by washing the dough preparation under water, and then centrifugation. This process removes soluble fibres and starch fractions and recovers the insoluble protein fraction that is mainly constituted of two fractions defined as follows according to their solubility in aqueous alcohols: soluble gliadins and insoluble glutenins [12], which are balanced with equal amounts.

Gliadins, which are monomeric proteins (intrachain disulfide bonds) with relatively low molecular weight (30 to 100 kDa), contribute to dough viscosity and extensibility,

* Correspondence: Emmanuelle.apper-bossard@tereos.com
Tereos Syral, Z.I. Portuaire, 67 390, Marckolsheim, France

whereas glutenins, which are polymeric proteins (intra- and interchain disulfide bonds) with high molecular weight (100 kDa to more than 10,000 kDa), are both cohesive and elastic, being responsible for tenacity (resistance to deformation) and elasticity (Figure 1). Gliadins can be qualified as plasticiser for glutenins.

A pellet binder

Due to its visco-elasticity, VWG can act as a pellet binder in extruded fish feed to partially replace starch or indigestible binders [8]. Indeed, the ability of fish to hydrolyse starch in the intestine and to regulate blood glucose concentration when the digestible carbohydrate level is high varies among species and is generally rather low [13], related to the enzymatic digestive capacity [14]. For example, the Atlantic salmon capacity to hydrolyse starch is low. Furthermore, administering glucose results in persistent hyperglycaemia in rainbow trout, common carp, red sea bream, yellowtail, and catfish [13], suggesting these species are not able to regulate glycaemia when fed with high level of digestible carbohydrates. That is why the starch amount is generally kept low in diets for fish, limiting its use as a technological binder.

Upon hydration, mixing, shearing, and heating (effects induced by the feed preparation, which is mostly done by extrusion-cooking), gliadins and glutenins interact in the dough through forces of various natures linked to their compositions: non-covalent bonds (hydrogen, ionic, and hydrophobic bonds) and covalent disulfide bonds [12] (Figure 1). Thus, gluten forms a strong cohesive network to entrap the other ingredients, providing adapted physical characteristics to the pellet in term of binding: improvement of the pellet hardness and pellet durability index (properties defined in Sorensen 2012). On our pilot extrusion line (Application Centre Tereos Syral, Marckolsheim, France), the production of two different salmon formulations containing 10 or 20% VWG (replacement of Soy Protein Concentrate (SPC) with VWG to switch from one formulation to the other) emphasised the importance of this trend: hardness and pellet durability index increase

from 35 N to 48 N and from 97% to 98%, respectively, with increasing VWG. Similarly, a higher breaking force was induced by incorporating VWG compared to fish meal and SPC [15]. Moreover, VWG water insolubility reduces pellet breakdown [11], which can be interesting in cases where water stability must be high: for shrimp feeds (long residence time in water before eating) and in farms transporting the pellets from the weighing cell to the fish cages with water.

A protein and amino acid source

The protein/energy ratio recommendation is higher for fish diets than for terrestrial vertebrates like pigs or poultry. Indeed, their basal energy requirements are lower than those of terrestrial vertebrates, due to the aquatic mode of life, poikilothermy, and ammonotelism [16]. As a result, the relative proportion of dietary protein in fish feed is higher than in terrestrial vertebrate feed. Dietary protein is then the major component of formulated fish feed. It is necessary to get both ingredients high in Crude Protein (CP) and high-quality proteins. Besides its technological properties, VWG has interesting nutritional values for fish feeds and is high in CP. The CP content is 80%, which is higher than in FM, SPC, Soybean Meal (SBM) or Pea Protein Concentrate (PeaPC) in which it represents between 44 and 72% [17].

Regarding Essential Amino Acids (EAA), VWG is rather low in lysine, tryptophan, and arginine (Figure 2). Thus, a dietary supplementation with lysine in fish feed high in VWG is necessary. As example, while body of salmon contains 9.3 g lysine/100 g CP, VWG contains only 1.5 to 1.7 g/100 g CP [18]. Several experiments showed VWG can successfully replace a large part of FM when diets are supplemented with free lysine in salmonids [19,20]. A dose-effect relationship study involving VWG and lysine in rainbow trout showed better growth performance was obtained when FM was replaced with up to 50% VWG and when 0.29% to 0.58% lysine was added [20].

Beside the low amount of lysine, VWG contains a relatively high concentration of sulfur-containing amino acids,

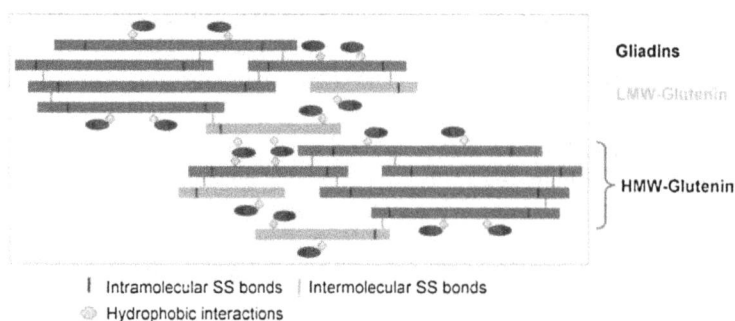

I Intramolecular SS bonds | Intermolecular SS bonds
◎ Hydrophobic interactions

Figure 1 Representation of the gluten visco-elastic network and the place occupied by the gliadin and glutenin fractions within its structure.

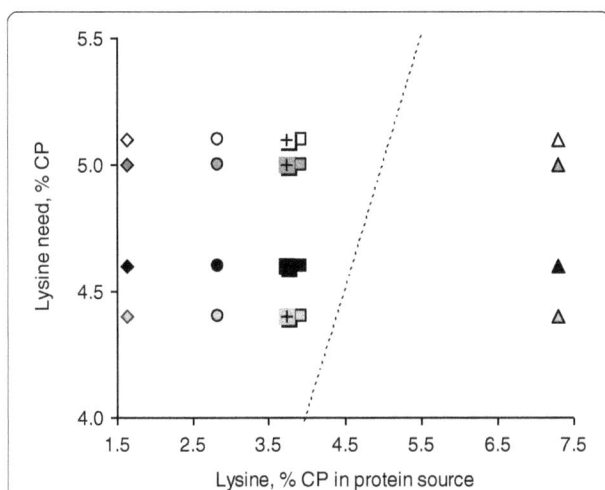

Figure 2 Lysine requirements of several species of fish compared to content in different protein sources expressed as % of Crude Protein (CP). White symbols: Nile tilapia requirement; dark grey symbols: Atlantic salmon requirement; black symbols: shrimp requirement; light grey symbols: European sea bass requirement. ◊: Vital Wheat Gluten; o: Soybean Meal; +: Pea Protein Concentrate; □: Soy Protein Concentrate; Δ: Fish Meal. Dotted line: requirement of fish = content in protein source.

due to the numerous di-sulfur bounds (1.8% CP of methionine and 2.6% CP of cysteine), whereas PP sources are generally low in sulfur-containing amino acids (Figure 3). As examples, SBM and SPC respectively contain 1.4 and 1.3 g/100 g CP of methionine and 1.3 and 1.4 g/100 g CP of cysteine, which is lower than fish requirements. Furthermore,

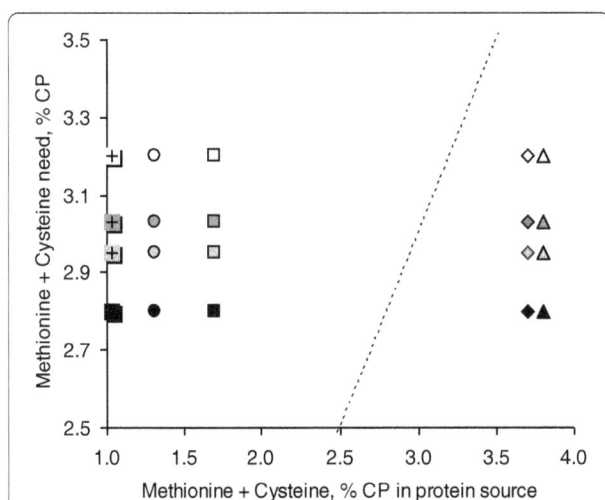

Figure 3 Sulfur amino-acid requirements of several species of fish compared to content in different protein sources, expressed as % of Crude Protein (CP). White symbols: Nile tilapia requirement; dark grey symbols: Atlantic salmon requirement; black symbols: Asian sea bass requirement; light grey symbols: European sea bass requirement. ◊: Vital Wheat Gluten; o: Soybean Meal; +: Pea Protein Concentrate; □: Soy Protein Concentrate; Δ: Fish Meal. Dotted line: requirement of fish = content in protein source.

VWG is high in leucine, with about 7.9 g/100 g CP. Leucine is considered as the main amino acid triggering muscle protein synthesis and inhibiting proteolysis in mammals [21]. In particular, leucine can stimulate the PI3K-Akt-TOR pathway [22]. This mechanism may also occur in fish. Indeed, it is shown that in different species, amino acids regulate TOR signalling pathway [23]. Furthermore, supplementing media containing 0.6 mM leucine with an additional 2.5 mM leucine reduced rates of protein degradation of rainbow trout primary myocytes by 8% [24].

Regarding non-essential amino acids, VWG is high in glutamine: from 35 to 40% CP. Glutamine is a major substrate for all rapidly proliferating cells and plays an important role in maintaining intestinal trophicity [25] (Table 1). Also, glutamine is one of the most important energy substrate of enterocytes. Several studies recently demonstrated free glutamine significantly increases enterocyte and microvilli length of catfish gut [26], hybrid striped bass [27], and juvenile hybrid sturgeon [28] (Table 1). Besides these effects on gut morphology, glutamine also constitutes a major substrate for immune cells, thus modulating immune response [25] (Table 1). The serum non-specific immunity of juvenile hybrid sturgeon is modulated whereas lysozyme activity is increased in hybrid striped bass by adding glutamine in diets [27,29]. Moreover, glutamine also plays a role in eliminating free radicals as it acts as a precursor for the glutathione synthesis [30]. Such effects are reported for juvenile hybrid sturgeon [29] and hybrid striped bass [27]. Glutamine has proven to stimulate muscle synthesis in terrestrial vertebrates [31], but such results are not available for fish. However, dietary glutamine supplementation increases growth performance in juvenile hybrid sturgeon fed with a SPC-based diet [28], and in hybrid striped bass fed with a FM and SBM-based diet [27] (Table 1). Nevertheless, to our knowledge, the effects of VWG on immune response or on antioxidant status, compared to other PP sources, have not been investigated whereas its high amount of glutamine should be explored.

Wheat gluten in diets of fish
Highly digestible
VWG is a highly digestible protein source for several species (Table 2). Apparent digestibility of 99% for CP when feeding a diet with 92.7% CP from VWG and 1.45% lysine is obtained in rainbow trout [33], whereas the apparent CP digestibility of VWG was estimated to approximate 100% [34]. In shrimp (*Litopenaeus vannamei*), the apparent CP digestibility is estimated at 98% [35], whereas it is 100% in Nile tilapia [17]. The apparent CP digestibility of VWG is higher than the apparent CP digestibility of numerous protein sources derived from plants and animals (Table 2). An inclusion of 15% VWG in a FM-based diet results in increasing CP digestibility in juvenile Nile tilapia. The same result is observed when around 30% VWG

Table 1 Effects of adding glutamine on growth performance, gut structure and activity, and immune parameters in different species of fish

Species	Physiologic status	Trial period (d)	Proteins in diets	Glutamine (Gln% in diet)	Main results	Reference
Jian carp *Cyprinus carpio*	Juvenile iBW[1] = 7.78 g	80	35% FM[2]; 20.5% Rice Gluten Meal; CP[3] = 33% Substitution of glycine by Gln	0, 0.4, 0.8, 1.2, 1.6, 2.0	*Performance* Percent weight gain, feed efficiency, and productive protein linearly increase until 1.2% and reach a plateau afterwards. *Gut morphology and activity* Intestine fold height, and gut length and weight linearly increase until 1.2% and reach a plateau afterwards. Protease, lipase, AKP, and Na,K-ATPase activities linearly increase until 1.2% and reach a plateau afterwards in the 3 segments of gut.	[32]
Hybrid sturgeon *Acipencer schrenckii × Huso dauricus*	Juvenile iBW = 22.38 g	56	20% SBM[2]; 10% Blood Meal; 20% FM; 8% Corn Protein; CP = 44.8% Substitution of SPC[2] by HWG[2] (0 to 5%) or by Aln-Gln (1%)	0, 0.63* 0, 0.3, 0.6, 0.9, 1.2, 1.5**	*Performance* Relative Growth Rate and feed efficiency linearly increase from 0 to 1.2% of Gln in HWG and increase with free Gln. *Gut morphology and activity* Intestine fold height, and gut length and weight linearly increase from 0 to 1.2% of Gln in HWG and increase with free Gln. Protease, amylase, lipase, and Na,K-ATPase activities linearly increase until 1.2% of Gln in HWG and increase with free Gln in the 3 segments of gut. *Immune response and antioxidant status* Glutathione peroxydase, glutataione, and superoxide dismutase concentrations linearly increase whereas malondialdehyde concentrations decrease until 1.2% Gln in HWG and with free Gln in intestine, hepatopancreas, and muscle. Modulation of serum non-specific immune parameters according to a quadratic response with 0 to 1.5% Gln in HWG. Linearly increase of serum concentrations of C3 and C4 with addition of free Gln.	[28,29]
Hybrid striped bass *Morone chrysops × Morone saxatilis*	Juvenile iBW = 4.12 g	56	34.6% Menhaden FM; 29.6% SBM; CP = 45% Substitution of glycine by Gln	0, 1, 2	*Performance* Specific Growth Rate and feed efficiency increase with 1% of Gln, not with 2%. *Gut morphology* No change in pyloric caeca. Enterocyte and microvillus heights in distal intestine and fold length in the proximal intestine increase with Gln. *Immune response* Higher serum lysozyme activity with 1% Gln. Increased extracellular superoxide anion production with 1% and 2% Gln.	[27]

Table 1 Effects of adding glutamine on growth performance, gut structure and activity, and immune parameters in different species of fish (*Continued*)

				Performance	
Channel catfish *Ictalurus punctatus*	Juvenile iBW = 6.1 g	70	13.4% casein; 3.76% gelatine; 7.1% of amino-acid premix; CP = 28%	0, 0.5, 1, 1.5, 2, 3	[26]
				No significant effect on growth performance. Increase of lipids in carcasses with 2 and 3% Gln.	
				Gut morphology Mucosal fold length, enterocyte height and microvilli height increase until 2% Gln in the 3 segments of intestine.	

[1]iBW: initial Body Weight; [2]FM: Fish Meal, SBM: Soybean Meal, SPC: Soy Protein Concentrate, HWG: Hydrolysed Wheat Gluten [3]CP: Crude Protein.
*Inclusion of 1% Aln-Gln; **: inclusion of 0 to 5% Hydrolysed wheat gluten.

Table 2 Apparent digestibility coefficient (ADC) of crude protein (CP) of diets with different protein sources depending on the inclusion rate and the species

Species	Physiological status	Protein source[1]	Protein source in diet (%)	ADC of CP in diet or of ingredient (%)	Reference
Coho salmon	Growing	VWG	30[2]	98.4	[34]
		SBM	100	96.4	
		Herring FM		96.6	
		Deboned meal		96.7	
		CGM		95.2	
		VWG		99.6	
		SBM		93.0	
		Herring FM		94.7	
		Deboned meal		95.1	
		CGM		91.9	
Rainbow trout	Growing	VWG	30[2]	98.6	[34]
		SBM	100	95.5	
		Herring FM		96.6	
		Deboned meal		97.2	
		CGM		97.4	
		VWG		100	
		SBM		90.1	
		Herring FM		94.6	
		Deboned meal		96.7	
		CGM		97.3	
Australian silver perch	Juvenile	VWG	29.7	99.8	[37]
		SBM		94.8	
		Australian FM		92.3	
		SDBM		90.2	
		CGM		95.4	
Gilthead sea bream	Growing	VWG	100[3]	96	[39]
		SPC		92	
		FM		86	
		CGM		90	
Nile tilapia	Juvenile	VWG	15	92.6	[40]
		SBM		90.5	
		DWD		88.9	
		FM		89.7	
Atlantic cod	Juvenile	VWG	29.8	99.9	[36]
		SBM		92.3	
		SPC		98.6	
		PeaPC		89.8	
		CGM		86.3	
		Herring FM		93.3	

[1]VWG: Vital Wheat Gluten; SBM: Soybean Meal; CGM: Corn Gluten Meal; SDBM: Spray-Dried Blood Plasma; SPC: Soy Protein Concentrate; PeaPC: Pea Protein Concentrate; FM: Fish Meal.
[2]Digestibility coefficients are given for both the diet with 30% protein source and for each protein source.
[3]Digestibility coefficients are indicated for each protein source.

are added in a herring FM-based diet in Atlantic cod [36] or Australian silver perch [37] (Table 2). In salmonids, the apparent CP digestibility of diets rises with increasing inclusion rate of VWG [8,20,38], (Figure 4). In particular, in Atlantic salmon, the Apparent Digestibility (AD) of CP linearly increases from 88.6% with a FM-based diet to 93.6% when including 50% CP as VWG, and by extrapolating these results it was 100% with 100% CP from VWG [8].

In these species, the apparent availability of EAA is also very high. In shrimp (*Litopenaeus vannamei*), the apparent availability of EAA from VWG is similar to or higher than the one of FM, except for lysine and cysteine. VWG has been reported to have equivalent or higher value with respect to apparent digestibility of protein and to apparent absorption of amino acids when compared to FM [8,34]. Specifically, the apparent digestibility of CP and apparent absorption of amino acids significantly increased with higher proportion of CP from VWG, except alanine and lysine [8] (Figure 5). The trend for a lower apparent absorption coefficient of lysine may be explained by a deviation between true and apparent lysine digestibility with increasing VWG, due to a relative higher excretion of endogenous lysine with high level of VWG.

The digestibility of energy from VWG is also higher than the digestibility of energy from FM in Atlantic salmon as well as in gilthead sea bream and Atlantic cod [8,39]. In contrast with these results, a decrease of energy and fat digestibility occurs when FM is replaced with SBM

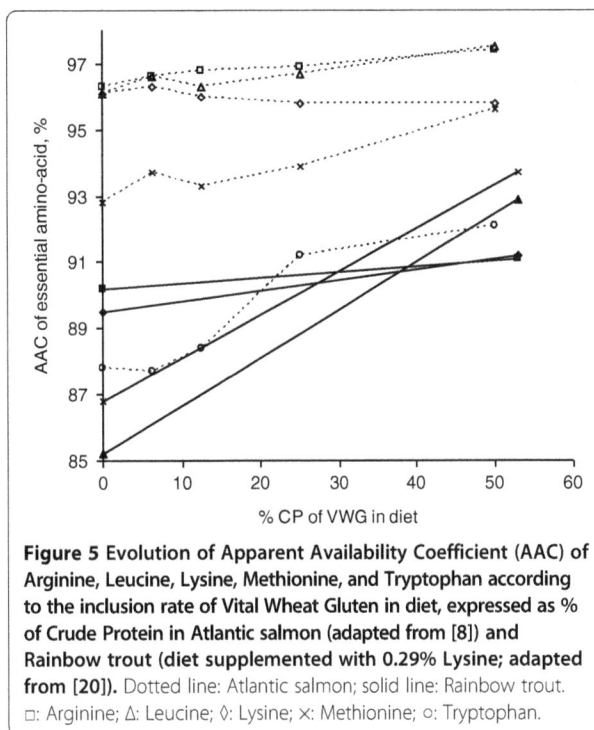

Figure 5 Evolution of Apparent Availability Coefficient (AAC) of Arginine, Leucine, Lysine, Methionine, and Tryptophan according to the inclusion rate of Vital Wheat Gluten in diet, expressed as % of Crude Protein in Atlantic salmon (adapted from [8]) and Rainbow trout (diet supplemented with 0.29% Lysine; adapted from [20]). Dotted line: Atlantic salmon; solid line: Rainbow trout. □: Arginine; Δ: Leucine; ◊: Lysine; ×: Methionine; o: Tryptophan.

in Atlantic salmon [41] or by SPC and Corn Gluten Meal (CGM) in sea bream [39]. These differences may be attributed to a relative higher content of carbohydrates in soy protein ingredients or CGM compared to VWG. Carbohydrates are less available as an energy source to carnivorous fish for which the natural diet -and thus digestive metabolism- is based on lipids and proteins. VWG possesses both high gross energy and protein content and availability; it is then possible that a part of protein from VWG is used as an energy source in these species.

Phosphorus (P) is an essential component of fish diets. Not only it affects hard tissues but also intermediary metabolism and in turn, feed conversion ratio. In the same time, as the aquaculture industry grows, there are concerns regarding P in effluents. Thus, it is important to formulate diets containing adequate P for fish growth and ensuring minimal level of P in effluent. P bioavailability in PP sources is widely variable because of the presence of ANF. P availability is higher in VWG than in FM, SBM, SPC, and CGM for rainbow trout and coho salmon [34]. The content of ANF, such as phytic acid, is a major factor restricting the use of SBM and SPC in diets of fish [8]. The major vegetable part of P (60 to 70%) is bound to phytic acid, leading to a decrease in P availability in fish [42]. This phytic acid decreases the availability of cations such as zinc, magnesium, and calcium [43,44] or protein [42]. This problem does not appear for VWG, in which P content is relatively low but is highly available because of the absence of phytic acid.

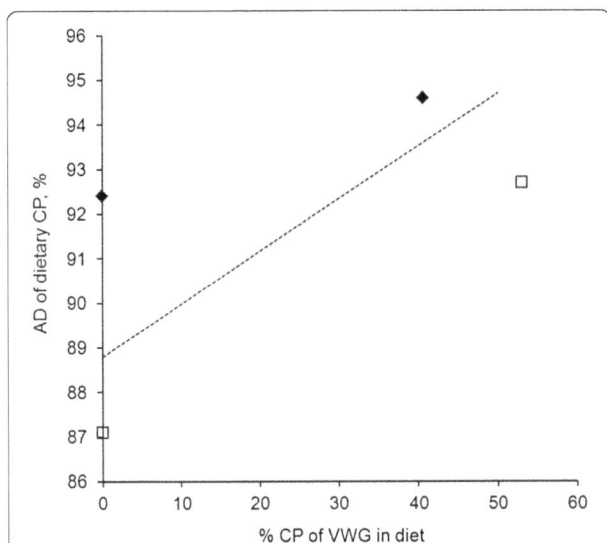

Figure 4 Evolution of Apparent Digestibility (AD) of dietary Crude Protein (CP) according to the inclusion rate of Vital Wheat Gluten in diet, expressed as% of CP in Rainbow trout (adapted from [20]), European sea bass (adapted from [32]), and Atlantic salmon (adapted from [8]). ◆: European sea bass; ———: Atlantic salmon (linear regression: $AD_{CP} = 88.8 + 0.118 \times G$, G being protein from VWG expressed as% of CP); □: Rainbow trout.

Low amount of fibres and no Antinutritional factors

The replacement of FM with other PP sources may result in increasing the level of fibres in diets [45]. Shrimp and fish do not have a high capacity to digest dietary fibres, and a high dietary level of fibres reduces digestibility and utilisation of other nutrients, acting as antinutritional factors (ANF) in carnivorous, as well as in herbivorous species. For instance, the level of dietary fibres must be less than 10% in rainbow trout otherwise leading to decreased growth and dry matter digestibility [46]. In tilapia, reducing weight gain is observed when fed diets with 10% soluble as well as insoluble fibres compared to a control diet [47]. The negative effect of dietary fibres on digestibility of dry matter or CP and growth is also demonstrated in common carp [48], Atlantic salmon [49], or European sea bass [50]. Such results have not been described when adding VWG in fish feed. Compared to other PP sources, VWG does not contain a high level of fibres. The amount of fibres in VWG averages 0.5 to 1% and is comparable to the amount found in FM. It is more than 10-fold lower than the level of fibres in SPC or SBM that averages 4.5% and 7.0%, respectively [17].

Additionally to increasing fibres, replacing FM by PP sources results in higher ANF content in diets. Such factors are defined as endogenous compounds of ingredients that may reduce feed intake, growth, nutrients digestibility and utilisation [51]. ANF are involved in the aetiology of diseases like enteritis, low lipid or protein digestibility and diarrhoea. These substances are numerous but the more relevant ones are phytic acids, enzyme inhibitors, lectin, saponins, phytoestrogens, phytosterols, and oligosaccharides. Due to their detrimental effects on zootechnical performance and nutrient utilisation, ANF must be taken into account when using high inclusion rates of PP for replacing FM [51].

The negative effects of ANF present in PP are different according to the type of PP, the process of production, and the dietary inclusion rate. Indeed, it appears that SBM induce histological and functional changes in gastro-intestinal tract (enteritis, increased susceptibility to bacterial infections, changes in absorptive cells ...) in salmonids [52,53]. Inconsistent results are described for PeaPC: they can induce enteritis in Atlantic salmon [54] whereas no change on digestibility and mineral availability of rainbow trout was observed when replacing 50% FM by PeaPC [54]. Several experiments testing SPC reported no change on growth when comparing to a control, even if some modifications of gut structure were noticed [54]. Compared to other PP sources, no action similar to ANF has been described when VWG was used as FM replacement in different species [7].

Finally, due to the absence of fibres and ANF combined with a high amount of glutamine, adding VWG in

diets may represent an opportunity in the current context in which PP are more and more present in diets. Indeed, diets with less than 5-10% FM are now formulated and used in practical diets for carnivorous fish. Such diets contain a mixture of several PP which have various nutritive properties. This strategy may lead to use complementary properties of different PP for balancing diets and diluting ANF.

Growth performance

Because of its high digestibility and its absence of actions typical of ANF, replacement of a large proportion of FM with VWG results in similar growth performance and fish composition whatever the species are (Table 3). In Rainbow trout, VWG successfully substitutes more than 50% FM providing diets are supplemented with lysine without affecting protein and lipid composition of the pooled carcasses [9,20]. Furthermore, another experiment shows the inclusion of 14.5% VWG in diets does not adversely affect the flavour of fillets [55]. In Atlantic salmon, the replacement of 35% FM with VWG without supplementing by lysine results in similar final body weight and growth [8]. These authors estimate the replacement of FM with VWG without amino acid supplementation can go up to 50% based on the amount and the availability of lysine in VWG and on the requirement of fish. In European sea bass, substituting more than 50% FM with VWG in feeds does not impair palatability, growth performance, and nitrogen-energy retention [10]. In gilthead sea bream, the use of 88% CP from VWG not only replaced successfully FM but also produced better growth and feed conversion ratio, probably related to higher protein and energy intake of fish [39]. Such results are not obtained with CGM and SPC diets: feed efficiency and weight gain decrease when compared to FM (Table 4). In Nile tilapia fed with diets differing by their protein sources, the highest growth is reported for VWG, FM, and Soybean extract diet. In the same experiment, fish body composition is not significantly affected by diet ingredients [40] (Table 4). In shrimp, only few results are available but it seems the replacement of up to 20% marine protein with VWG does not modify feed efficiency [56].

Health

Even if the proportion of PP increases in fish feeds, and despite removing most of the indigestible carbohydrates, it still remains challenging when using plant protein concentrates at very high levels to replace up to 80% FM. Indeed, there are profound effects on metabolism, immunity, and gut health. In Rainbow trout and gilthead sea bream, protease activity is reduced with higher content of PP source until 100% FM are replaced [57]. A total replacement of FM with a mixture of PP exerts a

Table 3 Effects of replacement of fish meal (FM) with vital wheat gluten (VWG) on growth performance, N and P losses, and whole body composition of several species of fish

Species	Duration (d.)	FM (%CP¹)	VWG (%CP)	Other PP² (%CP)	CP in diet (%)	Lys. Supp. (% diet)	iBW (g)	Main results	Reference
Atlantic salmon	126	100	0	No	41.3	No	952	Similar growth - No modification in the mucosa of the posterior intestine.	[8]
		65	35	No	39.6	No	956		
Nile tilapia	49	88	0	No	32.5	No	56.4	Similar growth with VWG as with FM. Higher non faecal N losses with VWG than with FM. Lower faecal P losses than with FM. Same protein, fat and phosphorus body composition with VWG as with FM.	[40]
		58.3	32.8	No	34.7	No	55.6		
Rainbow trout	60	50	23.7	No	30.3	1.40	50	Similar growth with FM diet and when FM is totally replaced with a mixture of VWG and crystalline amino acids.	[9]
		0	77.0	No	31.6	2.50	50		
Rainbow trout	65	72	0	SBM³: 23;	45.5	No	23.7	Better growth and feed efficiency with VWG and 0.29% Lys than with the FM diet. Greater apparent net protein utilisation and protein efficiency ratio. No modification of protein and lipid content in carcasses.	[20]
		31	53	WM³: 5SBM: 6; WM: 5	46.6	0.29	23.7		
European sea bass	96	100	0	No	49.9	No	23.9	Weight gain slightly lower in fish fed with 70% VWG and no change in feed conversion ratio. Same faecal, retention and metabolic losses of N.	[10]
		50	50	No	49.7	0.20	23.9		
		30	70	No	49.9	0.80	23.9		
Sea bream	85	95	0	No	44.8	No	41.6	Significantly higher weight gain and feed efficiency with VWG than with FM. Same productive protein value and higher energy retention value with VWG than with FM. Same protein, lipid, energy, and phosphorus body composition with VWG as with FM.	[39]
		0	88	No	45.1	2.10	41.2		

¹CP: Crude Protein; ²PP: Plant Protein; ³SBM: Soybean Meal; WM: Wheat Middlings.

Table 4 Effects of Vital Wheat Gluten (VWG) or other Plant Protein (PP) sources on growth performance of different fish species

Species	Duration, d.	Tested Protein[1]	Protein, %CP[2]	%CP in diet	Lys. Supp. (% diet)	iBW[3] (g)	Main results	Reference
Nile tilapia	49	FM	88	32.5	No	56.4	Highest feed efficiency and growth for FM, VWG, and SBE. No significant modifications in body composition for dry matter, protein, fat, P, and energy. Highest ash body content for DWD, SCP, and FM.	[40]
		VWG	32.8	34.7	No	55.6		
		SBM	23.9	30.5	No	55.6		
		SBE	34.2	35.7	No	54.3		
		DW	18.5	27.5	No	60.2		
		SCP	20.9	29.1	No	54.8		
Sea bream	85	FM	95	44.8	No	41.6	Better weight gain with VWG, before FM and before SPC and CGM. Same productive protein value and higher energy retention value with VWG as with FM.	[39]
		VWG	88	45.1	2.10	41.2		
		SPC	98	43.9	No	41.7		
		CGM	88	45.8	2.10	39.2		

[1]CP: Crude Protein; [2]FM: Fish Meal; VWG: Vital Wheat Gluten; SBM: Soybean Meal; SBE: Soybean Extract; DW: Duckweed; SCP: single-cell protein; SPC: Soy Protein Concentrate; CGM: Corn Gluten Meal; [3]iBW: initial Body Weight.

positive antioxidant effect by enhancing the glutathione metabolism in gilthead sea bream [58]. It also modulates immune system, notably by decreasing complement levels in blood [58]. The effects of high dietary concentrations of soybean proteins on the non-specific immune response are also reported in rainbow trout with inconsistent results [59,60]. Gut and liver morphology is also modified when replacing FM with up to 50% PP in gilthead sea bream and rainbow trout [58,61]. All these results are complex and not always consistent. Furthermore, quantifying the relative effect of each PP in order to balance a diet high in PP remains difficult since current knowledge is still limited. Further investigations are needed to establish the effects of each PP and their interrelationships.

In this field, there is evidence in terrestrial animals showing that VWG is an interesting source of proteins for maintaining gut health and stimulating the immune system. In piglets, weaning is a highly stressful period characterised by an immediate and transient drop in feed intake, by modifications of various aspects of small intestine morphology and function (villi atrophy, drop in enzyme activity, local inflammation); and may lead to intestinal disorders and diarrhoea [62]. As for fish feed, replacement of animal proteins is investigated in starter feed. Replacing 6% FM or 6% Whey Protein Concentrate, considered as good protein sources for piglets, with 6% SBM or Rice Protein Concentrate results in diminishing villi height [63]. On the contrary, replacing 10% FM with 9.5% VWG does not negatively affect villi height, which is even numerically higher with VWG than with FM, with a similar feed intake [64]. These authors conclude that the positive effect of VWG on villi height may be attributed to the high amount of glutamine rather than feed intake. In fish, to our knowledge, very few data are available. Nevertheless, the replacement of 15% FM by SBM leads to pathological changes in the distal intestine of Atlantic salmon while the replacement of 35% of FM by

VWG does not result in such damages, , probably related to the absence of ANF and to the high content of glutamine in VWG [8]. Furthermore, replacing 5% SPC with 5% Hydrolysed Wheat Gluten (HWG) results in improved performance, gut structure and activity, and immune response in juvenile hybrid sturgeons [28,29]. The results obtained with 3 to 5% HWG are similar to the results obtained when adding 1% glutamine, so that the authors conclude the effect of HWG on health may be due to the high amount of glutamine. The determination of the glutamine effect in native *vs* hydrolysed Wheat Gluten has not been investigated in fish and leads to inconsistent results in terrestrial animals, with either difference [64] or not [65]. These data suggest VWG can be a PP source limiting the negative impact on fish health when fed with high PP levels.

Lipid metabolism

Lipid metabolism may be affected when fish are fed with high PP levels. It is important to take into account lipid metabolism in farmed fish since these factors may affect flesh quality. Adding high levels of PP sources in diets modulates lipid metabolism pathways when comparing to FM diets in several species [66]. Plasma cholesterol decreases in European sea bass (replacement up to 95% FM [67], in gilthead sea bream (replacement up to 100% FM; [58]) and in whole fillet of large rainbow trout (replacement up to 100% FM; [68]) when fish are fed with high amount of a blend composed of several PP sources. Also, a modification of fatty acid desaturation or elongation, with a strong increase of PUFA n-6 and a slight decrease of PUFA n-3 in fillets of large rainbow trout as in muscle of European sea bass are described [68,69]. Recently, a different lipid distribution between dorsal muscle and liver was observed when European sea bass is fed with up to 70% CP with different PP sources without modifying overall adiposity and mesenteric fat [69].

Several mechanisms are proposed to explain such effects. First, there is evidence from different animal models indicating the effect of varying the dietary protein source on fat content could depend on changes in the activity of liver lipogenic enzymes. In rats, the differential effect of dietary proteins on plasma cholesterol level is mainly associated with sulfur amino acids including in the protein [70]. More generally, differences in the amino acid composition of diets including plant *vs* animal proteins have been mentioned as a major factor in the mechanisms modulating *de novo* fatty acid synthesis [71]. Secondly, an increased PP source in diets is related to increased starch level so that both effects are confounded. A significant correlation between starch, glycaemia, and the activities of such lipogenic effects are already underlined in sea bream [72]. Thirdly, some ANF like isoflavones contained in soy proteins are shown to affect lipid metabolism and lower plasma cholesterol [73].

The effects of PP sources on lipid metabolism seem to be various, regarding the properties of PP source itself, i.e. the amino acid composition, the content of starch and the ANF. A dietary plant-protein substitution affects hepatic metabolism of rainbow trout but that the metabolic effects of PP replacement in fish feed varied with PP source [74]. In the different experiments previously cited, a blend of several PP sources was tested so that it is not possible to conclude which PP is the most involved compared to the others on the modulation of lipid metabolism and to define potential interactions between PP. Interestingly, when compared extruded or pelleted wheat gluten-based diets in European sea bass, no clear postprandial patterns of plasma cholesterol were discernible and cholesterol with VWG diet was only slightly, but not significantly, lower than plasma cholesterol of fish fed with FM diet [38]. Effects of up to 70% VWG as replacement of FM, a blend of VWG and pea, and a blend of VWG and SBM have been compared [69]. The plasma cholesterol and triacylglycerol concentrations were not significantly modified by VWG or VWG and pea but decreased with the blend of VWG and SBM, suggesting soy proteins play a major role in decreasing cholesterol and triacylglycerols in blood. However, VWG diets as soy-protein diets were found to decrease PUFA (n-3) and to increase PUFA (n-6) in muscle. These results underline PP have probably different routes of action on lipid metabolism.

Conclusion

The use of VWG in fish feed offers different opportunities in term of technological properties. Furthermore, because of its nutritional properties, the replacement of a large part of FM with VWG is well-established and does not adversely affect feed intake, growth, feed and nutrient efficiency, and overall composition of different species of fish. Nowadays, research is refining and is moving towards understanding the impact of PP-based diets on fish health and metabolism. The knowledge of the pattern of each PP has to be completed by the knowledge of the effects of several PP in a same diet, their possible interactions, in order to optimise the formulation and the choice of ingredients. Indeed, using the correct mixture of PP provides not only the possibility of limiting damages, but also offers an interesting possibility to enhance antioxidant status and to modulate the immune response. In this way, the use of VWG in a diet containing high amount of PP may contribute to secure diets and to preserve health of fish.

Competing interests
The authors declare that they have no competing interests.

Authors' contributions
EAB, AF, AW and FR participated in the sequence alignment and drafted the manuscript. All authors read and approved the final manuscript.

References
1. Olsen RL, Hasan MR: **A limited supply of fishmeal: Impact on future increases in global aquaculture production.** *Trends in Food Sci Technol* 2012, **27**:120–128.
2. Tacon AGJ, Hasan MR, Metian M: **Demand and supply of feed ingredients for farmed fish and crustaceans: trends and prospects.** *FAO Fisheries Aquaculture Tech Paper* 2011, **564**:87. Rome.
3. Mambrini M, Roem AJ, Carvèdi JP, Lallès JP, Kaushik S: **Effects of replacing fish meal with soy protein concentrate and of DL-methionine supplementation in high-energy, extruded diets on the growth and nutrient utilization of rainbow trout, Onchorhynchus mykiss.** *J Anim Sci* 1999, **77**:2990–2999.
4. Carter CG, Hauler RC: **Fish meal replacement by plant meals in extruded feeds for Atlantic salmon, Salmo salar L.** *Aquaculture* 2000, **185**:299–311.
5. Bransden MP, Carter CG, Nowak BF: **Effects of dietary protein source on growth, immune function, blood chemistry and disease resistance of Atlantic salmon (Salmo salar L.) parr.** *Anim Sci* 2001, **73**:105–113.
6. Zhang Y, Overland M, Shearer KD, Sorensen M, Mydland LT, Storebakken T: **Optimizing plant protein combinations in fish meal-free diets for rainbow trout (Onchorhynchus mykiss) by a mixture model.** *Aquaculture* 2012, **360–361**:25–36.
7. Tusche K, Arning S, Wuertz S, Susenbeth A, Schulz C: **Wheat gluten and potato protein concentrate — Promising protein sources for organic farming of rainbow trout (Oncorhynchus mykiss).** *Aquaculture* 2012, **344–349**:120–125.
8. Storebakken T, Shearer KD, Baeverfjord G: **Digestibility of macronutrients, energy and amino acids, ansorption of elements and absence of intestinal enteritis in Atlantic salmon, Salmo salar, fed diets with wheat gluten.** *Aquaculture* 2000, **184**:115–132.
9. Rodehutscord M, Mandel S, Pack M, Jabobs S, Pfeffer E: **Free amino acids can replace protein-bound amino acids in test diest for studies in rainbow trout (Oncorhynchus mykiss).** *J Nutr* 1995, **125**:956–963.
10. Tibaldi E, Tulli F, Piccolo G, Guala S: **Wheat gluten as a partial substitute for fish meal protein in sea bass (D. labrax) diets.** *Ital J Anim Sci* 2003, **2**:613–615.
11. Day L, Augustin MA, Batey IL, Wrigley CW: **Wheat-gluten uses and industry needs.** *Trends Food Sci Technol* 2006, **17**:82–90.
12. Wieser H: **Chemistry of gluten proteins.** *Food Microbiol* 2007, **24**(2):115–119.
13. Wilson RP: **Utilization of dietary carbohydrate by fish.** *Aquaculture* 1994, **124**(1–4):67–80.
14. Stone DAJ: **Dietary Carbohydrate Utilization by Fish.** *Rev Fisheries Sci* 2003, **11**(4):337–369.
15. Draganovic V, GootbVan der AJ, Boomb R, Jonkersa J: **Assessment of the effects of fish meal, wheat gluten, soy protein concentrate and feed moisture on extruder system parameters and the technical quality of fish feed.** *Anim Feed Sci d Technol* 2011, **165**:238–250.

16. Kaushik SJ, Seiliez I: **Protein and amino acid nutrition and metabolism in fish: current knowledge and future needs.** *Aquacult Res* 2010, **41**(3):322–332.

17. NRC: *Nutrient requirements of fish and shrimp.* Washington: The National Academies Press; 2011.

18. Wilson RP, Cowey CB: **Amino acid composition of whole body tissue of rainbow trout and Atlantic salmon.** *Aquaculture* 1985, **48**(3–4):373–376.

19. Pfeffer E, Kinsinger S, Rodehutscord M: **Influence of the proportion of poultry slaughter by-products and of untreated or hydrothermically treated legume seeds in diets for rainbow trout (Onchorhyncus mykiss) on apparent digestibilities of their energy and organic components.** *Aquacult Nutr* 1995, **1**:111–117.

20. Davies SJ, Morris PC, Baker RTM: **Partial substitution of fish meal and full-fat soya bean meal with wheat gluten and influence of lysine supplementation in diets for rainbow trout, Oncorhyncus mykiss (Walbaum).** *Aquacult Res* 1997, **28**:317–328.

21. Li P, Mai K, Trushenski J, Wu G: **New developments in fish amino acid nutrition: towards functional and environmentally oriented aquafeeds.** *Amino Acids* 2009, **37**(1):43–53.

22. Nicklin P, Bergman P, Zhang B, Triantafellow E, Wang H, Nyfeler B, Yang H, Hild M, Kung C, Wilson C, Myer VE, MacKeigan JP, Porter JA, Wang YK, Cantley LC, Finan PM, Murphy LO: **Bidirectional transport of amino acids regulates mTOR and Autophagy.** *Cell* 2009, **136**:521–534.

23. Seiliez I, Gabillard JC, Skiba-Cassy S, Garcia-Serrana D, Gutie'rrez J, Kaushik SJ, Panserat S, Tesseraud S: **An in vivo and in vitro assessment of TOR signaling cascade in rainbow trout (Oncorhynchus mykiss).** *Am J Regul Integ Comp Physiol* 2008, **295**:R329–R335.

24. Cleveland BM, Weber GM: **Effects of insulin-like growth factor-I, insulin, and leucine on protein turnover and ubiquitin ligase expression in rainbow trout primary myocytes.** *Am J Regul Integ Comp Physiol* 2010, **298**:R341–R350.

25. Verlhac-Trichet V: **Nutrition and immunity: an update.** *Aquacult Res* 2010, **41**:356–372.

26. Pohlenz C, Buentello AMB, GatlinIII DM: **Free dietary glutamine improves intestinal morphology and increases enterocyte migration rates, but has limited effects on plasma amino acid profile and growth performance of channel catfish Ictalurus punctatus.** *Aquaculture* 2012, **370–371**:32–39.

27. Cheng Z, Gatlin DM III, Buentello A: **Dietary supplementation of arginine and/or glutamine influences growth performance, immune responses and intestinal morphology of hybrid striped bass (Morone chrysops × Morone saxatilis).** *Aquaculture* 2012, **362–363**:39–43.

28. Qiyou X, Qing Z, Hong X, Chang'an W, Dajiang S: **Dietary glutamine improves growth performance and intestinal digestion/absorption ability in young hybrid sturgeon.** *J Appl Ichthyol* 2011, **27**:721–726.

29. Zhu Q, Xu QY, Wang CA, Sun DJ: **Dietary glutamine supplementation improves tissue antioxidant status and serum non-specific immunity of juvenile hybrid sturgeon.** *J Appl Ichthyol* 2011, **27**:715–720.

30. Wu G: **Intestinal mucosal amino acid catabolism.** *J Nutr* 1998, **128**(8):1249–1252.

31. Wu G, Bazera FW, Davisa TA, Jaegere LA, Johnsone GA, Kima SW, Knabea DA, Meiningerc CJ, Spencera TE, Yina YL: **Important roles for the arginine family of amino acids in swine nutrition and production.** *Livest,Sci* 2007, **112**(1–2):8–22.

32. Yan L, Qiu-Zhou X: **Dietary glutamine supplementation improves structure and function of intestine of juvenile Jian carp (Cyprinus carpio var. Jian).** *Aquaculture* 2006, **256**:389–394.

33. Pfeffer E, Al-Sabty H, Haverkamp R: **Studies on lysine requirements of rainbow trout (Oncorhynchus mykiss) fed wheat gluten as only source of dietary protein.** *J Anim Physiol Anim Nutr* 1992, **67**(1–5):74–82.

34. Sugiura SH, Dong FM, Rathbone CK, Hardy RW: **Apparent protein digestibility and mineral availabilities in various feed ingredients for salmonid feeds.** *Aquaculture* 1997, **159**:177–202.

35. Lemos D, Lawrence AL, Siccardi AJ III: **Prediction of apparent protein digestibility of ingredients and diets by in vitro pH-stat degree of protein hydrolysis with species-specific enzymes for juvenile Pacific white shrimp Litopenaeus vannamei.** *Aquaculture* 2009, **295**(1–2):89–98.

36. Tibbetts SM, Milley JE, Lall SP: **Apparent protein and energy digestibility of common and alternative feed ingredients by Atlantic cod, Gadus morhua.** *Aquaculture* 2006, **261**:1314–1327.

37. Allan GL, Parkinson S, Booth MA, Booth MA, Stone DAJ, Rowland SJ, Frances J, Warner-Smith R: **Replacement of fish meal in diets for Australian silver perch, Bidyanus bydianus: I. digestibility of alternative ingredients.** *Aquaculture* 2000, **186**:293–310.

38. Robaina L, Corraze G, Aguirre P, Blanc D, Melcion JP, Kaushik SK: **Digestibility, postprandial ammonia excretion and selected plasma metabolites in European sea bass Dicentrarchus labrax/ fed pelleted or extruded diets with or without wheat gluten.** *Aquaculture* 1999, **179**:45–56.

39. Kissil GW, Lupatsch I: **Successful replacement of fishmeal by plant proteins in diets for the gilthead seabream, Sparus aurata L.** *Israeli J Aquacult* 2004, **56**(3):188–199.

40. Schneider O, Amirkolaie AK, Vera-Cartas J, Eding EH, Schrama JW, Verreth JAJ: **Digestibility, faeces recovery, and related carbon, nitrogen and phosphorus balances of five feed ingredients evaluated as fishmeal alternatives in Nile tilapia, Oreochromis niloticus L.** *Aquacult Res* 2004, **35**:1370–1379.

41. Refstie S, Storebakken T, Roem AJ: **Feed consumption and conversion in Atlantic salmon (Salmo salar) fed diets with fish meal, extracted soybean meal or soybean meal with reduced content of oligosaccharides, trypsin inhibitors, lectins and soya antigens.** *Aquaculture* 1998, **162**(3–4):301–312.

42. Storebakken T, Shearer KD, Roem AJ: **Availability of protein, phosphorus and other elements in fish meal, soy protein concentrate and phytate-treated soy protein concentrate based diets to Atlantic salmon, Salmo salar.** *Aquaculture* 1998, **161**:365–379.

43. Denstadli V, Skrede A, Krogdahl A, Sahlstrømd S, Storebakken T: **Feed intake, growth, feed conversion, digestibility, enzyme activities and intestinal structure in Atlantic salmon (Salmo salar L.) fed graded levels of phytic acid.** *Aquaculture* 2006, **256**(1–4):365–376.

44. Fredlund K, Isaksson M, Rossander-Hulthénc L, Almgrena A, Sandberga AS: **Absorption of zinc and retention of calcium: dose-dependent inhibition by phytate.** *J Trace Elems Med Biol* 2006, **20**(1):49–57.

45. Gatlin DM III, Barrows FT, Brown P, Dabrowski K, Gaylord TG, Hardy RW, Herman E, Hu G, Krogdahl A, Nelson R, Overturf K, Rust M, Sealey W, Skonberg D, Souza EJ, Stone D, Wilson R, Wurtele E: **Expanding the utilization of sustainable plant products in aquafeeds: a review.** *Aquacult Res* 2007, **38**(6):551–579.

46. Hilton JW, Atkinson JL, Slinger SJ: **Effect of increased dietary fiber on the growth of rainbow trout (Salmo gairdneri).** *J Can Sci Halieutiques Aquat* 1983, **40**(1):81–85.

47. Shiau SY, Kwok CC, Chen CJ, Hong HT, Hsieh HB: **Effects of dietary fibre on the intestinal absorption of dextrin, blood sugar level and growth of tilapia, Oreochromis niloticm × O. aureus.** *J Fish Biol* 1989, **34**(6):929–935.

48. Hossain MA, Focken U, Becker K: **Evaluation of an unconventional legume seed, Sesbania aculeata, as a dietary protein source for common carp, Cyprinus carpio L.** *Aquaculture* 2001, **198**(1–2):129–140.

49. Refstie S, Svihusb B, Shearerc KD, Storebakken T: **Nutrient digestibility in Atlantic salmon and broiler chickens related to viscosity and non-starch polysaccharide content in different soyabean products.** *Anim Feed Sci Technol* 1999, **79**(4):331–345.

50. Dias J, Huelvan C, Dinis M, Metailler R: **Influence of dietary bulk agents (silica, cellulose and a natural zeolite) on protein digestibility, growth, feed intake and feed transit time in European seabass (Dicentrarchus labrax) juveniles.** *Aquat Living Resour* 1998, **11**(4):219–226.

51. Krogdahl A, Penn M, Thorsen J, Refsie S, Bakke AM: **Important antinutrients in plant feedstuffs for aquaculture: an update on recent findings regarding responses in salmonids.** *Aquacult Res* 2010, **41**:333–344.

52. McKellep Bakke AM, Penn M, Mora Salas P, Refstie S, Sperstad S, Landsverk T, Ringo E, Krogdahl A: **Effects of dietary soyabean meal, inulin and oxytetracycline on intestinal microbiota and epithelial cell stress, apoptosis and proliferation in the teleost Atlantic salmon (Salmo salar L.).** *Brit J Nutr* 2007, **97**:699–713.

53. Dimitroglou A, Reynolds P, Ravnoy B, Johnsen F, Sweetman JW, Johansen J, Davies SJ: **The effect of Mannan Oligosaccharide supplementation on Atlantic salmon smolts (Salmo salar L.) fed diets with high levels of plant proteins.** *J Aquacult Res Dev* 2011, **S1**:011.

54. Penn MH, Bendiksen EA, Campbell P, Krogdahl A: **High level of dietary pea protein concentrate induces enteropathy in Atlantic salmon (Salmo salar L.).** *Aquaculture* 2011, **310**:267–273.

55. Skonberg D, Hardy RW, Barrows RT, Dong FM: **Color and flavour analyses from farm-raised rainbow trout (Onchorhyncus mykiss) fed low-phosphorus feeds containing corn or wheat gluten.** *Aquaculture* 1998, **166**:269–277.

56. Molina-Poveda C, Morales ME: **Use of a mixture of barley-based fermented grains and wheat gluten as an alternative protein source in practical diets for Litopenaeus vannamei (Boone).** *Aquacult Res* 2004, **35**:1158–1165.

57. Santigosa E, Sánchez J, Medale F, Kaushik SK, Pérez-Sánchez J, Gallardo MA: **Modifications of digestive enzymes in trout (Oncorhynchus mykiss) and**

sea bream (Sparus aurata) in response to dietary fish meal replacement by plant protein sources. *Aquaculture* 2008, **282**:68–74.

58. Sitjà-Bobadilla A, Peña-Llopisa S, Gómez-Requenia P, Medale F, Kaushik SK, Pérez-Sánchez J: Effect of fish meal replacement by plant protein sources on non-specific defence mechanisms and oxidative stress in gilthead sea bream (Sparus aurata). *Aquaculture* 2005, **249**:387–400.

59. Rumsey GL, Siwicki AK, Anderson DP, Bowser PR: Effect of soybean protein on serological response, non-specific defense mechanisms, growth, and protein utilization in rainbow trout. *Vet Immunol Immunopathol* 1994, **41**:323–339.

60. Burrells C, Williams PD, Southgate PJ, Crampton VO: Immunological, physiological and pathological responses of rainbow trout (Oncorhynchus mykiss) to increasing dietary concentrations of soybean proteins. *Vet Immunol Immunopathol* 1999, **72**:277–288.

61. Barrows FT, Gaylord TG, Stone DAJ, Smith CE: Effect of protein source and nutrient density on growth efficiency, histology and plasma amino acid concentration of rainbow trout (Oncorhynchus mykiss Walbaum). *Aquacult Res* 2007, **38**:1747–1758.

62. Lallès JP, Boudry G, Favier C, Le Floch N, Luron I, Montagne L, Oswald IP, Pié S, Piel C, Sève B: Gut function and dysfunction in young pigs: physiology. *Anim Res* 2004, **53**:301–316.

63. Yun JH, Kwon IK, Lohakare JD, Choi JY, Yong JS, Zheng J, Cho WT, Chae BJ: Comparative efficacy of plant and animal protein sources on the growth performance, nutrient digestibility, morphology and caecal microbiology of early-weaned Pigs. *Asian-Aust J Anim Sci* 2005, **18**:1285–1293.

64. Blasco M, Fondevila M, Guada JA: Inclusion of wheat gluten as a protein source in diets for weaned pigs. *Anim Res* 2005, **54**:297–306.

65. Vente-Spreeuwenberg MAM, Verdonk JMA, Koninkx JFJG, Beynen AC, Verstegen MW: A. Dietary protein hydrolysates vs. the intact proteins do not enhance mucosal integrity and growth performance in weaned piglets. *Livest Prod Sci* 2004, **85**:151–164.

66. Piccolo G, Centoducati G, Bovera F, Marrone R, Nizza A: Effects of mannan oligosaccharides and inulin on sharpsnout seabream (Diplodus puntazzo) in the context of partial fish substitution by soybean meal. *Ital J Anim Sci* 2013, **12**(1):133–138.

67. Kaushik SJ, Covès D, Dutto G, Blanc D: Almost total replacement of fish meal by plant protein sources in the diet of a marine teleost, the European seabass, Dicentrarchus labrax. *Aquaculture* 2004, **230**:391–404.

68. De Francesco M, Parisi G, Medale F, Lupia P, Kaushik SK, Polia BM: Effect of long-term feeding with a plant protein mixture based diet on growth and body/fillet quality traits of large rainbow trout (Oncorhynchus mykiss). *Aquaculture* 2004, **236**:413–429.

69. Messina M, Piccolo G, Tulli F, Messina CM, Cardinaletti G, Tibaldi E: Lipid composition and metabolism of European sea bass (Dicentrarchus labrax L.) fed diets containing wheat gluten and legume meals as substitutes for fish meal. *Aquaculture* 2013, **376–379**:6–14.

70. Sugiyama K, Ohkawa S, Muramatsu K: Relationship between amino acid composition of diet and plasma cholesterol level in growing rats fed a high cholesterol diet. *J Nutr Sci vitaminol* 1986, **32**:413–422.

71. Dias J, Alvarez MJ, Arzel J, Corraze G, Diez A, Bautist JM, Kaushik SJ: Dietary protein source affects lipid metabolism in the European seabass (Dicentrarchus labrax). *Comp Biochem Physiol* 2005, **142**:19–31.

72. Meton I, Mediavilla D, Caseras A, Canto E, Fernandez F, Baanante IV: Effect of diet composition and ration size on key enzyme activities of glycolysis–gluconeogenesis, the pentose phosphate pathway and amino acid metabolism in liver of gilthead sea bream (Sparus aurata). *Bi J Nutr* 1999, **82**:223–232.

73. Kaushik SJ, Cravedi JP, Lalles JP, Sumpter J, Fauconneau B, Laroche M: Partial or total replacement of fish meal by soybean protein on growth, protein utilization, potential estrogenic or antigenic effects, cholesterolemia and flesh quality in rainbow trout, Oncorhynchus mykiss. *Aquaculture* 1995, **133**:257–274.

74. Vilhelmsson OT, Martin SAM, Medale F, Kaushik SJ, Houlihan DF: Dietary plant-protein substitution affects hepatic metabolism in rainbow trout (Oncorhynchus mykiss). *Br J Nutr* 2004, **92**:71–80.

6

Indo-Pacific bottlenose dolphin (*Tursiops aduncus*) habitat preference in a heterogeneous, urban, coastal environment

Nardi Cribb[1*], Cara Miller[1,2,3] and Laurent Seuront[1,4,5]

Abstract

Background: Limited information is available regarding the habitat preference of the Indo-Pacific bottlenose dolphin (*Tursiops aduncus*) in South Australian estuarine environments. The need to overcome this paucity of information is crucial for management and conservation initiatives. This preliminary study investigates the space-time patterns of habitat preference by the Indo-Pacific bottlenose dolphin in the Port Adelaide River-Barker Inlet estuary, a South Australian, urbanised, coastal environment. More specifically, the study aim was to identify a potential preference between bare sand substrate and seagrass beds, the two habitat types present in this environment, through the resighting frequency of recognisable individual dolphins.

Results: Photo-identification surveys covering the 118 km^2 sanctuary area were conducted over 2 survey periods May to August 2006 and from March 2009 to February 2010. Sighting frequency of recognisable individual Indo-Pacific bottlenose dolphins established a significant preference for the bare sand habitat. More specifically, 72 and 18% of the individuals sighted at least on two occasions were observed in the bare sand and seagrass habitats respectively. This trend was consistently observed at both seasonal and annual scales, suggesting a consistency in the distinct use of these two habitats.

Conclusions: It is anticipated that these results will benefit the further development of management and conservation strategies.

Keywords: Bottlenose dolphin, Conservation, estuaries, Photo-identification, Management, Habitat type

Background

Cetacean habitats are typically heterogeneous, comprising a mosaic of patches which differ in their biological and physical properties [1]. Understanding the space-time movement patterns and distribution of organisms within their environments can provide insight into the preference of specific areas [1]; information considered essential in the development of management and conservation initiatives [2]. In this context, bottlenose dolphins (*Tursiops* spp.) are no exception. They occur globally in both temperate and tropical waters [3,4], and are common in coastal waters, in particular estuaries, over a wide range of habitat types, such as seagrass beds, sandy substrates and reefs [5-8]. The occurrence of bottlenose dolphins in different habitats illustrates the ecological plasticity and adaptability of this species [2,9-11]. This highlights the need to understand at the individual and population level the key habitat types and locations they preferentially frequent [12]. This is especially critical for populations frequenting coastal environments, which are increasingly impacted by anthropogenic activities, such as tourism, chemical and noise pollution, habitat degradation, commercial and recreational fisheries and aquaculture [13-19], thus making them more susceptible to threats [20,21].

The Indo-Pacific bottlenose dolphin (*Tursiops aduncus*) is a prime example of a coastal dolphin species with many populations throughout the Indo-Pacific region [22], and more specifically Australia, where they are found in a range of coastal environments such as bays, gulfs, lagoons and estuaries that are often highly urbanised [8,16,23,24]. However, little is still known about this species habitat preference in estuarine locations [16].

* Correspondence: nardi.cribb@flinders.edu.au
[1]School of Biological Sciences, Flinders University, Box 2100, Adelaide, SA 2001, Australia
Full list of author information is available at the end of the article

In South Australian waters, *T. aduncus* is a known resident, especially in the Port Adelaide River – Barker Inlet estuary, where animals have been recorded year-round over the past 18 years [25]. This area supports a small population of approximately 30 resident individuals as well as visiting non-regular transient animals [25,26]. Field observations indicate no other marine mammals, specifically delphinids, living in direct sympatry with this population. Fur seals and sea lions were, however, punctually observed hauled out within the study site. The Port Adelaide River – Barker Inlet estuary is situated in close vicinity to the city of Adelaide, hence it is highly urbanised and subjected to a variety of anthropogenic activities such as industrial and sewage pollution, recreational and commercial vessel traffic, dredging, urban development and habitat degradation [19,27-31]. As a result this area was proclaimed the Adelaide Dolphin Sanctuary (ADS) in 2005 in order to protect both the resident dolphins and their habitat [32].

Baseline habitat information is, however, still scarce and limited to the presence of bottlenose dolphins being independent of environmental features [8]. This potentially limits the development and implementation of effective conservation and management strategies, hence the long term-survival of this population. This also stresses the need to further understand and monitor the preference of habitats within this area at both the seasonal and annual scales, and to identify potential areas of high occurrence of specific individuals. In this context, the objective of this paper was to use photo-identification to assess whether recognisable individuals were consistently sighted in the same benthic habitat type at both seasonal and annual scales.

Methods

Study site
The ADS is situated in the north-eastern region of Gulf St Vincent (GSV), South Australia (Figure 1), located 15 km northwest of Adelaide. This area is characterised by high biodiversity and has both considerable commercial fisheries value and biological significance [33]. The sanctuary area which includes the Port Adelaide River – Barker Inlet estuary and the coastal waters extending northwards out into GSV covers an area of 118 km^2. In the absence of a map of the benthic habitat in the ADS, we conducted a preliminary sampling survey to assess the nature of the benthic habitat, which showed that the ADS supports two main benthic habitat types that may be used by dolphins (Figure 1). The northern part of the sanctuary extending into the open, unsheltered waters of Gulf St Vincent is characterised by the presence of seagrass beds (predominantly *Posidonia*, *Zostera* and *Heterozostera* sp.; [30,34]. No seasonal fluctuations in seagrass coverage were observed. In contrast, the southern area of the sanctuary consists of

shallow sheltered estuarine waters and narrow channels, bordered by mangrove forest, which are essentially devoid of vegetation such as seagrass and attached algae and consist predominantly of bare sand [35]. There is a distinct separation between these two habitat types from the mouth of the estuary out into gulf waters due to the presence of a seasonal sand bars, which constantly change the dynamics of the environment. Water depths in both habitat types range from 0.5 to 6 m, they increase in depth ranging from 10 to 17 m in the dredged shipping channel of the Port Adelaide River.

Data collection
Photo-identification data from the ADS were collected between the 5 May and 30 August 2006 and 6 March 2009 to 6 February 2010 (Table 1) following the same methodology. Survey transects were designed to provide both even and representative coverage of the sanctuary and the two benthic environment types found here. Specifically, four transects were used to survey the area (Figure 1). Surveys were always conducted at steady speed of 12 knots aboard either a 6 m rigid-hulled inflatable vessel powered by a 70 HP outboard engine, or a 5 m vessel powered by 70 HP outboard motor and were carried out at a Beaufort Sea state of less than 3, under daylight conditions, between 7:30 am and 3:00 pm and fluctuating tidal conditions. Whilst on transect a constant watch for dolphins was maintained by two observers who scanned the water with the naked eye ahead and to 90° either side of the transects. As boat access was limited in the estuary due the presence of exposed intertidal mud flats not accessible by dolphins and seasonal sand bars, sighting visibility was restricted to 200 m either side of the transect. Upon sighting an individual or group of dolphins (i.e. all animals within a 100 m radius of each other; [12]) the survey effort was ceased to record the time of the sighting and the number of dolphins present. The vessel was then moved as close to the location of the initial sighting as possible to determine the benthic environment type and record the GPS location. Benthic environment type was determined by visual analysis, as the bottom was visible due to the shallow nature and good water clarity. Note that in waters deeper than 10 m the bottom was not visible from the surface. Specifically in the dredged shipping channels of the Port Adelaide River, preliminary benthic sampling consistently showed the benthos to be devoid of vegetation. The benthic environment type was therefore defined within the study area by the presence of seagrass or bare sand. Once the benthic environmental data was recorded the vessel approached the individual or group and it was then endeavoured to photograph as many of the dorsal fins of the animals present as possible [36,37]. A Canon EOS 350D digital SLR with a 75–300 zoom lens was used to take all photographs. Encounters (i.e. an interaction with an individual

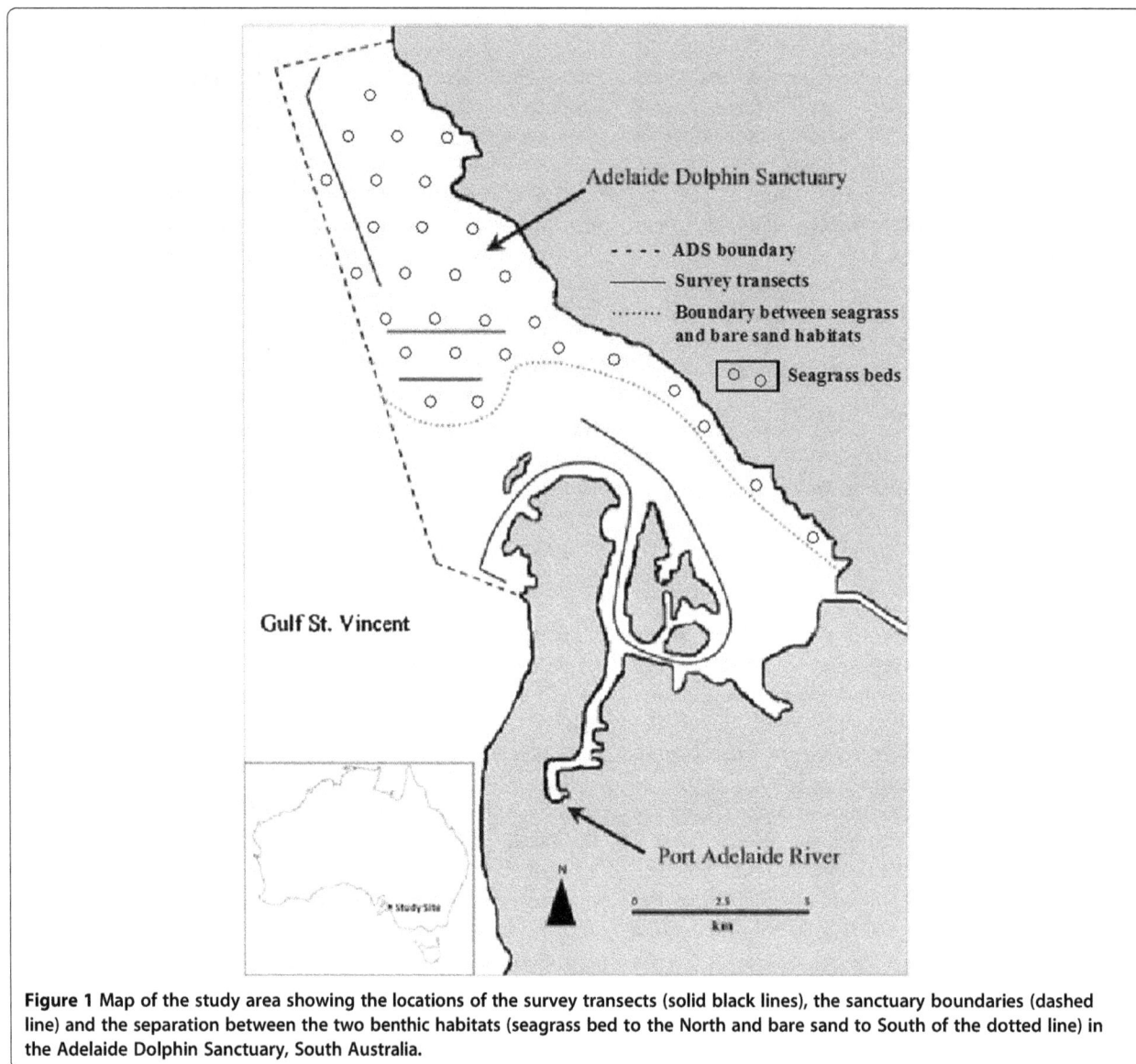

Figure 1 Map of the study area showing the locations of the survey transects (solid black lines), the sanctuary boundaries (dashed line) and the separation between the two benthic habitats (seagrass bed to the North and bare sand to South of the dotted line) in the Adelaide Dolphin Sanctuary, South Australia.

or dolphin group; [12]) were restricted to a maximum period of 20 minutes in order to attempt to minimise disturbance to the group or until all individuals in the group were photographed. The vessel then returned to the transect and continued until the transect was completed or all of the study area had been surveyed.

Table 1 Number of survey days shown as a function of both season and photo-identification survey periods

Season	Photo-Identification Survey Periods	
	Survey Period 2006	Survey Period 2009 – 2010
Spring	1	4
Summer	2	5
Autumn	3	6
Winter	4	8

Photo-identification analysis

Photo-identification of bottlenose dolphins relies on the matching of distinctive dorsal fin features, such as nicks and notches present on both the trailing and leading edges of the fin, and tip [36,37]. Photographs were assessed for photographic quality (e.g. focus, clarity, contrast, angle, portion of the fin visible and the percentage of picture filled by the fin) and graded according to quality (excellent, average, poor) using Adobe *Photoshop Elements 5.0* imaging software. Only those photographs considered to be of excellent quality were included in the analysis. Poor quality photographs were always discarded from the analysis. Photographs were checked systematically against each other to develop a master catalogue of recognisable individuals and to determine the number of re-sights. The individuals not matched with animals previously recorded were given a unique identification number and added to the catalogue.

Data analysis

The statistical package PWAS for Windows, version 18, was used for all statistical analysis. As the data failed to meet the assumptions of normality (Kolmgorov-Smirnov test, $p < 0.05$), non-parametric tests were therefore used to make comparisons between data sets. In order to explore the habitat preference of bottlenose dolphins in the ADS the resighting frequency of individuals (i.e. the sighting frequency of recognisable individuals seen at least on two or more occasions) was estimated for each benthic habitat type. Resighting frequencies were also assessed to identify potential habitat preference between seasons, defined as spring (September - November), summer (December - February), autumn (March - May) and winter (June - August), and years. Additionally, the resighting frequencies were examined to identify habitat preferences on an individual level. Sighting frequencies between habitats were compared using the χ^2 test [38]. Specifically, our survey equally covered the two habitat types; hence we compared the observed habitat preference frequencies to theoretical frequencies (50% – 50%).

Results

Survey and photo-identification effort

Twenty two survey days were completed during the two study periods (Table 1). An individual or group of dolphins were sighted on 126 occasions, which resulted in a total of 1602 photographs, and 502 of excellent quality used in the analysis. Although surveys were conducted on different tidal regimes, no effect of tide on the frequency of dolphin occurrence was ever observed. Note, however, that the microtidal regime [39] of the Adelaide Dolphin Sanctuary (and more generally in South Australian gulfs) is unlikely to affect the dynamics of bottlenose dolphins in contrast to megatidal areas such as Aberdeen harbour [40]. A total of 75 distinct individuals were identified based on permanent dorsal fin markings ranging from tip nicks to trailing and leading edge notches. The 75 distinct individuals photographed during the study were sighted between 1 and 8 times. Forty nine of these individuals (65.3%) were sighted on only one occasion. In contrast, 21 (28%) individuals were sighted on two or three occasions and only 5 (6.7%) were sighted on 4 or more occasions (Figure 2).

Habitat preference

The survey effort equally covered the two habitat types. Bottlenose dolphins were observed throughout the study area over both habitat types. However, the majority of sightings (i.e. 76%, n = 96) was concentrated in the bare sand habitat (χ^2 test, df = 1, $p < 0.05$; Figure 3A). A clear seasonal (Figure 3B) and inter-annual (Figure 3C) preference for one of the two habitat types was also observed, with individuals consistently sighted in the bare sand habitat over the four seasons. However, seagrass preference increased from 0 and 10% in winter and spring to 27 and 34% summer and autumn (Figure 3B). The preference for the bare sand habitat was consistent throughout the 3 years of the study (Figure 3C), suggesting that bare sand is the preferred habitat type used by bottlenose dolphins in this area.

Individual habitat preference

Recognisable individuals sighted in the ADS on two or more occasions showed a preference for habitat type. Twenty six dolphins were sighted on 2 or more occasions, and 18 of them (69%) were consistently resighted in the same habitat over time. Only 8 individuals (31%) were sighted both over bare sand and seagrass beds

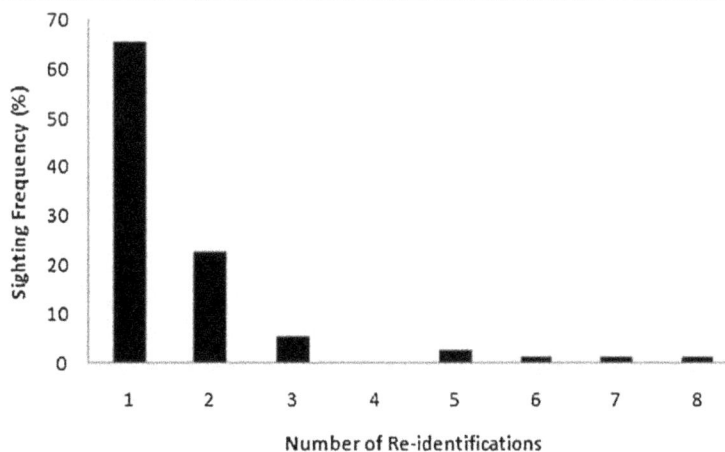

Figure 2 Sighting frequencies for individual dolphins identified in the Adelaide Dolphin Sanctuary in 2006 and between 2009 and 2010.

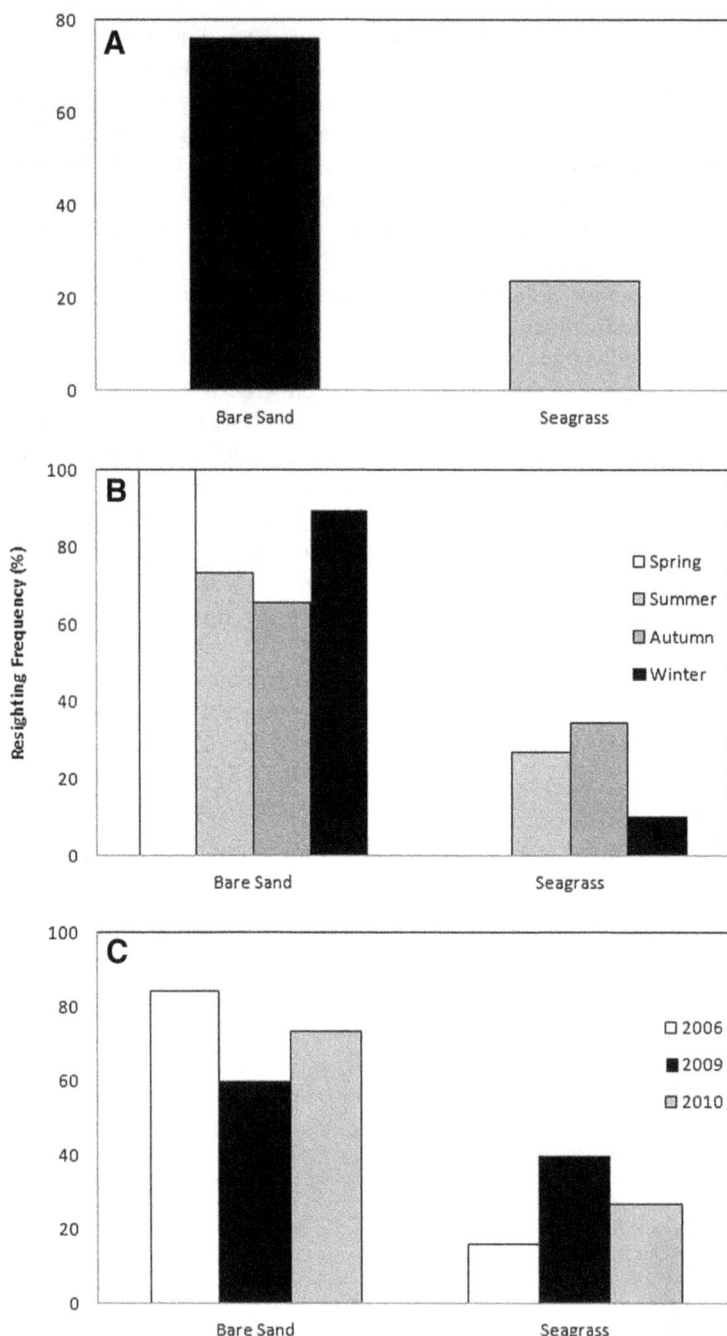

Figure 3 Sighting frequency of recognisable dolphins in the Adelaide Dolphin Sanctuary in relation to habitat type (bare sand and seagrass) over the duration of the whole study (A), and as a function of the season (spring: white; summer; light grey; autumn: dark grey; winter: black; B) and the year (2006: white; 2009: black; 2010: grey; C).

(Figure 4A). Additionally, from the 18 animals consistently sighted in the same habitat, 13 (72.2%) and 5 (27.8%) were respectively predominantly (χ^2 test, df = 1, p < 0.05) resighted in the bare sand and seagrass habitats over time (Figure 4B).

Discussion

Indo-Pacific bottlenose dolphin habitat preference in the ADS

Our observations of dolphin presence and significantly higher sighting frequency in the bare sand habitat (76%;

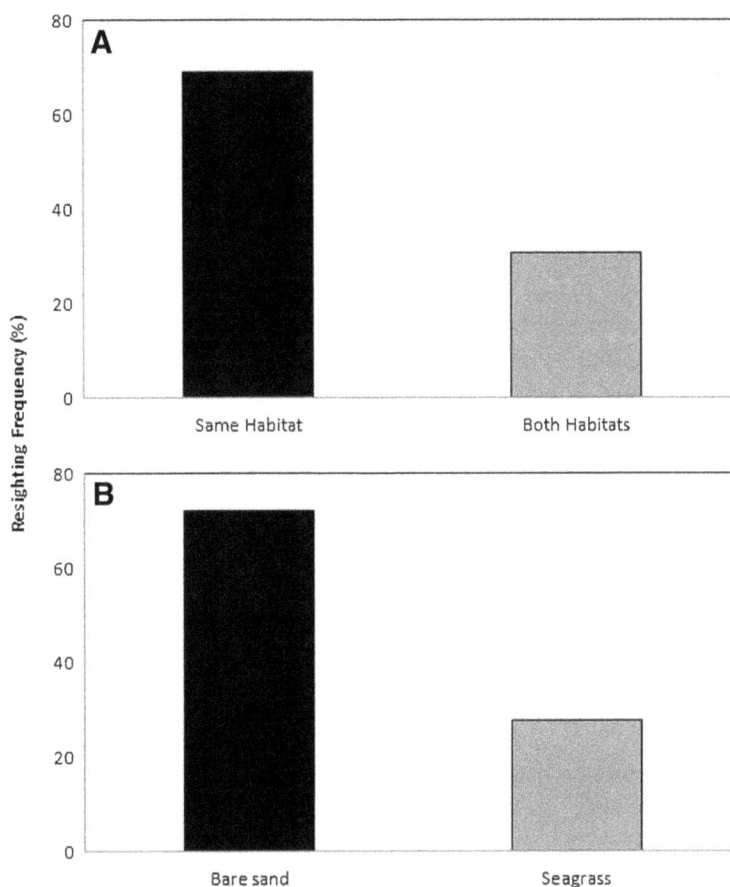

Figure 4 Resighting frequency of (A) individuals consistently sighted in the same habitat or sighted in both habitat types, and (B) only sighted in the same habitat as a function of habitat type.

Figure 3A) at both the seasonal and annual scales (Figure 3B,C) and the significantly higher resighting frequency in the same habitat (69%; Figure 4A) are consistent with the previously reported regular occurrence and preference of bottlenose dolphins in one habitat over another [1,6,7,41,42]. While further work is needed to specifically address this issue, our results suggest the presence of a seasonal pattern in habitat preference with an increase in dolphin frequency in the seagrass habitat in autumn and summer (Figure 3B). Seasonal shifts and variations in habitat preference by bottlenose dolphins have also been observed in other locations such as the San Luis Pass (Texas, USA; [43]), the Moray Firth (Scotland; [9]) and the Hauraki Gulf (New Zealand; [44]). However, the occurrence of nearly one-third of the individuals (31%; Figure 4A) over both the bare sand and the seagrass habitats may indicate that a non-negligible proportion of the *T. aduncus* occurring in the ADS have enough behavioural flexibility to use the seagrass beds found in the open waters of Gulf St. Vincent as well as the sheltered waters found in the inner estuarine part of the ADS (Figure 1). More specifically,

respectively 72 and 28% of the resighted individuals were observed over the bare sand and the seagrass habitats (Figure 4B). This suggests that the bare sand habitat may be a core area for this population, in contrast to previous work stressing the vital role of seagrass beds for bottlenose dolphins [41,42]. However, further investigation into the behavioural budget of this population is needed to determine how and why these habitats differ in their importance and use.

Estuaries as important dolphin habitats
Our observations of higher dolphin frequency in the bare sand habitat of the Adelaide Dolphin Sanctuary (ADS) further support the importance of estuarine waters for this species [12,16,45-49]. This may be linked to the overall nature of estuaries and their potential for high productivity and prey abundances [50]. Bare sand substrates may also provide a less complex habitat than seagrass in which to feed, particularly as seagrass beds impair their ability to echolocate to find prey [51]. In addition,, the consistent high occurrence of individuals at the seasonal and annual scales in the shallow and sheltered waters of the bare sand

habitat (Figure 3B,C) may also be related to threat avoidance, as bottlenose dolphin habitat preference is influenced by shark predation [52]. Specifically, in South Australian waters, dolphins are considered the primary prey of white sharks [53]. Although occasional, the white shark (*Carcharodon carcharias*) and the bronze whaler (*Carcharodon brachyurus*) both frequent the ADS [26]. Despite the relatively low occurrence of sharks in the ADS compared to other locations such as Sarasota (Florida), Moreton Bay (Queensland) and Shark Bay (Western Australia) [26,54-56], one dolphin observed during the study had, a large healed scar on the leading edge of its dorsal fin (Figure 5A). This scar is likely the result of a shark and not other sources such as boat strike, entanglement or other dolphins due to its distinct crescent-shape which contrasts with the deeper penetrating laceration caused by boat strikes and entanglements (Figure 5B), [56]. This suggests that predation may be a potential influencing factor for the high frequency of dolphin sightings in shallow and sheltered waters characterising the bare sand habitat. The bare sand habitat may hence provide a suitable haven from predators, in contrast to the open environment characterising the seagrass habitat.

On the influence of sex and social structure on habitat preference

The frequency of the same individuals within the same habitat over time (Figure 4) may be linked to other factors such as social organisation and association patterns [40]. Specifically, bottlenose dolphin habitat preference has been explained by the home range of individuals and the social strategies which individuals or different sexes adopt [1,57]. It has been suggested that protected, shallow and narrow waterways which are geographically further from the open ocean such as the bare sand environment in the present work (see Figure 1), generally promote limited movement patterns and therefore some degree of site fidelity [47,58]. In contrast, individuals found in open

habitats have more extensive home ranges and a lesser degree of site fidelity [47,58]. The individuals resighted consistently over time in the bare sand habitat may hence potentially represent resident individuals, and therefore those sighted on fewer occasions in the seagrass habitat may be transients. Additionally, this sighting frequency may be related to foraging or social specific strategies of male and females. Females have smaller home ranges and frequent more areas which provide a higher concentration of resources, such as estuaries that are important for reproduction and calving and the avoidance of predators [47,59]. In contrast, males cover wider ranges than females which has been attributed to female breeding cycles and accessibility [45,59]. As a consequence, the animals sighted consistently in the bare sand habitat might be females utilising local resources, whilst those sighted on fewer occasions in the seagrass may be males searching for females.

Conclusion

Our results show that bottlenose dolphins in the Adelaide Dolphin Sanctuary occur predominantly in a bare sand habitat. The consistent occurrence and resighting of individuals at both the seasonal and annual scale clearly highlight the importance of the sheltered, bare sand habitat for this population. With a paucity of information available on dolphin habitat due to a lack of monitoring and research in this area, these results provide critical information, which can improve conservation and management strategies previously implemented in the ADS [32]. Specifically, it is recommended to monitor future trends in dolphin spatial and temporal habitat preference, as initiated here through photo-identification surveys. Additionally, due to the presence and potential growth of anthropogenic activities in the vicinity of the ADS, it is critical to understand the details of the seasonal patterns of habitat preference and social activities of bottlenose dolphins that will ultimately help in objectively establishing restricted access to specific core locations of the Adelaide Dolphin Sanctuary.

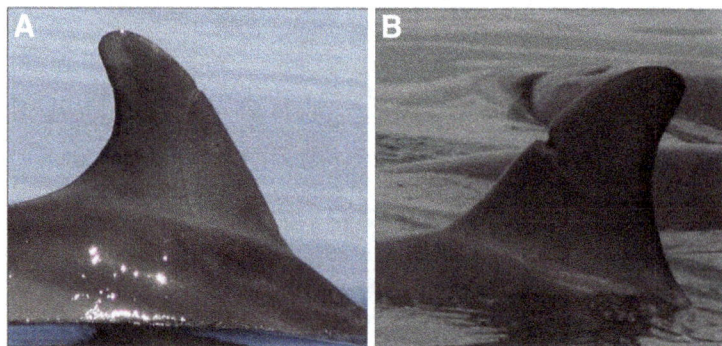

Figure 5 Examples of both natural (A) and anthropogenic (B) injuries observed on Indo-Pacific bottlenose dolphins photographed in the Adelaide Dolphin Sanctuary. Natural injuries considered to be inflicted by shark bites are crescent in shape, whilst in contrast those inflicted by anthropogenic causes (e.g. boat strike, entanglements) are usually deeper penetrating 'slash like' lacerations.

We also stress that the approach presented here may be applicable to other anthropogenically impacted coastal environments, where the identification of dolphin habitat preferences might have critical conservation and management implications. Finally, as the driving mechanisms influencing bottlenose dolphin habitat preferences may differ depending on the intrinsic properties of their environment, such as the nature of anthropogenic activities, coastal geomorphology and bottom topography, further studies are needed to understand habitat choice on both local and global scales.

Competing interests
The authors declare that they have no competing interests.

Authors' contributions
NC and LS designed the research. NC conducted field work and data analysis and wrote the final draft of the paper. CM contributed to previous drafts of the manuscript. LS assisted with statistical analysis and data interpretation and contributed to previous drafts of the manuscript. All authors contributed to the planning and design of the study and have read and approved the final manuscript.

Acknowledgements
Rob Laver, Verity Gibbs, Brett Williams and staff from the Department of Environment and Natural Resources are acknowledged for providing resources and support in the collection of data. Thank you to Clayton McCloud, Coraline Chapperon, Stephanie Duong, Peter Cribb, and Mike Bossley for their contribution to support in the field. We also thank William Allen for providing comment on an earlier draft of this manuscript. This research was financially supported under Australian Research Council's Discovery Projects funding scheme (projects number DP0664681 and DP0988554), Marine Innovation South Australia (MISA), the Department of Environment and Heritage and the Flinders Collaborative Research Grant Scheme (FCRGS) from the Faculty of Science and Engineering, Flinders University. Professor Seuront is the recipient of an Australian Professional Fellowship (project number DP0988554). Field work was done under ethics approval from Flinders University (Project E269).

Author details
[1]School of Biological Sciences, Flinders University, Box 2100, Adelaide, SA 2001, Australia. [2]Faculty of Science, Technology and the Environment, University of the South Pacific, Laucala campus, Suva, Fiji. [3]The Whale and Dolphin Conservation Society International, Chippenham, UK. [4]South Australian Research and Development Institute, Aquatic Science, West Beach, SA 5022, Australia. [5]Centre National de la Recherche Scientifique, Laboratoire d'Océanologie et de Géosciences, UMR LOG 8187, Université des Sciences et Technologies de Lille, Station Marine, Wimereux 62930, France.

References
1. Ballance LT: Habitat use patterns and ranges of the bottlenose dolphin in the Gulf of California, Mexico. Marine Mammal Sci 1992, 8:262–274.
2. Bearzi G, Azzellino A, Politi E, Costa M, Bastianini M: Influence of seasonal forcing on habitat use by bottlenose dolphins Tursiops truncatus in the North Adriatic Sea. Ocean Sci J 2008, 43:175–182.
3. Leatherwood S, Reeves RR: The Sierra Club handbook of whales and dolphins. San Francisco: Sierra Club Books; 1983.
4. Connor RC, Wells RS, Mann J, Read AJ: The bottlenose Dolphin: Social Relationships in a fission-fusion society. In Cetacean Societies: Field Studies of Dolphins and Whales. Edited by Mann J, Connor RC, Tyack PL, Whitehead H. Chicago: The University of Chicago Press; 2000:1–34.
5. Hansen MT, Defran RH: The behaviour and feeding ecology of the Pacific coast bottlenose dolphin, Tursiops truncatus. Aquatic Mammals 1993, 19:127–142.
6. Grigg E, Markowitz H: Habitat use by bottlenose dolphins (Tursiops truncatus) at Turneffe Atoll, Belize. Aquatic Mammals 1997, 23:163–170.
7. Allen MC, Read AJ, Gaudet J, Sayigh LS: Fine-scale habitat selection of foraging bottlenose dolphins Tursiops truncatus near Clearwater, Florida. Mar Ecol Prog Ser 2001, 222:253–264.
8. Cribb N, Miller C, Seuront L: Assessment of bottlenose dolphin (Tursiops aduncus) habitat characteristics in the estuarine waters of the Adelaide Dolphin Sanctuary, South Australia. J Mar Ani Ecol 2008, 1:1–3.
9. Wilson B, Thompson PM, Hammond PS: Habitat use by bottlenose dolphins: seasonal distribution and stratified movement patterns in the Moray Firth, Scotland. J Appl Ecol 1997, 34:1365–1374.
10. Shane SH: Residence patterns, group characteristics, and association patterns of bottlenose dolphins near Sanibel Island, Florida. Gulf Mexico Sci 2004, 22:1–12.
11. Sargeant BL, Wirsing AJ, Heithaus MR, Mann J: Can environmental heterogeneity explain individual foraging variation in wild bottlenose dolphins (Tursiops sp.)? Behav. Ecol Soc 2007, 61:679–688.
12. Ingram SN, Rogan E: Identifying critical areas and habitat preferences of bottlenose dolphins Tursiops truncatus. Mar Ecol Prog Ser 2002, 244:247–255.
13. Watson-Capps JJ, Mann J: The effects of aquaculture on bottlenose dolphin (Tursiops sp.) ranging in Shark Bay, Western Australia. Biol Conserv 2005, 124:519–526.
14. Bedjer L, Samuels A, Whitehead H, Gales N, Mann J, Connor R, Heithaus M, Watson-Capps J, Flaherty C, Krützen M: Decline in relative abundance of bottlenose dolphins exposed to long-term disturbance. Conserv Biol 2006, 20:1791–1798.
15. Wright AJ, Soto NA, Baldwin AL, Bateson M, Beale CM, Clark C, Deak T, Edwards EF, Fernández A, Godinho A, Hatch LT, Kakuschke A, Lusseau D, Martineau D, Romero ML, Weilgart LS, Wintle BA, Notobartolo-di-Sciara G, Martin V: Do marine mammals experience stress related to anthropogenic noise? Int J Comp Psychol 2007, 20:274–316.
16. Fury CA, Harrison PL: Abundance, site fidelity and range patterns of Indo-Pacific bottlenose dolphins (Tursiops aduncus) in two Australian subtropical estuaries. Mar Freshwater Res 2008, 59:1015–1027.
17. Lukoschek V, Chilvers BL: A robust baseline for bottlenose dolphin abundance in coastal Moreton Bay: a large carnivore living in a region of escalating anthropogenic impacts. Wildl Res 2008, 35:593–605.
18. Lavery TJ, Kemper CM, Sanderson K, Schultz CG, Coyle P, Mitchell JG, Seuront L: Heavy metal toxicity of kidney and bone tissues in South Australian adult bottlenose dolphins. Mar Environ Res 2009, 67:1–7.
19. Seuront L, Cribb N: Fractal analysis reveals pernicious stress levels related to boat presence and type in the Indo-Pacific Bottlenose dolphin, Tursiops aduncus. Physica A 2011, 390:2333–2339.
20. Stockin KA, Lusseau D, Binedell V, Wiseman N, Orams MB: Tourism affects the behavioural budget of the common dolphin Delphinus sp. in the Hauraki Gulf, New Zealand. Mar Ecol Prog Ser 2008, 355:287–295.
21. Wright AJ, Deak T, Parsons ECM: Concerns related to chronic stress in marine mammals. Inter Whaling Commission 2009, 1–7. SC/61/E16.
22. Ross GJB: Review of the conservation status of Australia's smaller whales and dolphins. Canberra: Australian Government; 2006:124.
23. Möller LM, Allen SJ, Harcourt RG: Group characteristics, site fidelity and seasonal abundance of bottlenose dolphins Tursiops aduncus in Jervis Bay and Port Stephens, south-eastern Australia. Aust. Mammal 2002, 24:11–21.
24. Kemper CM, Harcourt R, Gibbs S, Miller C, Wright A: Estimating population size of 'at risk' bottlenose and common dolphins in Spencer Gulf and Gulf St Vincent. Canberra: South Australia. Consultancy report to Department of the Environment and Heritage; 2006.
25. Kemper CM, Bossley M, Shaughnessy P: Marine mammals of gulf St vincent and investigator straight and backstairs passage. In Natural History of Gulf St Vincent. Edited by Shepherd S, Bryars S, Kirkegaard I, Harbison P. Adelaide: Royal Society of South Australia; 2008:339–352.
26. Steiner A, Bossley M: Some reproductive parameters of an estuarine population of Indo-pacific bottlenose dolphins (Tursiops aduncus). Mar Mammal Sci 2008, 2:34–63.
27. Edyvane K: Pollution: The death knell of our mangroves? SAFISH 1991, 16:47.
28. Edyvane K: Conserving marine biodiversity in South Australia - Part 2 - Identification of high conservation value in South Australia. Final Report No 39. Adelaide, South Australia: South Australian Research and Development Institute, Department of Primary Industries and Resources; 1999:328.

64   Aquatic Ecosystem Management

29. Edyvane K: *Environmental management review of the Barker Inlet and Port Estuary Environs*, Report to AGC Woodward-Clyde PTY LTD and Dames and Moore NRM; 2000:21.

30. Connolly RM: **A comparison of fish assemblages from seagrass and unvegetated areas of a Southern Australian estuary.** *Aust J Mar Freshwater Res* 1994, **45**:1033–1044.

31. Bryars S: *An inventory of important coastal fisheries habitats in South Australia.* South Australia: Fish Habitat Program, Primary Industries and Resources; 2003:1000.

32. *Adelaide Dolphin Sanctuary Act*; http://www.legislation.sa.gov.au/LZ/C/A/ADELAIDE%20DOLPHIN%20SANCTUARY%20ACT%202005.aspx.

33. Tanner J, Fairhead A, Marsh R, Miller D: *Environmental assessment of the dredge site at Outer Harbour: Berth 8 Option.* SARDI Aquatic Sciences: Report for GHD Pty Ltd; 2003.

34. Bloomfield A, Gillanders B: **Fish and invertebrate assemblages in seagrass, mangrove, saltmarsh, and nonvegetated habitats.** *Estuaries* 2005, **28**:63–77.

35. Jones GK, Baker JL, Edyvane K, Wright GJ: **Nearshore fish community of the Port River-Barker Inlet Estuary, South Australia. I. Effect of thermal effluent on fish community structure, and distribution and growth of economically important fish species.** *Mar Freshwater Res* 1996, **47**:785–799.

36. Würsig B, Würsig M: **The photographic determination of group size, composition, and stability of coastal porpoises (*Tursiops truncatus*).** *Science* 1997, **198**:755–756.

37. Würsig B, Jefferson TA: **Methods of photo-identification for small cetaceans.** In *Individual Recognition of Cetaceans: Use of Photo-Identification and other Techniques to Estimate Population Parameters.* Edited by Hammond PS, Mizroch SA, Donovan GP. Cambridge: Report of the International Whaling Commission, Special Issue 12; 1990.

38. Zar JH: *Biostatistical Analysis.* Upper Saddle River, NJ: Prentice Hall; 1996.

39. Tomczak M, Godfrey JS: *Regional Oceanography: An Introduction.* Oxford, GB: Pergamon; 1994.

40. Sini MI, Canning SJ, Stockin KA, Pierce GJ: **Bottlenose dolphins around Aberdeen harbour, north-east Scotland: a short study of habitat utilization and the potential effects of boat traffic.** *J Mar Biol Assoc UK* 2005, **85**:1547–1554.

41. Shane SH: **Behaviour and ecology of the bottlenose dolphin at Sanibel Island, Florida.** In *The Bottlenose Dolphin.* Edited by Leatherwood S, Reeves RR. San Diego: Academic; 1990:245–261.

42. Barros N, Wells RS: **Prey and feeding patterns of resident bottlenose dolphins (*Tursiops truncatus*) in Sarasota Bay, Florida.** *J Mammal* 1998, **79**:1045–1059.

43. Maze KS, Wursig B: **Bottlenose dolphins of San Luis Pass, Texas: Occurrence patterns, site-fidelity, and habitat use.** *Aquatic Mammals* 1999, **25**:91–103.

44. Berghan J, Algie KD, Stockin KA, Wiseman N, Constantine R, Tezanos-Pinto G, Mourao F: **A preliminary photo-identification study of bottlenose dolphin (*Tursiops truncatus*) in Haruaki Gulf, New Zealand.** *New Zeal J Mar Fresh Res* 2008, **42**:465–472.

45. Wells RS, Scott MD, Irvine AB: **The social structure of free-ranging bottlenose dolphins.** In *Current Mammalogy*, Volume 1. Edited by Genoways HH. New York: Plenum Press; 1987:247–305.

46. Wilson B, Hammond PS, Thompson PM: **Estimating size and assessing trends in a coastal bottlenose dolphin population.** *Ecol Appl* 1999, **9**:288–300.

47. Gubbins C: **Use of home ranges by resident bottlenose dolphins (*Tursiops truncatus*) in a South Carolina estuary.** *J Mammal* 2002, **83**:178–187.

48. Zolman ES: **Residence patterns of bottlenose dolphins (*Tursiops truncatus*) in the Stono River estuary, Charleston County, South Carolina. U.S.A.** *Mar Mammal Sci* 2002, **18**:879–892.

49. Irwin LJ, Wursig B: **A small resident community of bottlenose dolphins, *Tursiops truncatus*, in Texas: Monitoring recommendations.** *Gulf of Mexico Science* 2004, **22**:13–21.

50. Moyle PB, Cech JJ Jr: *Fishes: An introduction to ichthyology.* New Jersey: Prentice Hall Inc; 1982.

51. Nowacek DP: **Acoustic ecology of foraging bottlenose dolphins (*Tursiops truncatus*), habitat-specific use of three sounds types.** *Mar Mammal Sci* 2005, **21**:587–602.

52. Heithaus MR, Dill LM: **Food availability and tiger shark predation risk influence bottlenose dolphin habitat use.** *Ecology* 2002, **83**:480–491.

53. Bruce BD: **Preliminary observations on the biology of the white shark, *Carcharodon carcharias*, in South Australian waters.** *Aust Mar Freshwater Res* 1992, **43**:1–11.

54. Urian KW, Wells RS, Scott MD, Irvine AD, Read AJ, Hohn AA: *When the shark bites: An analysis of shark bite scars on wild bottlenose dolphins (*Tursiops truncatus*) from Sarasota Florida.* Monaco: The World Marine Mammal Conference; 1998.

55. Corkeron PJ, Morris RJ, Bryden MM: **A note on healing of large wounds in bottlenose dolphins, *Tursiops truncatus*.** *Aquatic Mammals* 1987, **13**:96–98.

56. Heithaus MR: **Shark attacks on bottlenose dolphins (*Tursiops aduncus*) in Shark Bay, Western Australia: Attack rate, bite scar frequencies, and attack seasonality.** *Mar Mammal Sci* 2001, **17**:526–539.

57. Wells R: **The role of long term study in understanding the social structure of a bottlenose dolphin community.** In *Dolphin societies: discoveries and puzzles.* Edited by Pryor K, Norris KS. Berkeley: University of California Press; 1991:199–225.

58. Defran RH, Weller DW: **Occurrence, distribution, site fidelity and school size of bottlenose dolphins (*Tursiops truncatus*) off San Diego, California.** *Mar Mammal Sci* 1999, **15**:366–380.

59. Quintana-Rizzo E, Wells RS: **Resighting and association patterns of bottlenose dolphins (*Tursiops truncatus*) in the Cedar Keys, Florida: insights into social organization.** *Can J Zool* 2001, **79**:447–456.

7

Community solar salt production in Goa, India

Kabilan Mani, Bhakti B Salgaonkar, Deepthi Das and Judith M Bragança[*]

Abstract

Traditional salt farming in Goa, India has been practised for the past 1,500 years by a few communities. Goa's riverine estuaries, easy access to sea water and favourable climatic conditions makes salt production attractive during summer. Salt produced through this natural evaporation process also played an important role in the economy of Goa even during the Portuguese rule as salt was the chief export commodity. In the past there were 36 villages involved in salt production, which is now reduced to 9. Low income, lack of skilled labour, competition from industrially produced salt, losses incurred on the yearly damage of embankments are the major reasons responsible for the reduction in the number of salt pans.

Salt pans (*Mithagar or Mithache agor*) form a part of the reclaimed waterlogged *khazan* lands, which are also utilised for aquaculture, pisciculture and agriculture. Salt pans in Goa experience three phases namely, the ceased phase during monsoon period of June to October, preparatory phase from December to January, and salt harvesting phase, from February to June. After the monsoons, the salt pans are prepared manually for salt production. During high tide, an influx of sea water occurs, which enters the reservoir pans through sluice gates. The sea water after 1–2 days on attaining a salinity of approximately 5°Bé, is released into the evaporator pans and kept till it attains a salinity of 23 - 25°Bé. The brine is then released to crystallizer pans, where the salt crystallises out 25 - 27°Bé and is then harvested.

Salt pans form a unique ecosystem where succession of different organisms with varying environmental conditions occurs. Organisms ranging from bacteria, archaea to fungi, algae, etc., are known to colonise salt pans and may influence the quality of salt produced.

The aim of this review is to describe salt farming in Goa's history, importance of salt production as a community activity, traditional method of salt production and the biota associated with salt pans.

Keywords: Salt pan, Goa, Estuary, Community, *Khazan*, Tidal influx, India, Salt production

Background

Goa, together with Daman and Diu, was a province under the Portuguese rule from 1510 and referred to as *Estado da India*. Goa was annexed by India on 19th December 1961 and liberated from the Portuguese rule [1]. Solar salt production in Goa had been an important activity through its history.

Goa experiences a tropical monsoon climate with hot summers followed by long monsoons from June to October. Goa has 9 rivers, most of them forming estuaries, the major being river Mandovi and Zuari. These rivers experience high tidal influx during summers and therefore the salinity varies during monsoon (2–3°Bé) and non-monsoon times (4–5°Bé) [2]. Various factors such as favourable climatic conditions and easy accessibility to sea water have aided salt production through natural evaporation in Goa.

Today, solar salt production has become a declining industry due to low income generated, competition from industrially produced iodized salt, yearly damage and repairs of the embankments and pollution. Currently there are 9 villages producing salt each having a few operational salt pans.

Historical background

Solar salt production in Goa, described as a traditional village industry, has been practiced for the past 1,500 years by various communities [3,4]. Since most of the rivers form estuaries and experience tidal influxes, salt production was started mainly in the coastal villages. Salt served as an important trade commodity too, playing an important role in the economy of Goa. The salt

* Correspondence: judith@goa.bits-pilani.ac.in
Department of Biological Sciences, BITS PILANI, K K Birla Goa Campus, Zuarinagar, Goa 403 726, India

produced in the Goan salt pans was considered to be of superior quality and was exported to Burma, Thailand and other Asian countries [5]. With the Portuguese colonization of Goa in 1510, the salt production gained a huge momentum because of the increased demand for consumption. Portuguese cuisine required surplus salt and it was also used in balancing the hull of ships on sailing. With the maritime dominance of Portuguese, salt produced in Goa was exported even to the Middle Eastern countries. Salt was thus the major export commodity of the 'Estado da India' through the Mormugao port [6-11].

Due to the 1878 Anglo – Portuguese Treaty, the British monopolised the salt production in Goa and resold it to the Portuguese. After the British took over the salt pans, the average quota of salt for an individual was reduced from 14.5 kg to 6.5 kg. This had forced many people to reduce the intake of salt which in turn resulted in hyponatremia [12]. With the annexion of Goa by India in 1961, the salt production through natural evaporation faced severe decline, which continues till date. Competition from iodized salt, availability of salt at less cost from other states and lack of skilled labour forced the salt pan operators to abandon the salt pans and look for other employment opportunities [13].

Areas where salt is produced

Goa has an area of 3,702 sq. km and lies between 14°54′ to 15°48′ North and 73°41′ to 74°26′ East, with a coastline of about 110 km. The Arabian Sea borders Goa on the west, Maharashtra to the north and Karnataka to the east and south. Goa consists of 443 villages with a population of 1.2 million. A former union territory, Goa was added as the 25th state of Indian union on 30th May, 1987. It is divided into 2 districts, North Goa and South Goa together consisting of 12 talukas [14]. Goa receives an annual rainfall of about 280 to 480 cm, most of it during the months of June to October. The salterns receive high intensity sunlight and strong winds, making salt production a successful activity in Goa only during the summer. Salt production was concentrated around in 36 villages mainly in the four talukas Pernem, Bardez, Tiswadi and Salcete. These villages lie on the estuaries of the Terekhol, Chapora, Baga, Mandovi, Zuari and Sal rivers (Figures 1 and 2). Currently, the number of salt producing villages has drastically reduced to 9 and the total area under current salt production is about 2,978 ha [15]. Because of the landscape and ownership of lands, all the salt pans in Goa come under the category of a small scale production, which is less than 4.04 ha and, owned by the private sector. In 1876, Goa's salt production was about 44 kton and in 1961, it reduced to 31 kton. In 2011, Goa's total salt production was a mere 2.1 kton. This is very low when compared with India's total salt production in 2011–2012, which was about 22179 kton [16].

India is the third largest salt producer only after China and USA. At the time of independence, India's annual salt production was 1900 kton making it an importer from United Kingdom and Adens. But within a short span of time, India filled the gap in supply chain and became an exporter of salt. The main contribution comes

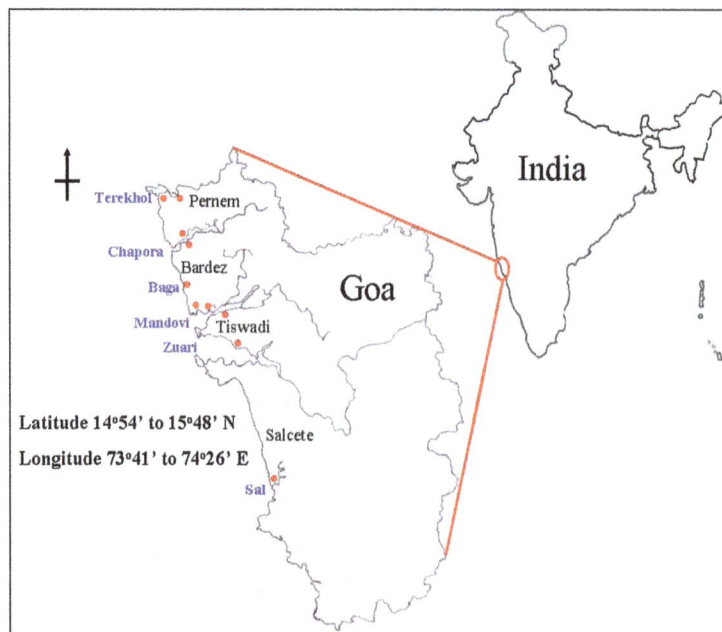

Figure 1 Map of Goa showing the four salt producing talukas (Pernem, Bardez, Tiswadi and Salcete) of Goa (located on the west coast of India) and their location on different river estuaries (Terekhol, Chapora, Baga, Mandovi, Zuari and Sal).

Figure 2 Salt production at Siridao located in Tiswadi taluka of Goa. Precipitated salt is heaped and kept for drying at the corners of the crystallizer pans.

from the states of Gujarat, Rajasthan and Tamil Nadu, which is about 90% of the country's total production [16]. Salt production methods vary widely among the different salt producing states. Apart from sea water, sub-soil brine, lake brine and rock salt deposits are also being used for salt production [17]. Private sector plays a major role in salt production, contributing 90.3%. Most of the salt pans have iodisation plants located nearby for fortifying the salt with iodine and iron. Looking at the scenario of other Indian salt producing states, Goa's salt production fails to meet its local demand.

Communities involved in salt production

A unique organizational structure called *comunidade*, headed by a hereditary descendant, involves in governing villages and regulating the agricultural activities in Goa. Each village constitutes a *comunidade* and has its own rules depending on the local customs. This is one of the oldest administrative setup, which is in existence for the past thousand years and has been recognized by the constitution. In the past, *comunidade* was responsible for reclaiming the waterlogged lands (*khazans*) along the coasts and making them suitable for agricultural activities, aquaculture, pisciculture and salt production. Therefore all the activities in these *khazan* lands were regulated by *comunidade* including salt production. The income obtained from these *khazan* lands were utilized for community development activities [18].

Five communities are involved in the salt production. They are *Mithgaudas, Gauddos, Bhandaris, Agris* and *Agers* [4,19]. They either own the salterns or are employed by one of these communities. The salt making art was pioneered by the ancestors of the *mithgauda* community known as '*Shamans*' [20]. The *mithgauda* is a subdivision of the *gauda / govada* community mainly settled in Corgaon and Agarwada region of Pernem taluka. They are believed to have migrated from the

konkan belt of Maharashtra [19]. Even though the procedure for producing salt followed by all the communities is the same; with some minor variations such as the collection of salt crystals from crystallizer pans. This indicates the evolution of salt production process within a community over a period of time. In the past salt production alone was the only source of income for the people of these communities. With the increase in profits and the importance emphasised by Portuguese on salt production, many people who had lands accessible to seawater started salt production. Salt pans also provided employment opportunities for the migrant workers from the neighbouring states. [4,20].

Process of salt production

Khazan lands of Goa are reclaimed mesohaline agricultural lands in the estuarine regions. At most of the places these *khazan* lands are surrounded by a thick lush of mangrove vegetation. The salinity and tidal influx is regulated by embankments (dykes/bunds/*mero*) and sluice gates (*manos*) [21]. These sluice gates are symbols of rich cultural heritage and engineering skills. *Khazans* are described as contour controlled, topo-hydro engineering, agro-economic and agro-ecological sustainable productive systems. These *khazan* lands are utilized for agriculture, pisciculture and salt production [22,23].

The salt pans *(Mithache Agor or Mithagar)* in Goa experience three phases; namely the monsoons ceased phase, the salt pan preparatory phase and the salt harvesting phase [2]. These salt pans are located in close proximity to the sea or may be located on the estuaries of a river. During monsoons, salt pans lie submerged in rain water and therefore abandoned or utilized for aquaculture for breeding fishes, shrimps and prawns [24].

a) Preparative phase (December to January)

The preparation of salt pans is carried out from December to January. The previous embankments (dykes/bunds) that were damaged due to the monsoon are repaired. Rain water / sea water from the salterns is drained using motor pumps. Once the water has been completely drained, the preparation of salt pan beds begins. The beds are ploughed, levelled by stamping and/ using a device called '*saalon*'. The extra clay is raked onto the walls of the bunds. *Saalon* has a long bamboo stick, approximately of 4 m in length, attached to a circular wooden base. During this process the borders of different pans are also constructed with clay.

The salterns consists of three distinct pans namely; reservoir pans (*tapovanim / tapounni*), evaporator pans (*podshing*) and crystallizer pans (*pikechem agor*) [25]. All the pans are inter-connected through an opening at the corners. Reservoir pan is used for receiving the sea water during tidal influxes and is connected to many evaporator pans. Crystallizer pans in turn is fed by the evaporator pans. The dimension of reservoir pan is 18–20 × 10–12 m, while that of the evaporator pan is of 18–20 × 6–8 m. In some salt pans the reservoir pan may be of the same size as evaporator pan; however the main difference is the depth of these two pans. Reservoir pans are around 20 in. deep, maintaining sea water level of upto 15–18 in. deep while the evaporator pans are 10 in. deep in which water is filled for upto 5 in.. The reservoir pan is twice or thrice the size of the crystallizer pan. The dimension of crystallizer pan is 6.5–8 × 4–5 m and the depth is of same as that of the evaporator pan. The brine level in the crystallizer pan is maintained at a maximum level of 3 in.. The size of evaporator pans plays a critical role in the production of salt. Bigger the size of evaporator pan, better the production of salt *(personal communication)*.

The reservoir pan is connected with the creek or canals, supplying seawater, during tidal influxes, through a sluice gate (*Manos*). Sluice is made of wood and the gates are made up of clay mixed with hay. This helps regulating the flow of water during the monsoon rains and tidal fluctuations. It helps in the controlled release of seawater into the reservoir pan during high tide and prevents the backflow of water during the low tide, thereby maintaining the level of water in the reservoir pan. Algal growth occurs in these pans which is harvested regularly and used as fertilizers for coconut and cashew plantations. The reservoir pans are also used for pisciculture especially for breeding salt water fishes, during the months of October to December [26].

Once the salinity of seawater in the reservoir pans is around 5°Bé, it is released to the first evaporator pan. Calcium carbonate ($CaCO_3$) starts precipitating at a salinity around 5°Bé [27] in the reservoir pan and completely precipitates in the first evaporator pan. Once the brine attains salinity around 13–15°Bé, it is released from the first evaporator pan to the second evaporator pan. In the second evaporator pan, calcium sulphate ($CaSO_4$) crystallizes in the form of gypsum. These precipitates form a hard crust at the bed of the evaporator pans. The brine, now having a salinity around 23–25°Bé, is released from the second evaporator pan to the crystallizer pan. Sodium chloride (NaCl) crystallizes around 27°Bé, first as flakes which float on the surface (*sai*) and then settle at the bottom of the pan (Figure 3). The brine in the crystallizer pan appears to be frothing due to the crystallisation of salt. The workers monitor the salinity of each pan by tasting the brine.

During the preparatory phase, seawater in the evaporator and crystallizer pan is allowed to stand and stirred time and again for about 20-25 days using a teeth shaped tool called '*danto*'. *Danto* has a long stick approximately 4 m in length, attached to a tool with teeth like projections [27]. The fed water is allowed to evaporate completely and the pans are fed again. This is done for removing the extra clay, which in turn will be raked

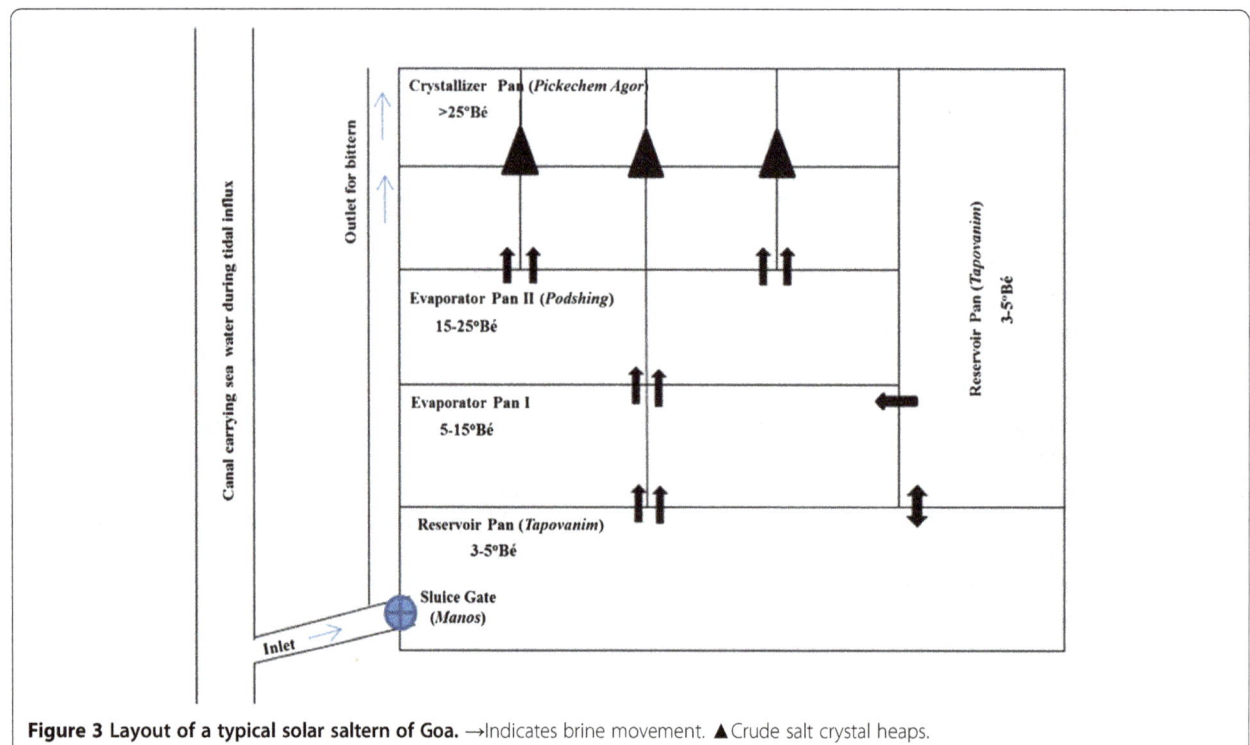

Figure 3 Layout of a typical solar saltern of Goa. →Indicates brine movement. ▲Crude salt crystal heaps.

onto the walls of the pan, thus further setting the salt pan beds. Most of the salterns will have a layout as described in Figure 3, however depending on the total area of the solar saltern, smaller salterns may not have separate evaporator pans.

Once the beds are set fresh brine is released from the reservoir to the evaporator pan and finally to the crystallizer pan for salt crystallization. It takes upto 10 days for the salt to crystallize during the first harvesting. Some crude salt (approximately 50–60 kg) is sprinkled over the crystallizer pans for aiding the salt crystallisation. This is repeated two to three times. When the salt crystallises, it is left on the pan unharvested for the initial three to four times. This is done for hardening the beds and making the beds uniform for further harvesting. The first salt harvested, contains lots of impurities mainly suspended clay particles and is brown to grey in colour [4,25].

b) Salt harvesting phase (February to May)
When the salt pans are completely prepared, the peak salt harvesting season begins, usually from mid-February and lasts till end of May or early June, depending on the monsoons.

The brine in the evaporator pan having a salinity of 23–25 °Bé is released to the crystallizer pan every morning. The NaCl crystallizes out at around 27°Bé in the crystallizer pan and is harvested in the evening on a daily basis. The salt crystals are harvested with the wooden rake, 'foyem' (a long stick of approximately 4 m in length, attached to a wooden rectangular block of 50–70 × 15–20 cm) and piled as small heaps at the intersection of the pans [2]. In the Pernem taluka, the salt crystals are heaped at the centre of the bunds. If the brine is kept for a longer time, salts of magnesium and potassium will co-precipitate, making the salt bitter and unfit for consumption. In such situations, the entire brine is let off in the drain and the process restarted. Once sodium chloride precipitates out, the remaining brine rich in magnesium and potassium (bittern/mother liquor) is drained off in to the canals *(personal communication)* (Additional file 1: Figure S1).

The harvested salt is further washed with the concentrated brine solution to remove impurities if any and allowed to drain on the bunds itself. It is then transferred with bamboo baskets into a large heap off the salt pans before transferring it to a store house. The salt produced is transported to the local market through pickup vans.

The salt produced in Goa is of two grades: (i) Fertilizer grade and (ii) Preservative / Consumption grade. The salt harvested initially (for a month) will be of fertilizer grade because of the impurities of mud and clay contained. It is mostly used as manure or as soil conditioner

and termite repellent for coconut trees. This salt is also used for the salting or preserving dry fish. The salt produced after the fine setting of the bed is the white consumption grade salt. This is used as a brine solution for pickling raw mangoes and in cuisine. The salt produced in different areas may render unique taste owing to the soil texture [25,26,28,29].

Studies on biodiversity of solar salterns
Solar salterns are extreme environments that act as a niche for organisms which thrive over a range of extreme salinities, temperatures, pH, nutrient concentrations, oxygen availability, water activity and solar radiation. Microorganisms which survive in such high salinities are known as halophilic extremophiles, which include bacteria, archaea and fungi. Solar salterns are man-made ecosystems showing diversity of microbial populations during different stages of salt crystallisation [30-34]. These microorganisms play a vital role in the recycling of materials (nutrients and other substances) in saltern ecosystem and therefore, important members of biogeochemical cycles [35,36]. Since the estuarine regions are situated in the vicinity of thickly populated areas, they experience pollution from anthropogenic activities. Adding to this, barges carrying mineral ores pollute the rivers and estuarine regions with crude oil and heavy metals. As a consequence of this, the brine gets concentrated along with the metals [37-41]. Diversity studies on bacteria indicated the presence of members belonging to genera *Aeromonas, Pseudomonas, Vibrio, Desulfobacter, Desulfovibrio, Desulfococcus* and *Chromohalobacter*. These bacteria play an important role in the cycling of substances within the saltern ecosystem [35,36,42,43]. Culture dependent haloarchaeal diversity studies in the salterns of Goa indicated that *Halococcus sp.* are the dominant haloarchaeal member during less saline conditions. During salt harvesting phase, members belonging to the genera *Halococcus, Halorubrum, Haloarcula* and *Haloferax* have been identified [2]. Culture independent studies reported the presence of novel archaeal members belonging to phylum *Crenarchaeota* and *Euryarchaeota* in various salterns of Goa [44]. Fungal communities in salt pans of Goa were dominated by members belonging to the genera *Aspergillus* and *Penicillium* such as *Aspergillus versicolor, A. wentii, A. candidus, A. penicilloides A. flavus, A. sydowii, Penicillium chrysogenum, P. corylophilum, P. griseofulvum, Eurotium amstelodami* and *Hortaea werneckii* [45].

For any ecosystem to be successful, transport of energy across the food web should be regulated. Algae act as the sole producer in the salt pans producing energy by trapping sunlight. They provide food for crustaceans like *Artemia sp.*, for birds to feed on [46]. The dominant components of phytoplankton community in Goan salt pans were reported to be the members of *Cyanophyceae,*

Chlorophyceae, Bacillariophyceae and *Dinophyceae*. The various species and their distribution showed strong correlation with the change in salinity. The salt pans contained algal members belonging to *Pediastrum* sp., *Oedogonium* sp., *Cladophora* sp., *Tetraselmis* sp., *Spirulina* sp. and *Spirogyra* sp. *Dichotomosiphon salina* and *Enteromorpha flexuosa*. Filamentous algae such as *Oscillatoria* sp. and *Phormidium* sp. and diatoms such as *Pleurosigma* sp., *Navicula* sp., *Chaetoceros* sp., *Amphora* sp., *Coscinodiscus* sp., *Surirella* sp. and *Nitzschia* sp. were common in all pans during less saline conditions but the growth of *Dunaliella* sp. were found to be dominating with the increase in salinity. More than 90% of the micro zooplankton in salt pans is dominated by ciliates viz., *Fabrea salina, Brachionus* sp., *Tintinnids, Indomysis* sp. and *Artemia* sp. *F. salina* were observed between December and May but they were seen to be absent during the monsoon period. The microzooplankton diversity and species richness was found to be poor in the Goan salt pans [47-52]. The salt pans represent a specific ecosystem which depends not only on the salinity but also on the temperature. It permits the growth of only a few adapted organisms which can survive the extreme variations in the environmental conditions.

Since most of the salt pans of Goa lie in estuarine regions, the plants typical of mangrove vegetation dominate the surrounding areas [53]. Vegetation found surrounding the salt pans and bordering the creeks in rows of 4–8 are *Rhizophora mucronata* and *Avicennia officinalis*. Few trees of *Bruguiera gymnorrhiza, Sonneratia acida, Kandelia rheedei, Excoecaria agallocha* are also present along with shrubs of *Acanthus ilicifolius* and *Bruguiera parviflora*. Among these species *R. mucronata* and *S. acida* has shown tolerance to salinity till 30°Bé [54]. These plants provide additional reinforcements for the embankments by checking the floods during rainy season.

Salt pans don't serve as nesting grounds for birds because of the continuous human disturbances. However a thick lush of mangroves provide shelter for many resident birds and migrating birds. However salt pans are excellent feeding grounds because of the availability of planktons like *Artemia sp.* [55]. Apart from the brine shrimps, fishes can also attract a large number of piscivorous birds. Birds like Indian Cormorant *Phalacrocorax fuscicollis*, Marsh sandpiper *Tringa stagnatilis*, Little Stint *Calidris minuta*, Jungle Myna *Acridotheres fuscus*, Plovers like *Charadrius dubius, Charadrius alexandrinus, Pluvialis fulva, Vanellus indicus*, Egrets like *Casmerodius albus, Egretta garzetta, Egretta gularis* and grey heron *Ardea cinerea* have been sighted regularly on the salt pans and they feed mostly on the crustaceans and fishes in and around the salt pans. Since, Goa is located in the Central Asian – Indian Flyway, the salt pans provide an attractive source of food for the migratory birds. Birds like pintail duck *Anas acuta*, northern shoveller *Anas clypeata*, storks like *Anastomus oscitans, Ciconia episcopus* and *Leptoptilos javanicus*, redshanks *Tringa tetanus*, sandpipers like *Actitis hypoleucos, Calidris ferruginea, Limicola falcinellus, Xenus cinereus* were found in the regions surrounding the salt pans in winters. Birds that were recorded in summer included brown-headed gulls *Larus brunnicephalus* and terns like *Gelochelidon nilotica, Sterna acuticauda* and *Chlidonias hybrida* [56,57]. Salt pans are shallow water bodies making it easy for birds to feed on the benthic communities and thereby regulating the growth of benthic invertebrates. Salt pans thus provide an excellent demonstration of energy flow through the food web.

Conclusions

Salt production has played an irreplaceable role in the lives of Goan people. It is important to take measures for preventing the loss of these unique ecosystems. Apart from producing salt, it can be employed for pisciculture, aquaculture and agriculture. Awareness of the proper physicochemical and biological management of the salt pans can lead in the co-cultivation of algae like *Dunaliella sp.*, shrimps, etc. and could be used as natural fermenters for large scale cultivation of halophilic archaea, along with salt harvesting. Salt pans are good model for studying the ecological succession of organisms ranging from microbes to avifauna.

Salt making process is simple but extremely laborious. The workers working in these salt pans experience fatigue as they have to put in 8–12 h of work under the merciless sun, many times with a bending posture. The workers collect the salt with bare hands and walk about bare foot to maintain cleanliness. Although, saline water has been used as a therapy to ease arthritic pain, excess of exposure to salt leads to skin rashes. Reflection of the sun or UV rays off the salt also leads to a blinding effect [58]. The workers should be provided with gumboots, gloves and goggles to protect them from the high solar radiation. Simple machinery could be introduced for mechanised salt collection and piling.

Embankments should be constructed strong enough for withstanding the floods during rainy season and it should be checked at frequent time intervals for any leakages. The present government is now encouraging salt farming in traditional salt pans through the announcements of subsidies or financial support schemes [23]. Cooperative societies can be established for procuring the salt produced from the manufacturers and distributing it in the markets. The salt pans could also be encouraged for ecotourism. It is of utmost importance to protect the existing salt pans and thereby protect their contribution to the global ecological scenario.

Additional file

Additional file 1: Figure S1. Traditional salt harvesting during various stages of salt production (a-d) in Goa with the different tools (f and e) employed during the process and the use of the area for pisciculture (g).

Competing interests
The authors declare that they have no competing interests.

Authors' contributions
All the authors conducted field visits to various saltpans, interviewed various salt pan workers and conducted extensive literature survey. All authors have written, read and approved the final manuscript.

Acknowledgements
KM and BBS acknowledge Council of Scientific and Industrial Research, India (CSIR) for award of Senior Research Fellowship (09/919(0017)/2012-EMR-I) / (09/919(0016)/2012-EMR-I). The authors thank Ms. Bilva Salgaonkar for sketching the figures. The authors are grateful to Prof. Nicholas A. Korovessis, Director of Hellenic Saltworks S.A. and the other anonymous reviewers for their valuable suggestions in greatly improving this manuscript.

References
1. Pinho V: *"Snapshots" of Indo-Portuguese History*, Volume I. Panaji: Vasco Pinho; 2007.
2. Mani K, Salgaonkar BB, Braganca JM: Culturable halophilic archaea at the initial and final stages of salt production in a natural solar saltern of Goa, India. *Aquat Biosyst* 2012, 8:15.
3. Furtado I, Fernandes CFE: Traditional Salt Production in Goa- India. Enriches Diverse Microbial Resource. In *Proceedings of the 9th International Symposium on Salt 4–6 September 2009*, Volume A. Edited by Zuoliang S. Beijing: Gold Wall Press; 2009:781–786.
4. Sequeira R: *Mitagars of Goa (A Sociological study of a Community in Transition). PhD thesis*. Goa University, Department of Sociology; 2009.
5. Pinto C: At the Dusk of the Second Empire: Goa – Brazil Commercial Links, 1770–1826. *Purabhilekh Puratatva* 1990, 7:1770–1825.
6. Pinto C: Goa-Based Overseas and Coastal Trade 18th and 19th Centuries. In *Goa through the ages and economic history*, Volume II. Edited by De Souza TR. New Delhi: Concept Publishing Company; 1989:176–212.
7. Scammell GV: England, Portugal and the Estado da India c. 1500–1635. *Mod Asian Stud* 1982, 16:177–192.
8. Lobo ADV: The salt industry. *Goa Today* 1967.
9. Gracias JBA: O Caminho de Ferro e Porto de Mormugao. In *O Oriente Portugues*, Volume II. Edited by Tipografia R; 1940:334–335.
10. Pereira A: *History of MPT*. Goa: Mormugao Port Trust; 1981.
11. Souza GB: Imperial foundations: the Estado da India and Macao. In *The Survival of Empire: Portuguese Trade and Society in China and the South China Sea 1630–1754*. Cambridge: Cambridge University Press; 2009:12–29.
12. Nagvenkar HT: *Salt and Goan Economy (A Study of Goa's Salt Trade in the 19th and 20th centuries during the Portuguese). PhD thesis*. Goa University: Department of Economics; 1999.
13. Almeida JC: A profile- economic and financial features of Goa. In *Techno – economic survey of Goa, Daman and Diu.*: Goa Economic Development Corporation; 1985.
14. Xavier PD: *Goa, a Social History, 1510–1640*. Panaji: Rajhauns Vitaran; 1993.
15. Panigrahy S, Singh TS, Patel JG, Murthy TVR, Inamdar A: *National Wetland Atlas: Goa*. Ahmedabad: Space Applications Centre (ISRO); 2009.
16. *Salt Department: Annual Report 2011-2012*. Jaipur: Ministry of Commerce & Industry; 2012.
17. Dravid SK: *Development of salt industry in India*. Jaipur: Upma Prakashan; 1972.
18. De Souza TR: *Goa Through the Ages: An economic history*, Volume 2. New Delhi: Concept Publishing Company; 1990.
19. Singh KS, Shirodkar PP, Mandal HK: In *People of India: Goa*, Volume XXI. Edited by Singh KS. Bombay: Anthropological Survey of India, Popular Prakashan; 1993.
20. Kamat N: Reviving the salt industry in Goa. *Navhind Times* 2005.
21. Sonak S, Kazi S, Abraham M: *Khazans in troubled waters*. New Delhi: TERI Press; 2005:54.
22. Kamat N: History of *Khazan* land management in Goa: ecological, economic and political perspective. In *Seminar on History of agriculture in Goa*. Goa University; 2004.
23. *Goa Government Budget Speech*; www.goa.gov.in/pdf/speech12-13.pdf.
24. Rubinoff JA: Pink Gold: Transformation of Backwater Aquaculture on Goa's Khazan Lands. In *Economic and Political Weekly*, Volume 36; 2001:1108–1114.
25. Kamat N: *Santa Cruz: The Ecological Mysteries of a Goan Village*; http://www.goa-world.com/goa/about_goa/santac.htm.
26. Fernandes TEC: *Salt pan ecology and its impact on community structure of halophilic Archaea. PhD thesis*. Goa University: Department of Microbiology; 2006.
27. Korovessis NA, Lekkas TD: Solar Saltworks' wetland function. *Global NEST J* 2009, 11:49–57.
28. Parrikar R: *Salt of the Earth*; www.parrikar.com/blog/2011/05/27/salt-of-the-earth/.
29. Khedekar V: Salt of the land. *Navhind Times* 2011.
30. Salgaonkar BB, Kabilan M, Nair A, Sowmya G, Braganca JM: Interspecific interactions among members of family halobacteriaceae from natural solar salterns. *Probiotics Antimicrob Proteins* 2012, 4(2):98–107.
31. Raghavan TM, Furtado I: Occurrence of extremely halophilic archaea in sediments from the continental shelf of west coast of India. *Curr Sci* 2004, 86:1065–1067.
32. Braganca JM, Furtado I: Isolation and characterization of haloarchaea from low-salinity coastal sediments and waters of Goa. *Curr Sci* 2009, 96:1182–1184.
33. Sequeira F: *Microbiological study of salt pans of Goa. Master thesis*. Goa University: Department of Microbiology; 1992.
34. Braganca JM: Microbiology of salt pans. *Navhind Times* 2005.
35. Kerkar S, Loka Bharathi PA: Stimulation of sulphate reducing activity at salt saturation in the saltern of Ribandar, Goa, India. *Geomicrobiol J* 2007, 24:101–110.
36. Kerkar S, Loka Bharathi PA: G model revisited: seasonal changes in the kinetics of sulfate reducing activity in the saltern of Ribander, Goa, India. *Geomicrobiol J* 2011, 28:187–197.
37. Braganca JM: *Uptake of arsenic and cadmium by halophilic archaebacteria. PhD thesis*. Goa University: Department of Microbiology; 2003.
38. Khandavilli S, Sequiera F, Furtado I: Metal tolerance of extremely halophilic bacteria isolated from estuaries of Goa, India. *Ecol Env Cons* 1999, 5:149–152.
39. Raghavan TM, Furtado I: Tolerance of an estuarine halophilic archaebacterium to crude oil and constituent hydrocarbons. *Bull Environ Contam Toxicol* 2000, 65:725–731.
40. Raghavan TM, Furtado I: Expression of carotenoid pigments of haloarchaeal cultures exposed to aniline. *Environ Toxicol* 2005, 20:165–169.
41. Aguiar R, Furtado I: Growth of Halobacterium strain R1 on sodium benzoate. In *Perspectives in Microbiology*. Edited by Kahlon RS. India: National Agricultural Technology Information Centre; 1996:78–79.
42. Kerkar S, Raiker L, Tiwari A, Mayilraj S, Dastager SG: Biofilm associated indole acetic acid producing bacteria and their impact in the proliferation of biofilm mats in solar salterns. *Biologia* 2012, 67:454–460.
43. Salgaonkar BB, Kabilan M, Braganca JM: Sensitivity of Haloarchaea to eubacterial pigments produced by Pseudomonas aeruginosa SB1. *World J Microbiol Biotechnol* 2011, 27:799–804.
44. Ahmed N, Johri S, Sultan P, Abdin MZ, Qazi GN: Phylogenetic Characterization of Archaea in Saltpan Sediments. *Indian J Microbiol* 2011, 51:132–137.
45. Nayak SS, Gonsalves V, Nazareth SW: Isolation and salt tolerance of halophilic fungi from mangroves and solar salterns in Goa-India. *Indian J Geomarine Sci* 2012, 41:164–172.
46. Rahaman AA, Jeyalakshui R: Integration of Artemia in Indian salt works economic opportunities. In *Proceedings of the 2nd International Conference on the Ecological Importance of Solar Saltworks (CEISSA2009): 26–29 March 2009*. Merida; 2009:51–56.
47. Rahman AA: Plankton communities in hypersaline waters of Indian solar salt work. In *Proceedings of 1st International Conference on the ecological importance of Solar Saltworks (CEISSA2006): 20–22 October 2006*. Santorin; 2006:20–22.

48. Preetha K, John L, Subin CS, Vijayan KK: **Phenotypic and genetic characterization of Dunaliella (Chlorophyta) from Indian salinas and their diversity.** *Aquat Biosyst* 2012, **8**:27.

49. Arora M, Chandrashekar AA, Leliaert F, Delany J, Mesbahi E: **Tetraselmis indica (Chlorodendrophyceae, Chlorophyta), a new species isolated from salt pans in Goa, India.** *Eur J Phycol*, in press.

50. Mustafa S, Nair VR, Govindan K: **Zooplankton community of Bhayandar and Thane salt pans around Bombay.** *Indian J Mar Sci* 1999, **28**:184–191.

51. Modassir Y, Ansari A: **Plankton community of the hypersaline salterns of Goa, India.** *Biol Forum* 2011, **3**:78–81.

52. Goswami SC: **Zooplankton ecology of the mangrove habitats of Goa.** In *Tropical ecosystems: Ecology and management.* Edited by Singh KP, Singh JS. New Delhi: Wiley Eastern; 1992:321–332.

53. Jagtap TG, Naik S, Nagle VL: **Assessment of Coastal Wetland Resources of Central West Coast, India, using LANDSAT Data.** *J Indian Soc Remote Sens* 2001, **29**:3.

54. Untawale AG, Parulekar AH: **Some observations on the ecology of an estuarine mangrove of Goa.** *Mahasagar* 1976, **9**:57–62.

55. Borges SD, Shanbhag AB: **Food Resource Partitioning among Water Birds Wintering on the Diwar Wetland in Mandovi Estuary of Goa, India.** In *Proceedings of Taal 2007; The 12th World Lake Conference;* 2008:124–130.

56. Borges SD: *Studies on the ecology of wader birds in the Mandovi estuary of Goa. PhD thesis.* Goa University: Department of Zoology; 2002.

57. *Real birder;* www.realbirder.com/Goa-Arpora.htm.

58. D'Souza G: **Tears of salt.** *Navhind Times* 2005, 3 July, 2005.

8

An annotated list of fish parasites (Isopoda, Copepoda, Monogenea, Digenea, Cestoda, Nematoda) collected from Snappers and Bream (Lutjanidae, Nemipteridae, Caesionidae) in New Caledonia confirms high parasite biodiversity on coral reef fish

Jean-Lou Justine[1]*, Ian Beveridge[2], Geoffrey A Boxshall[3], Rodney A Bray[3], Terrence L Miller[4], František Moravec[5], Jean-Paul Trilles[6] and Ian D Whittington[7]

Abstract

Background: Coral reefs are areas of maximum biodiversity, but the parasites of coral reef fishes, and especially their species richness, are not well known. Over an 8-year period, parasites were collected from 24 species of Lutjanidae, Nemipteridae and Caesionidae off New Caledonia, South Pacific.

Results: Host-parasite and parasite-host lists are provided, with a total of 207 host-parasite combinations and 58 parasite species identified at the species level, with 27 new host records. Results are presented for isopods, copepods, monogeneans, digeneans, cestodes and nematodes. When results are restricted to well-sampled reef fish species (sample size > 30), the number of host-parasite combinations is 20–25 per fish species, and the number of parasites identified at the species level is 9–13 per fish species. Lutjanids include reef-associated fish and deeper sea fish from the outer slopes of the coral reef: fish from both milieus were compared. Surprisingly, parasite biodiversity was higher in deeper sea fish than in reef fish (host-parasite combinations: 12.50 vs 10.13, number of species per fish 3.75 vs 3.00); however, we identified four biases which diminish the validity of this comparison. Finally, these results and previously published results allow us to propose a generalization of parasite biodiversity for four major families of reef-associated fishes (Lutjanidae, Nemipteridae, Serranidae and Lethrinidae): well-sampled fish have a mean of 20 host-parasite combinations per fish species, and the number of parasites identified at the species level is 10 per fish species.

Conclusions: Since all precautions have been taken to minimize taxon numbers, it is safe to affirm than the number of fish parasites is at least ten times the number of fish species in coral reefs, for species of similar size or larger than the species in the four families studied; this is a major improvement to our estimate of biodiversity in coral reefs. Our results suggest that extinction of a coral reef fish species would eventually result in the coextinction of at least ten species of parasites.

Keywords: Biodiversity, Coral reefs, Parasites, Coextinction, Lutjanidae, New Caledonia, South Pacific

* Correspondence: justine@mnhn.fr
[1]UMR 7138 Systématique, Adaptation, Évolution, Muséum National d'Histoire Naturelle, Case postale 51, 55, rue Buffon, 75231 Paris cedex 05, France
Full list of author information is available at the end of the article

Background

Parasites probably constitute the least known component of biodiversity in coral reefs, which are considered some of the most diverse ecosystems on the planet [1]. An early evaluation of parasite biodiversity of fish of the Great Barrier Reef (GBR) in Australia proposed a number of 20,000 parasites (all groups included) in the 1,000 fish species believed to exist in the area at this time; however, this evaluation, published as short papers [2,3] was based on very limited data. More reliable estimates are available for only two groups, the digeneans and monogeneans. Estimates were 2,270 digenean species in the 1,300 fish species of the GBR [4] and 2,000 monogenean species on the 1,000 fish species recorded around Heron Island, in the southern GBR [5].

An eight-year program allowed us to investigate the biodiversity of fish parasites off New Caledonia (South Pacific), the largest coral lagoon of the world. A compilation of available literature including a number of papers produced by this program [6] concluded that only 2% of fish parasite biodiversity was known in New Caledonia. Two subsequent comprehensive papers provided abundant, previously unpublished data and a compilation of already published information on two families of fish, the Serranidae (groupers) and the Lethrinidae (emperors) [7,8]. In this paper, we provide information about the parasites of the Lutjanidae, Nemipteridae and Caesionidae and compare our results with those already published for the other families.

Results

Results are presented as a host-parasite list (Appendix 1), a parasite-host list (Appendix 2) and a list of material deposited (Appendix 3). The number of host-parasite combinations (HPCs) and the number of species-level identified parasite – host-parasite combinations (SLIP-HPCs) found in each fish species are given in Table 1.

Discussion

Comments on each group

For brevity, in this section references to our own published papers on these fish families (available in Table 2 and Appendix 2) are kept to a minimum. For parasites, the "minimized number of taxa" is a cautious minimized evaluation in which all unidentified taxa in a group are counted as a single taxon [8].

Fish

In this paper, we group results from three families of fish, namely the Lutjanidae, Nemipteridae and Caesionidae. Clearly, most of the results concern the Lutjanidae but we included the two other families because they are closely related [9-11]. Modern molecular phylogenies are

available for the Lutjanidae [12-15] and confirm the close relationship of the Lutjanidae and Caesionidae.

According to the most recent survey [16], the Lutjanidae, Caesionidae and Nemipteridae include, respectively, 17, 4, and 5 genera and 108, 22 and 66 species, with a total of 26 genera, 196 species. The numbers of species in New Caledonia [17] are, respectively, 43, 13, and 16, with a total 72 species. In this work, we report parasitological results from 18 lutjanid species, 1 caesionid and 6 nemipterids; the total, 25 species, represents 34% of the species reported from New Caledonia, and 13% of the world number of species for the three families.

Diets of lutjanids and nemipterids off New Caledonia mainly comprise fish, crustaceans and occasionally molluscs [18], all of which can serve as intermediate hosts for parasites such as nematodes, digeneans and cestodes.

Most fishes included in this study are reef-dwelling; however, we also include several lutjanids (two species of *Etelis* and three species of *Pristipomoides)* which are deeper water fishes, collected from the outer slope of the barrier reef of New Caledonia [19]. These fishes provide data for a comparison of the parasitic fauna of coral-associated and deeper sea fishes.

As occurs often in the South Pacific, parasitologists have had to face problems with fish taxonomy [8,20-23]. *Pentapodus aureofasciatus* Russell, 2001, was first identified as *Pentapodus* sp. in the description of a nematode [24] but this was corrected later [25].

Isopoda

Adult isopods were rare and belonged to three families: Aegidae, Corallanidae and Cymothoidae. The single aegid, *Aega musorstom*, was found on a deep water lutjanid. Two cymothoids (*Anilocra gigantea* and *An. longicauda*) were found only on deep water lutjanids, but the single corallanid, *Argathona macronema*, was found on a coral dwelling lutjanid.

An. gigantea was already known from New Caledonia and was recorded from the branchial region of the deep water lutjanid *Etelis carbunculus* off "Banc de la Torche, au sud-est de la Nouvelle Calédonie" [26]. It was also recorded from the Pacific Ocean from the gills of *Epinephelus* sp. and *Pr. flavipinnis*, off Suva reefs, Suva, Fiji [27] and from the Indian Ocean from an unidentified host [28-30]. We found this species again on *Et. carbunculus*, but *Et. coruscans* and *Pr. filamentosus* are new host records. Interestingly, we did not collect this species from the branchial region or from the gills of the host fish, as reported by previous authors but on the anterior part of the body just behind the head. A female specimen of *An. gigantea* attached behind the head of *Pr. filamentosus* is illustrated by a colour photograph (Figure 1).

An. longicauda was already known from the Indian and the Pacific Oceans [30]. It was previously recorded

Table 1 Number of host-parasite combinations (HPCs) found in 24 species of caesionids, lutjanids and nemipterids in New Caledonia

Family and habitat	Fish species	Total	Gill	Abdo	Isop	Cope	Mono	Poly	Dige	Both	Tetr	Tryp	Nema	Other	Total
Caesionidae	Caesio cuning	8	4	8					2(0)	1(0)	1(0)		1(0)		5(0)
Lutjanidae, reef-associated	Aprion virescens	3	3	3			1(1)		2(2)						3(3)
	Lutjanus adetii	5	0	5					2(2)	1(0)			1(0)		4(2)
	Lutjanus argentimaculatus	4	3	3	1(0)	2(1)	5(3)		4(4)		1(0)				13(8)
	Lutjanus fulviflamma	17	11	11		2(1)	3(2)	1(0)	2(2)				1(0)		9(5)
	Lutjanus fulvus *	2	1	0			4(2)								4(2)
	Lutjanus gibbus *	2	2	1		1(1)	1(0)								2(1)
	Lutjanus kasmira	16	12	12		1(0)	6(0)		5(2)	1(0)	1(0)		1(0)		15(2)
	Lutjanus monostigma *	0	0	0	1(1)										1(1)
	Lutjanus quinquelineatus	12	0	6			5(3)		2(2)		1(0)		1(0)		9(5)
	Lutjanus rivulatus *	2	2	2			1(0)						1(0)	1(0)	3(0)
	Lutjanus russellii	6	0	1	1(0)		4(2)	1(0)	2(1)						8(3)
	Lutjanus vitta	42	19	31	1(0)		6(2)		5(3)		1(0)	2(2)	5(2)		20(9)
	Macolor niger *	2	2	1		3(1)	2(0)		1(0)						6(1)
Lutjanidae, deep-sea	Etelis carbunculus	16	5	3	1(1)	4(1)	1(1)		3(2)		1(0)	1(0)	2(1)	1(0)	14(6)
	Etelis coruscans	18	11	5	1(1)	4(1)	2(2)	1(0)	3(1)			1(0)	2(2)		14(7)
	Pristipomoides argyrogrammicus	20	14	18	1(1)	1(0)	3(1)		4(3)		1(0)	1(0)	2(1)		13(6)
	Pristipomoides auricilla *	2	2	2	1(0)		1(0)		1(1)		1(0)	1(0)	3(1)		8(2)
	Pristipomoides filamentosus	7	2	2	2(2)	1(1)	1(0)		2(0)				3(2)		9(5)
Nemipteridae	Nemipterus furcosus	239	111	160	1(0)		2(0)	1(0)	5(4)	1(0)	1(0)	6(6)	7(3)	1(0)	25(13)
	Pentapodus aureofasciatus	23	19	12	1(0)	1(0)	2(1)	1(0)	4(0)		1(0)		2(0)		12(1)
	Pentapodus nagasakiensis *	2	2	2					2(0)		1(0)		1(0)		4(0)
	Scolopsis bilineata	12	8	9	1(0)				2(1)				1(0)	1(0)	5(1)
	Scolopsis taenioptera *	3	2	3									1(0)		1(0)
	Total Caesionidae (1 species)	8	4	8					2(0)	1(0)	1(0)		1(0)		5(0)
	Partial total Lutjanidae, reef (13)	113	55	76	3(1)	10(4)	38(15)	2(0)	25(18)	2(0)	4(0)	2(2)	10(2)	1(0)	97(42)
	Partial total Lutjanidae, deep-sea (5)	63	34	30	6(5)	10(3)	8(4)	1(0)	13(7)	0(0)	3(0)	4(0)	12(6)	1(0)	58(25)
	Total Lutjanidae (18 species)	176	89	106	9(6)	20(7)	46(19)	3(0)	38(25)	2(0)	7(0)	6(2)	22(8)	2(0)	155(67)
	Total Nemipteridae (5 species)	279	142	186	3(0)	1(0)	4(1)	2(0)	13(5)	1(0)	3(0)	6(6)	12(3)	2(0)	47(15)
	Total (24 species)	463	235	300	12(6)	21(7)	50(20)	5(0)	53(30)	4(0)	11(0)	12(8)	35(11)	4(0)	207(82)

*: species with low sample size or only anecdotal collections, excluded from general calculations in Table 3. For each number: HPCs (SLIP-HPCs) i.e. number of host-parasite combinations, and, within parentheses, number of species level identified parasite – host-parasite combinations.

from Swains Reefs, Great Barrier Reef, Marion Reef, Australian Coral Sea, North West Shelf of Western Australia, Krakatua, Indonesia [31], Singapore and Poulo Condor, Vietnam [29,31], Ragay Gulf, Pasacao and Maribuyoc Bay, Bohol Island, Philippines [27]. This species has been reported from *Plectorhynchus goldmani*, *Diagramma picta* and *Priacanthus* sp. [31]. *Pr. argyrogrammicus* is a new host record and New Caledonia is a new geographical record.

Aega musorstom was already known from New Caledonia, in the vicinity of Western New Caledonia including the Coral Sea region of the Chesterfield Archipelago and the Loyalty Islands, at depths from 475 to 615 m [32]. Only one fish association was noted, "*Synagonopi* sp. 1" probably a species of *Synagrops* (Acropomatidae) [32]. *Pr. filamentosus* is a new host record.

Argathona macronema was already known from New Caledonia [7,33]. It was previously reported from

Table 2 List of 58 species identified at the species level with Latin binomial, with full authorities

Isopoda (4)	Aegidae: *Aega musorstom* Bruce, 2004
	Corallanidae: *Argathona macronema* (Bleeker, 1857) Monod, 1933
	Cymothoidae: *Anilocra gigantea* (Herklots, 1870) Schiœdte & Meinert, 1881
	Cymothoidae: *Anilocra longicauda* Schiœdte & Meinert, 1881
Copepoda (6)	Caligidae: *Caligus brevis* Shiino, 1954
	Dissonidae: *Dissonus excavatus* Boxshall, Lin, Ho, Ohtsuka, Venmathi Maran & Justine, 2008
	Hatschekiidae: *Hatschekia clava* Kabata, 1991
	Hatschekiidae: *Hatschekia tanysoma* Ho & Kim, 2001
	Lernaeopodidae: *Parabrachiella lutiani* (Pillai, 1968)
	Pennellidae: *Lernaeolophus sultanus* (H Milne Edwards, 1840) Heller, 1865
Monopisthocotylea (11)	Ancyrocephalidae: *Haliotrematoides lanx* Kritsky & Justine, 2009 in Kritsky, Yang & Sun, 2009
	Ancyrocephalidae: *Haliotrematoides longitubocirrus* (Bychowsky & Nagibina, 1971) Kritsky, Yang & Sun, 2009
	Ancyrocephalidae: *Haliotrematoides novaecaledoniae* Kritsky & Justine, 2009 in Kritsky, Yang & Sun, 2009
	Ancyrocephalidae: *Haliotrematoides patellacirrus* (Bychowsky & Nagibina, 1971) Kritsky, Yang & Sun, 2009
	Ancyrocephalidae: *Haliotrematoides potens* Kritsky & Justine, 2009 in Kritsky, Yang & Sun, 2009
	Ancyrocephalidae: *Haliotrematoides tainophallus* Kritsky & Justine, 2009 in Kritsky, Yang & Sun, 2009
	Capsalidae: *Benedenia elongata* (Yamaguti, 1968) Egorova, 1997
	Capsalidae: *Lagenivaginopseudobenedenia etelis* Yamaguti, 1966
	Capsalidae: *Pseudonitzschia uku* Yamaguti, 1965
	Capsalidae: *Trilobiodiscus lutiani* Bychowsky & Nagibina, 1967
	Diplectanidae: *Calydiscoides limae* Justine & Brena, 2009
Digenea (21)	Acanthocolpidae: *Pleorchis uku* Yamaguti, 1970
	Acanthocolpidae: *Stephanostomum uku* Yamaguti, 1970
	Cryptogonimidae: *Adlardia novaecaledoniae* Miller, Bray, Goiran, Justine & Cribb, 2009
	Cryptogonimidae: *Euryakaina manilensis* (Velasquez, 1961) Miller, Adlard, Bray, Justine, & Cribb, 2010
	Cryptogonimidae: *Euryakaina marina* (Hafeezullah & Siddiqi, 1970) Miller, Adlard, Bray, Justine, & Cribb, 2010
	Cryptogonimidae: *Metadena rooseveltiae* (Yamaguti, 1970) Miller & Cribb, 2008
	Cryptogonimidae: *Retrovarium manteri* Miller & Cribb, 2007
	Cryptogonimidae: *Retrovarium saccatum* (Manter, 1963) Miller & Cribb, 2007
	Cryptogonimidae: *Siphoderina hirastricta* (Manter, 1963) Miller & Cribb, 2008
	Cryptogonimidae: *Siphoderina ulaula* (Yamaguti, 1970) Miller & Cribb, 2008
	Cryptogonimidae: *Varialvus charadrus* Miller, Bray, Justine & Cribb, 2010
	Fellodistomatidae: *Tergestia magna* Korotaeva, 1972
	Hemiuridae: *Ectenurus trachuri* (Yamaguti, 1934) Yamaguti, 1970
	Lepocreadiidae: *Lepidapedoides kalikali* Yamaguti, 1970
	Monorchiidae: *Allobacciger macrorchis* Hafeezullah & Siddiqi, 1970
	Opecoelidae: *Hamacreadium mutabile* Linton, 1910
	Opecoelidae: *Macvicaria jagannathi* (Gupta & Singh, 1985) Bijukumar, 1997
	Opecoelidae: *Neolebouria blatta* Bray & Justine, 2009
	Opecoelidae: *Neolebouria lineatus* Aken'Ova & Cribb, 2001
	Sclerodistomidae: *Prosogonotrema bilabiatum* Vigueras, 1940
	Transversotrematidae: *Transversotrema borboleta* Hunter & Cribb, 2012
Trypanorhyncha (7)	Lacistorhynchidae: *Callitetrarhynchus gracilis* (Rudolphi, 1819) Pintner, 1931
	Lacistorhynchidae: *Floriceps minacanthus* Campbell et Beveridge, 1987
	Lacistorhynchidae: *Pseudolacistorhynchus heroniensis* (Sakanari, 1989) Palm, 2004
	Otobothriidae: *Otobothrium mugilis* Hiscock, 1954
	Tentaculariidae: *Nybelinia goreensis* Dollfus, 1960
	Tentaculariidae: *Nybelinia indica* Chandra, 1986

Table 2 List of 58 species identified at the species level with Latin binomial, with full authorities *(Continued)*

	Tentaculariidae: *Nybelinia queenslandensis* Jones & Beveridge, 1998
Nematoda (9)	Anisakidae: *Raphidascaris (Ichthyascaris) etelidis* Moravec & Justine, 2012
	Anisakidae: *Raphidascaris (Ichthyascaris) nemipteri* Moravec & Justine, 2005
	Camallanidae: *Camallanus carangis* Olsen, 1952
	Capillariidae: *Pseudocapillaria novaecaledoniensis* Moravec & Justine, 2010
	Cucullanidae: *Cucullanus bourdini* Petter & Le Bel, 1992
	Cucullanidae: *Dichelyne etelidis* Moravec & Justine, 2011
	Philometridae: *Philometra brevicollis* Moravec & Justine, 2011
	Philometridae: *Philometra mira* Moravec & Justine, 2011
	Trichosomoididae: *Huffmanela branchialis* Justine, 2004

This Table shows all binomial names of parasite taxa collected (SLIPs); since several names are extremely long, its main purpose is to lighten the other tables and the text. Authors involved in the description and combination of taxa for Isopoda: [28,29,32,167,168]; for Copepoda: [36,37,155,169-172]; for Monopisthocotylea: [45,49,66,67,156,173-175]; for Digenea: [71-74,76,77,79,84,86,87,89,97,176-180]; for Trypanorhyncha: [99,181-188]; for Nematoda: [24,157-161,189,190].

Epinephelus tauvina, *Diagramma cinerascens*, *Pseudolabrus* sp., *Trachichtodes affinis*, *Cromileptes altivelis*, *Lu. argentimaculatus*, *Plectropomus leopardus* and *Pl. maculatus* [33]. It was found again later on *Pl. leopardus* and in addition on *Pl. laevis* [7]. *Lu. monostigma* is a new host record.

Larval isopods belonged to the Gnathiidae. Gnathiids, found as praniza larvae, were collected on 6 species of nemipterids and lutjanids (5 reef-dwelling, 1 deep water). In New Caledonia, larval gnathiids were found on most fish families examined (serranids, lethrinids, lutjanids, nemipterids and many others). Adult isopods were found on serranids and lutjanids but not on lethrinids and nemipterids [7,8]. The biodiversity of larval gnathiids is hard to evaluate [34,35], but it is likely that several species are involved.

Figure 1 *Anilocra gigantea* (Isopoda, Cymothoidae), specimen MNHN Is6292, on the deep-sea lutjanid *Pristipomoides filamentosus*.

Copepoda

Fourteen taxa, including 6 identified at the species level, were found. Seven species of *Hatschekia* were distinguished but only two are known species, the other five (Figure 2) are not formally described. A total of 21 undescribed *Hatschekia* species has now been listed from New Caledonian fish ([7,8]; this paper). *Hatschekia tanysoma* was originally described from Kuwait Bay, from *Lu. fulviflamma* [36] and is reported here from the Pacific for the first time, but from the same host. In contrast *H. clava* was described from Heron Island from material collected from *Lu. carponotatus* (Richardson) (as *Lu. chrysotaenia*) [37].

The copepods belonged to five families, namely Caligidae, Dissonidae, Hatschekiidae, Lernaeopodidae and Pennellidae. Larvae and premetamorphic adults belonging to the Pennellidae were found on the deep-sea lutjanid *Et. coruscans*. The only adult member of this family found during eight years of sampling was a single female of *Lernaeolophus sultanus* (Figure 3) found on *Pr. filamentosus*. Pennellids are known to utilise two different hosts during their life cycle, either two different fish hosts or a pelagic mollusc and a final fish host [38]. However, the life cycle of no *Lernaeolophus* species has ever been elucidated so it is not possible to confirm whether the developmental stages found on *Pr. filamentosus* are those of *L. sultanus*. *L. sultanus* exhibits the lowest host specificity of any copepod parasite, occurring on 16 different host fishes in the Mediterranean [39].

Caligus brevis is reported here from two species of *Etelis*, *Et. carbunculus* and *Et. coruscans*, for the first time. This species was previously reported only from labrid hosts in Japanese [40] and New Zealand waters [41]. Ho & Lin (2004) suspected that *C. brevis* might be synonym of *Caligus oviceps* Shiino, 1952, but refrained from synonymizing them [42]. *C. oviceps* has already been reported from a lethrinid host (*Lethrinus haematopterus*

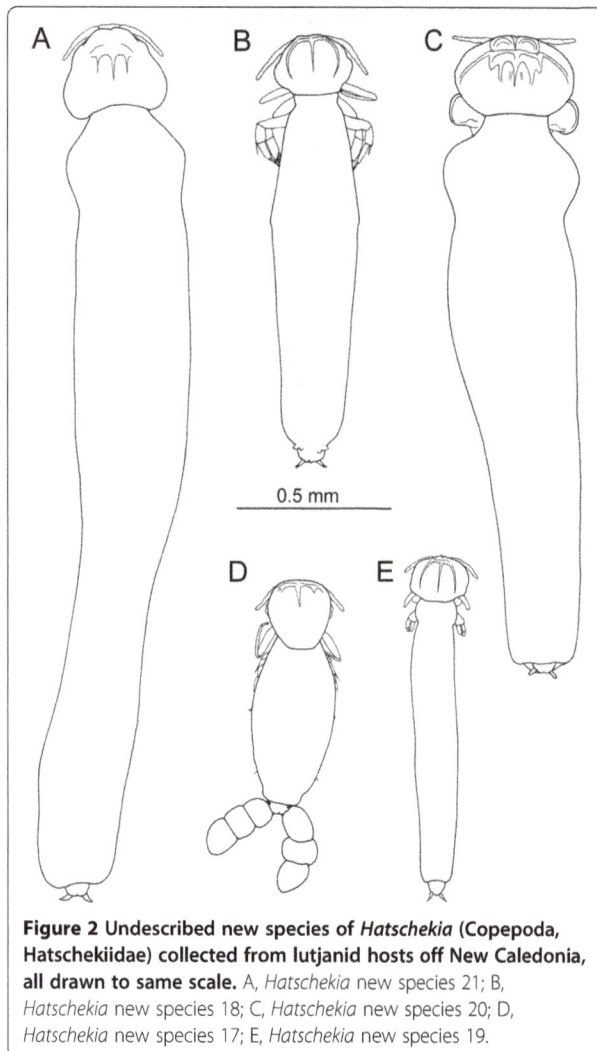

Figure 2 Undescribed new species of *Hatschekia* (Copepoda, Hatschekiidae) collected from lutjanid hosts off New Caledonia, **all drawn to same scale.** A, *Hatschekia* new species 21; B, *Hatschekia* new species 18; C, *Hatschekia* new species 20; D, *Hatschekia* new species 17; E, *Hatschekia* new species 19.

Temminck & Schlegel) but it has a broad range of hosts including species of Cheilodactylidae, Kyphosidae, Monacanthidae, Mullidae, Scaridae, and Siganidae [42].

All the copepods are from the gills; none was found on the skin. Insufficient sampling precludes interpretation of the absence of copepods from several of the fish species listed here; however, the absence of copepods on fork-tailed threadfin breams, *Ne. furcosus*, with 239 specimens examined at various seasons during eight years is certainly significant.

Monogenea

The minimized number of taxa for polyopisthocotyleans was 2 and for monopisthocotyleans was 23.

Polyopisthocotyleans were represented by *Allomicrocotyla* sp. on the deep-sea lutjanid *Et. coruscans*, and several records of unidentified microcotylids or other polyopisthocotylean families from coral-associated lutjanids and nemipterids. Polyopisthocotyleans on reef lutjanids were rare; similarly, polyopisthocotyleans were rare on *Lu. griseus* in the Gulf of Mexico [43]. The rarity of polyopisthocotyleans from 4 species of lutjanids off Heron Island on the Great Barrier Reef is apparent from the literature [44] and from unpublished observations by I.D. Whittington at the same locality.

Monopisthocotyleans included four families, the Ancyrocephalidae, Capsalidae, Diplectanidae and Gyrodactylidae.

Ancyrocephalids included a series of species of the recently described genus *Haliotrematoides* Kritsky, Yang & Sun, 2009, several species of *Euryhaliotrema* Kritsky & Boeger, 2002, which are still undescribed, and we also have records of various ancyrocephalids not attributed to a genus. Clearly, the lutjanids harbour an impressive ancyrocephalid radiation, with probably several species

Figure 3 *Lernaeolophus sultanus* (Copepoda, Pennellidae), specimen BMNH 2010.750, from *Pristipomoides filamentosus* off New Caledonia. Scale, each scale division = 1 mm.

on each fish species; some of these species seem to be strictly species-specific, others are found on up to 5 different host species [45]. These ancyrocephalids might be a threat for cultured snappers [46].

Diplectanids included several species of *Diplectanum* and one species of *Calydiscoides*. *Calydiscoides* species are numerous in lethrinids [8]; only one species was found in New Caledonia in the families studied here, in the nemipterid *Pentapodus aureofasciatus*, but none in members of *Scolopsis* and *Nemipterus*, which are known to harbour sometimes numerous species of *Calydiscoides* in other localities [44,47,48].

Four species are attributed here to *Diplectanum*. *D. opakapaka* Yamaguti, 1968 was described from *Pr. microlepis* and *Aphareus rutilans* off Hawaii [49] and *D. curvivagina* Yamaguti, 1968 was described from *Pr. sieboldii* and *Pr. auricilla* off Hawaii [49] and have rarely been recorded since [50]. *D. fusiformis* Oliver & Paperna, 1984, described from *Lu. kasmira* off Kenya [51] was recorded again from its type-host and *Lu. fulvus* in French Polynesia and Hawaii [52,53]. *D. spirale* Nagibina, 1976 described from *Lu. fulviflamma* [54] has apparently not been recorded since its original description. In the absence of a comparative examination of the type specimens, we prefer to keep our identifications of these four species as "cf." and we do not comment on their generic attribution.

Gyrodactylids were represented by a single specimen found on *Macolor niger*, which is a relatively difficult fish to catch and unfortunately additional specimens could not be obtained. It is the first gyrodactylid collected from a lutjanid. It should be noted that several coral-associated lutjanids were soaked for collection of skin monogeneans such as capsalids and digeneans such as transversotrematids, but no other gyrodactylid specimens were recovered.

Capsalid systematics is currently under reinvestigation (e.g. [55-57]) after Perkins et al. [58] demonstrated that the current morphological classification has limited congruence with a phylogenetic hypothesis based on three unlinked nuclear genes. Apparently homoplastic morphological features were highlighted throughout the molecular phylogeny [58]. This has entailed a reluctance to assign some taxa to genera until appropriate characters and generic and subfamilial definitions are better refined within a phylogenetic context. Hence 'identifications' of four taxa are herein provided only as Capsalidae sp. 6, 7, 13 and 17, differentiated phylogenetically by Perkins [59] using nuclear and mitochondrial markers, but from morphological 'identifications' made by I.D. Whittington.

Of the Capsalidae we report here, four from the gills were identified to species (*Benedenia elongata* from three lutjanid species; *Lagenivaginopseudobenedenia etelis* from *Et. coruscans*; *Pseudonitzschia uku* from *Aprion*

virescens; *Trilobiodiscus lutiani* from *Lu. argentimaculatus*). Two capsalid species recovered from gills of *Macolor niger* and *Ne. furcosus* remained unidentified. A species assigned only as a *Metabenedeniella* sp. was recovered from the pectoral fins (see [59]; probably a new species, I.D. Whittington unpublished) and the four capsalid species assigned only as Capsalidae sp. 6, 7, 13 and 17 [59] were recovered from the gills, fins, body washings, branchiostegal membranes and the head of their hosts (for details, see Appendix 1). Additional external sites, rarely examined in this study, received careful scrutiny only when I.D. Whittington visited Nouméa in October/November 2008. During his visit, thorough fish necropsies (four specimens of *Lu. vitta*; six specimens of *Ne. furcosus*; one specimen of *Lu. kasmira* [which was totally uninfected by capsalids]; two specimens of *Pr. argyrogrammicus*; one specimen of *Lu. argentimaculatus*) paid particular attention to external microhabitats known to support capsalids (e.g. [60]) followed by freshwater bathing of the same tissues to ensure recovery of parasites that may be cryptic due to camouflage or transparency (e.g. [61]). Dissections of fish at other times, when microhabitats other than gills that may harbour capsalids remained unstudied, seem likely to have appreciably underestimated the diversity of capsalid monogeneans from the caesionids, lutjanids and nemipterids examined.

Other factors in this study that contributed to an inability to assign capsalids to genera from gills, pelvic, pectoral and anal fins, body washings, branchiostegal membranes and head included small numbers of specimens recovered and the juvenile status of many capsalid individuals from *Ne. furcosus*. Adult specimens of *B. lutjani* from *Lu. carponotatus* from Heron Island on the Great Barrier Reef preferentially inhabit the branchiostegal membranes, and the pelvic fins was the site where protandrous parasites that possess a vagina may become inseminated [62,63]. Discovery of juvenile specimens of Capsalidae sp. 13 on pelvic and anal fins and branchiostegal membranes and small, recently matured adults on the branchiostegal membranes of *Ne. furcosus* suggests a similar migration and habitat partitioning for this taxon on the nemipterid.

Several of the capsalids reported in this study represent new records. *Metabenedeniella* species are previously reported from oplegnathids, serranids and haemulids [64]. Our report of a *Metabenedeniella* sp. from a lutjanid (Appendix 1) represents a new fish family and a new geographic record for this capsalid genus. *Benedenia elongata* was described as *Pseudobenedenia elongata* from a priacanthid, *Priacanthus boops*, and two lutjanids, *Pr. sieboldii* and *Arnillo auricilla* (now *Pr. auricilla*, see [65]), off Hawaii [49]. Although two specimens of *Pr. auricilla* were studied in the present investigation,

B. elongata was not recorded. It has, however, been recorded from three new host lutjanid species, *Et. carbunculus*, *Et. coruscans* and *Pr. argyrogrammicus*, each of which represents a new host and geographic record for *B. elongata*. Like deep water lutjanids, priacanthids can also occur in deep water and it seems as though *B. elongata* has relatively low host specificity among several deeper water fish species in these two families.

Lagenivaginopseudobenedenia etelis was originally described from *Et. carbunculus* off Hawaii by Yamaguti [66]. While 16 specimens of *Et. carbunculus* from New Caledonia were studied, *La. etelis* was not reported but we did record it from *Et. coruscans*, a new host and geographic record for this taxon (Appendix 1). As suggested for *B. elongata*, *La. etelis* may also exhibit relatively low host specificity and infect several deep-sea lutjanids. Further sampling may indicate whether *La. etelis* is specific to species of *Etelis* or whether this capsalid can also infect *Pristipomoides* species.

As far as we are aware, there have been no published reports of *Trilobiodiscus lutiani* since its original description [67] although it does occur on the type host, *Lu. argentimaculatus* in north Queensland (I.D. Whittington, unpublished). The present report of *T. lutiani* from New Caledonia is a new geographic record (Appendix 1).

For the same sampling limitations presented above, it is possible that specimens of Anoplodiscidae known from external surfaces of nemipterids on the Great Barrier Reef ([68]; I.D. Whittington, unpublished) may have been overlooked in the present study.

Digenea

The total minimized number of taxa was 33, with 21 species identified at the species level (SLIPs).

Eleven families were represented: Acanthocolpidae (2 SLIPs), Cryptogonimidae (12 spp, 9 SLIPs), Didymozoidae (2 unidentified adults, unknown number of species as unidentified larvae), Fellodistomatidae (1 SLIP), Hemiuridae (3 spp, 1 SLIP), Lecithasteridae (1 sp.), Lepocreadiidae (1 SLIP), Monorchiidae (1 SLIP), Opecoelidae (8 spp, 4 SLIPs), Sclerodistomidae (1 SLIP) and Transversotrematidae (1 SLIP).

The dominant digenean family is the Cryptogonimidae. Members of this family and of the Acanthocolpidae and the Didymozoidae utilise fishes as second intermediate hosts [69], indicating that this component of the lutjanid (and related families) diet is a major source of its digenean fauna. The other digenean families utilise a wide range of invertebrates, often crustaceans, as second intermediate hosts, and occasionally lepocreadiids and opecoelids also use fishes [69]. Some hemiuroids (e.g. *Lecithochirium*) have interpolated a third intermediate host, a fish, into their life-cycle [70]. Considering the importance of fishes in the diet and infection of lutjanids it

is, perhaps, surprising that we found no members of the common family Bucephalidae, which also utilises fishes as second intermediate hosts.

The cryptogonimids include one species, *Adlardia novaecaledoniae*, which is known only from New Caledonia [71]. Other identified cryptogonimids are reported more widely, from the Great Barrier Reef (GBR) (*Retrovarium manteri*), from Fiji (*R. saccatum*), from both these localities (*Siphoderina hirastricta*), from Hawaii (*Metadena rooseveltiae*), from Hawaii and China (*S. ulaula*) and from the Philippines and the GBR (*Euryakaina manilensis*) [72-76]. Two species also occur in the Indian Ocean. *Varialvus charadrus* occurs in the GBR and the Maldives and *E. marina* is reported from the GBR, the Bay of Bengal and Ningaloo Reef, Western Australia [77-79].

Both acanthocolpid species are reported from *Aprion virescens*, and are known from the Western Pacific, from Hawaii to the Great Barrier Reef [74,75,80,81].

The Opecoelidae is a difficult group, with many similar species described. Of the four SLIPs one, *Hamacreadium mutabile*, is a cosmopolitan parasite reported in many lutjanid species in the Atlantic, Indian and Pacific Oceans [82]. This is one of the few opecoelids known to utilise fishes as its second intermediate host [83]. Of the other species, *Neolebouria blatta* is reported only from New Caledonia, *N. lineatus* only from southern Western Australia, and *Macvicaria jagannathi* from the Bay of Bengal [84-88].

The Transversotrematidae includes a single species, which could be identified at the species level. Morphologically and biologically this form agrees with the variable species *Transversotrema borboleta*, reported from chaetodontids and lutjanids (including *Lu. kasmira*) from the northern and southern GBR [89]. The species includes 3 genotypes which are not partitioned to different host families, but only genotype 'G2' is reported in *Lu. kasmira*.

Didymozoidae include several records of unidentified juveniles from coral and deep-sea lutjanids and nemipterids; juvenile didymozoids are found in the intestine of most marine tropical fish [7,8,90,91] and the present records are not surprising. Adult didymozoids include a relatively abundant long filiform form found under the scales of the deep water *Et. carbunculus*.

Two of the other SLIPs are widespread in the Atlantic and Indo-West Pacific Region, i.e., *Ectenurus trachuri*, *Prosogonotrema bilabiatum* (see [92,93]). *E. trachuri* is mostly reported in carangids, but *P. bilabiatum* is a common parasite of lutjanids. *Allobacciger macrorchis* (also known as *Monorcheides macrorchis*) is reported mainly in the Indian Ocean, but also Japan [77,94,95]. *Lepidapedoides kalikali* is known only from Hawaii, Japan and Palau [74,96]. *Tergestia magna* is reported from emmelichthyid fishes from the waters off southern Australia [97].

Cestoda Bothriocephalidea and Tetraphyllidea

For these two cestode orders, only larvae were found, and species identification is not possible.

Bothriocephalideans were represented by larvae found in four species of fishes. They were especially abundant and highly prevalent in *Ne. furcosus*. White, flattish larvae, about 1 cm in length, were found in the abdominal cavity and in almost all organs of this fish, including the intestinal lumen. Sometimes larvae were visible when the fish was caught, because they were protruding in the region around the anus, with half of their body buried in the flesh. We interpret these larvae as bothriocephalideans due to the lack of morphological characters.

Tetraphyllideans include small forms found in the intestinal lumen of various fishes, both from coral-associated and deeper water fishes. It is possible that these include lecanicephalideans. A detailed morphological examination was not performed.

Cestoda Trypanorhyncha

Only larvae were found. All these trypanorhynchs have their adults parasitic in sharks and are probably transmitted to their final host by predation. Seven species of larval trypanorhynch cestodes, all from coral-associated fish, were identified at the species level on the basis of the armature of their tentacles.

Trypanoselachoidans [98] included four species, found as larvae in cysts in the abdominal cavity. The otobothriid *Otobothrium mugilis* was found only once in the nemipterid *Nemipterus furcosus*. The lacistorhynchids *Callitetrarhynchus gracilis*, *Floriceps minacanthus* and *Pseudolacistorhynchus heroniensis* were found in a lutjanid and a nemipterid; in New Caledonia, *C. gracilis* has already been recorded from 4 serranids and 1 lethrinid, *F. minacanthus* from 6 serranids and 1 lethrinid, and *Ps. heroniensis* from 7 serranids and 2 lethrinids [7,8] and show low host specificity at the larval stage. *C. gracilis* has already been found in *Lu. vitta* in Indonesia and *F. minacanthus* has already been found in *Ne. furcosus* in Indonesia [99]. *Ne. furcosus* is a new host record for *C. gracilis*, *Lu. vitta* is a new host record for *Ps. heroniensis* and *Ne. furcosus* is a new geographical and new host record for *O. mugilis*.

Trypanobatoidans [98] included three species of the tentaculariid genus *Nybelinia*, found free in the intestinal lumen of the nemipterid *Ne. furcosus*. *Ny. goreensis* has already been found in *Ne. furcosus* in Indonesia [99] and in New Caledonia in 2 lethrinids [8]. *Ne. furcosus* is a new geographical and new host record for *Ny. indica* and for *Ny. queenslandensis*.

It is likely that this high rate of records in *Ne. furcosus* is simply a consequence of its higher sampling (see also nematodes). In addition, cysts were repeatedly found in small numbers in the abdominal cavity of several deep water lutjanids. Although there is no direct evidence that these represent trypanorhynchs because they never contained larvae, the cysts (Figure 4), about 1 cm in length, were similar to sterile trypanorhynch cysts. We hypothesize that these larvae have a very long development time, perhaps related to the relatively cold deep water environment, and that the fish collected by us were too young to harbour fully developed larvae; deep water lutjanids have long life spans (30–40 years) [100].

Nematoda

The minimized number of taxa is 17, including 9 species identified at the species level.

Nematodes recorded belonged to eight families, the Anisakidae, Camallanidae, Capillariidae, Cucullanidae, Gnathostomatidae, Philometridae, Physalopteridae and Trichosomoididae.

The Anisakidae is represented by both larvae and adults. Larvae were encapsulated on the surface of organs or free in the lumen of the intestine; a few were occasionally identified at the genus level (*Anisakis* sp., *Hysterothylacium* sp., *Terranova* sp.), but most were identified only at the family level. Most fish species, whether coral-associated [7,8] or deep-sea, harbour these larvae, sometimes in very high numbers.

Adult anisakids included two newly described species of *Raphidascaris (Ichthyascaris)*, one from the coral-

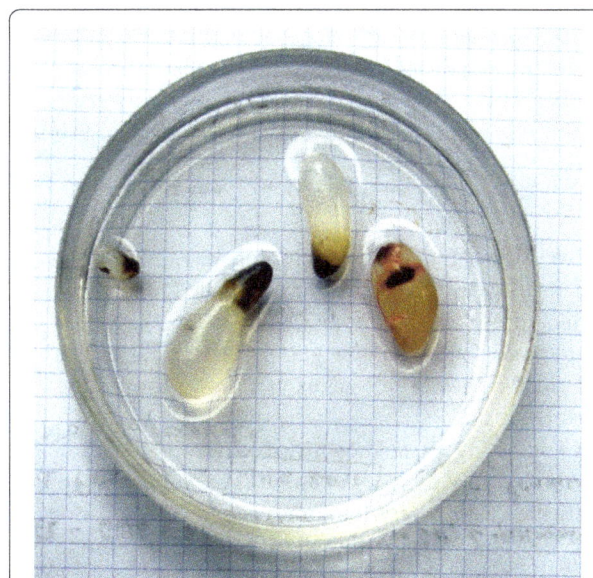

Figure 4 Sterile cysts from deep-sea lutjanids (here from *Pristipomoides auricilla*), interpreted as immature cysts of trypanorhynch cestodes. Scale, each square = 5 mm.

associated nemipterid *Ne. furcosus*, and one from the deep water lutjanid *E. coruscans*, and several additional specimens were found in other fish species and could not be identified at the species level.

Camallanids included one identified species, *Camallanus carangis*, and unidentified immature specimens.

Cucullanids were found only in deep water lutjanids, with *Cucullanus bourdini* in two species of *Pristipomoides*, and *Dichelyne etelidis* in two species of *Etelis*.

Unidentified gnathostomatids and physalopterids were occasionally found only in *Ne. furcosus* and probably illustrate the fact that this fish species was more intensively sampled that others, thus providing a number of records of parasites with low prevalence (see also trypanorhynchs).

Gonad-parasitic philometrids were found as two new species from *Lu. vitta*, but never in other fish species. Lutjanids and nemipterids are known as hosts of a few species of gonad-parasitic philometrids in other regions, such as the North Pacific, Indian and Atlantic Oceans [101,102].

Capillariids included a newly described species, *Pseudocapillaria novaecaledoniensis*, from a deep water lutjanid, but no other records were made.

Trichosomoidids included two species of *Huffmanela*, *H. branchialis* and *Huffmanela* sp., from the gills of two nemipterids. Interestingly, both species recorded in 2004, were never found again, despite intensive sampling of *Ne. furcosus*; their prevalence is probably very low, and their initial discovery in a small sample should be attributed to chance. Similarly, no *Huffmanela* species was recorded from more than 500 serranids examined [7], but a new species was described later from a serranid [103]. Tissue-dwelling trichosomoidids are characterised by two opposing features which probably balance each other out - very low prevalences and extremely high numbers (millions) of eggs in the few individual fish infected.

Hirudinea, 'Turbellaria' and Acanthocephala

Specimens of these three groups were rarely found.

Juvenile specimens of the piscicolid leech *Trachelobdella* sp. were found on the gills of a lutjanid and a nemipterid. Leeches of this genus are known from other lutjanids [104,105].

Cysts containing an unknown turbellarian were found rarely on the gills of *Ne. furcosus*; these were orange, abundant but with a very low prevalence, and were not studied in detail. Parasitic turbellarians are rarely found on coral reef fish [106,107].

An unidentified acanthocephalan was found once in the intestine of the deep water *Et. carbunculus*; this record constitutes a small addition to our very poor knowledge of the acanthocephalans of New Caledonia fish [6-8,108].

A numerical evaluation of parasite biodiversity in lutjanids and nemipterids

In presenting our results, we used the same methods as in previous similar papers of this series on serranids and lethrinids [7,8].

Our results (Appendices 1 and 2) include a number of parasite identifications, but the level of taxonomic identification varies greatly. Table 1 details the number of HPCs found in each fish species, and indicates how many fish specimens were examined; it was compiled by counting each parasitological record (i.e. each line in Appendix 1) as a host-parasite combination (HPC).

The number of HPCs differs from the actual number of different parasite species, for two reasons (a) a parasite species present in several hosts is counted as several HPCs; and (b) HPCs in Table 1 enumerate records which vary widely in systematic precision, and may designate, in a decreasing order of taxonomic precision:

- Species-level identified parasites (SLIPs); these have full binomial names, and we do not include 'cf' identifications within them.
- Species-level identified parasites which have not yet received a binomial (such as the numbered copepods *Hatschekia* sp. 17–23, *Diplectanum* 'cf.' species, and numbered capsalids). These represent valid, independent species but a comparison of the presence of these species in other localities or in other fish species will not be possible until the parasite species are formally identified, described and names are published.
- Parasite species identified at the generic level only, but which probably represent only a single species (examples: several digeneans).
- Parasite species identified at the generic level only, but for which we already know that they represent several species (example: several ancyrocephalid monogeneans).
- Parasite species identified at the family or higher level, for which we know that abundant biodiversity is hidden within this HPC. This includes unidentifiable larvae such as gnathiid isopods, anisakid nematodes, didymozoid digeneans and tetraphyllidean metacestodes. We estimate that these may represent a total of about 50–100 species.

Only species-level identified parasites with binomial (SLIPs) allow valid comparisons between localities and fish.

Table 1 includes all results, but some fishes were only studied superficially and their inclusion in further calculations would severely bias the results; for this reason,

Table 3 was constructed only from fish species of which at least several specimens have been studied. Table 3 also provides a comparison with the lethrinids and serranids of New Caledonia, based on previous results [7,8]; this comparison will be discussed below. Of course, we are aware that the number of fish studied is generally too low to provide truly significant results on parasite biodiversity, but at least these results allow comparison with other fish families studied using the same methods in the same area. Caesionids are not included in Table 3 because only a single species was involved.

For lutjanids (Table 3), the total number of HPCs was 131, and the number of different parasite species identified at the species level (SLIPs) was 40. For nemipterids (Table 3), the total number of HPCs was 42, and the number of SLIPs was 15. As usual, in addition,

indistinguishable larval taxa probably correspond to a high number (50–100?) of different species, but cannot be differentiated on the basis of morphological studies.

Table 3 includes evaluations of these numbers as means per species of fish. The main results for all lutjanids (reef-associated and deeper sea) were that 10.92 HPCs were found per fish species, with 3.33 SLIPs (identified with binomial) per fish species. For nemipterids, the results were 14.00 HPCs and 5.00 SLIPs per fish species.

A comparison between reef-associated and deeper-water lutjanids

Our results provide an opportunity to compare parasite biodiversity in reef-associated and deep water fishes. It is widely accepted that fishes in deeper waters have a lower

Table 3 Parasite biodiversity in lutjanids and nemipterids in New Caledonia for each parasite group, and a comparison with lethrinids and serranids

Family or group		Isop	Cope	Mono	Poly	Dige	Both	Tetr	Tryp	Nema	Other	Total
Lutjanidae, reef-associated	HPCs	2	6	30	2	24	2	4	2	9	0	81
(species: 8*; specimens: 105; gill: 48; abdomen: 72)	SLIPs	0	2	8	0	11	0	0	2	2	0	25
Expressed as mean per fish species:	HPCs mean	0.25	0.75	3.75	0.25	3.00	0.25	0.50	0.25	1.13	0.00	10.13
	SLIPs mean	0.00	0.25	1.00	0.00	1.25	0.00	0.00	0.25	0.25	0.00	3.13
Lutjanidae, deep-sea	HPCs	5	10	7	1	12	0	2	3	9	1	50
(species: 4*; specimens: 61; gill: 32; abdomen: 28)	SLIPs	3	2	2	0	4	0	0	0	4	0	15
Expressed as mean per fish species:	HPCs mean	1.25	2.50	1.75	0.25	3.00	0.00	0.50	0.75	2.25	0.25	12.50
	SLIPs mean	0.75	0.50	0.50	0.00	1.00	0.00	0.00	0.00	1.00	0.00	3.75
Lutjanidae, all	HPCs	7	16	37	3	36	2	6	5	18	1	131
(species: 12*; specimens: 166; gill: 80; abdomen: 100)	SLIPs	3	4	10	0	15	0	0	2	6	0	40
Expressed as mean per fish species:	HPCs mean	0.58	1.33	3.08	0.25	3.00	0.17	0.50	0.42	1.50	0.08	10.92
	SLIPs mean	0.33	0.33	0.83	0.00	1.25	0.00	0.00	0.17	0.50	0.00	3.33
Nemipteridae	HPCs	3	1	4	2	11	1	2	6	10	2	42
(species: 3*; specimens: 274; gill: 138; abdomen: 181)	SLIPs	0	0	1	0	5	0	0	6	3	0	15
Expressed as mean per fish species:	HPCs mean	1.00	0.33	1.33	0.67	3.67	0.33	0.67	2.00	3.33	0.67	14.00
	SLIPs mean	0.00	0.00	0.33	0.00	1.67	0.00	0.00	2.00	1.00	0.00	5.00
Lethrinidae	HPCs	9	21	53	4	59	1	7	10	21	3	188
(17 species: 17; specimens: 423; gill: 227; abdomen: 334)	SLIPs	0	7	11	0	13	0	0	6	5	0	42
Expressed as mean per fish species:	HPCs mean	0.53	1.24	3.12	0.24	3.47	0.06	0.41	0.59	1.24	0.18	11.06
	SLIPs mean	0.00	0.41	0.65	0.00	0.76	0.00	0.00	0.35	0.29	0.00	2.47
Serranidae	HPCs	20	53	97	0	76	4	13	35	37	2	337
(28 species, specimens: 540; gill: 394; abdomen: 275)	SLIPs	3	7	42	0	13	1	0	5	4	0	75
Expressed as mean per fish species:	HPCs mean	0.71	1.89	3.46	0.00	2.71	0.14	0.46	1.25	1.32	0.07	12.04
	SLIPs mean	0.11	0.25	1.50	0.00	0.46	0.04	0.00	0.18	0.14	0.00	2.68

HPCs, number of host-parasite combinations; SLIPs: number of species-level identified parasites (the same parasite species found in different hosts is counted a single time; hence differences with numbers of SLIP-HPCs in Table 1). Source of data for lutjanids and nemipterids: Table 1 and Appendix 2, this paper; for lethrinids and serranids, Table 3 in [8].
*: species with only anecdotal data (see Table 1) were excluded from these calculations.

parasitic diversity than surface fishes [109-112]. However, in comparative studies, fish species from the deep-sea are generally from orders (e.g. Gadiformes, Ophidiiformes, Notacanthiformes) which are different from the orders of surface fishes (e.g. Perciformes); in contrast, our study allows us to compare fish from the same family, the Lutjanidae, from both environments. Moreover, collection areas were very close and adjacent, with deeper-water fishes collected from just off the barrier reef along the outer slope, i.e. less than one kilometre away from the barrier reef and the lagoon [19]. Recently, a molecular study demonstrated that monogeneans of groupers tend to widen specificity when they infect fish from the outer slope, in comparison to lagoon fish where they are strictly specific [23].

Table 3 shows that the number of HPCs in reef-associated lutjanids was 10.13 per fish species, compared with 12.50 (123%) in deep-water lutjanids. The number of SLIPs in reef-associated lutjanids was 3.13 per fish species, compared with 3.75 (125%) in deeper water lutjanids. These figures suggest that parasite biodiversity was higher in deeper-water fishes than in reef-associated fishes, a highly unexpected result.

However, we identify four biases which diminish the validity of this comparison:

(a) Depth bias. Estimates of parasite biodiversity in deep-sea fish [109-112] generally concern fish from deeper seas (i.e. 1000 m vs 100–250 m) and from other fish orders than those studied here.

(b) Size bias. Most deeper water fish examined, especially *Etelis* spp., were large fish, usually in the metre range size, while reef lutjanids were smaller, usually 10–30 cm in length [113]. It is known that parasite biodiversity increases with the size of hosts [8,114,115].

(c) Monogenean bias. Coral-associated lutjanids harbour a number of small ancyrocephalid monogeneans, of which a large proportion has not been described yet [45].

(d) Human bias. The high number of parasites identified at the species level in deeper water fish is probably related to the scientific interest they elicit in scientists. Systematicians like to describe rare parasites from rarely examined fish! No particular direction was given to participants of this study to balance their descriptive effort between reef and deeper water fish.

A comparison with lethrinids and serranids

Data on parasite biodiversity, compiled using the same methods at the same location, are available for two other families [7,8].

Table 4 compares parasite biodiversity in four families of reef-associated fishes, the lutjanids and nemipterids (present study), and the serranids and lethrinids [7,8]. Table 4 also compares results for data compiled from fishes with variable sample size (anecdotal collections excluded) and for fishes with significant sampling (i.e. >25 individuals). For the latter, the numbers of HPCs for lutjanids and nemipterids per fish species are 20.00 and

Table 4 Parasite biodiversity in reef-associated families: lutjanids and nemipterids compared with lethrinids and serranids, and a calculation on all four families

Family or group		All data	Well-sampled		All data	Well- sampled
Lutjanidae, reef-associated	HPCs	81	20	HPCs mean	10.13	20.00
(All data: 8 species*; well-sampled: 1 species**)	SLIPs	25	9	SLIPs mean	3.13	9.00
Nemipteridae	HPCs	42	25	HPCs mean	14.00	25.00
(All data: 3 species*; well-sampled: 1 species**)	SLIPs	15	13	SLIPs mean	5.00	13.00
Lethrinidae	HPCs	188	89	HPCs mean	11.06	22.25
(All data: 17 species; well-sampled: 4 species)	SLIPs	42	38	SLIPs mean	2.47	9.50
Serranidae	HPCs	337	136	HPCs mean	12.04	19.43
(All data: 28 species; well-sampled: 7 species)	SLIPs	75	74	SLIPs mean	2.68	10.57
Four families of reef-associated fish	HPCs	648	270	HPCs mean	11.57	20.77
(All data: 56 species; well-sampled: 13 species)***	SLIPs	148	125	SLIPs mean	2.64	9.62

*: species with only anecdotal data (see Table 1) were excluded from these calculations.

** species sampled over 30: Lutjanidae, only *Lutjanus vitta* (n = 42); Nemipteridae, only *Nemipterus furcosus* (n = 239).

*** SLIPs for all four families are not a simple addition of SLIPs for each families, because a few taxa are shared. These include only the digenean *Stephanostomum japonocasum* (shared by 2 fish families), and the four trypanorhynchs *C. gracilis* (shared by 4 families), *F. minacanthus*, *Ps. heroniensis* and *Ny. goreensis* (shared by 2–3 families).

25.00, respectively, and the number of SLIPs are 9.00 and 13.00, respectively. These results are similar to previous results in serranids and lethrinids, in which HPCs in well-sampled fishes were 19.43 and 22.25, respectively, and SLIPs were 10.57 and 9.50, respectively. Results for all fishes (including species with low sample size, but excluding anecdotal collections) constitute about half of these figures.

In addition, Table 4 includes results obtained by pooling the data for all four families of fish. Results from 13 species of well-sampled reef-associated fishes (only n > 25 or n > 30 sampled individuals according to family), represented a sampling effort of almost 1,000 fishes (382 serranids, 329 lethrinids, 42 lutjanids, 239 nemipterids, total 992). The sampling effort and person-day time represented by these figures, which include collection, preparation, precise identification of parasite specimens, and curation in recognized collections, has no equivalent in the literature for reef-associated fish and all of their parasites (although similar efforts were devoted to digeneans only [4,116,117]). The number of HPCs is 20.77 per fish species, and the number of SLIPs is 9.62 per fish species. Many precautions were taken in these studies to *minimize* the number of taxa, and larval taxa (which are difficult to identify) certainly represent a significant additional biodiversity. We thus consider that the total result of *ten species of parasites per reef fish species* is a strict minimum, and that the real number is probably double or triple this estimate. All numbers here concern only macroparasites; the addition of Myxosporea, if they were known, would significantly enhance all results, with probably 2–3 additional parasite species per fish species [118,119].

Our published [6,81,120-124] and unpublished data on other families of fish suggest that this number of 10 species of parasites per fish species is generalizable to other families of fish, at least those with similar "average" size (10–40 cm), since parasite biodiversity depends upon the size of fish [4,8].

As found for serranids and lethrinids [7,8], the literature includes very few extensive lists of parasites from lutjanids and nemipterids in the Pacific. The checklist of parasites of Heron Island [44] includes a single lutjanid species in common with the present study, *Lu. fulviflamma*, with a single parasite record, and no nemipterid species in common. This precludes any biogeographical comparison for lutjanids and nemipterids, as was found for serranids and lethrinids [7,8].

Consequences for coextinction of parasites of coral reef fish

The word coextinction was coined by Stork and Lyal (1993) [125] to express that as a host species becomes extinct, so does one or more species of parasites, and was redefined by Koh et al. (2004) [126] in a slightly broader sense as "the loss of a species (the affiliate) upon the loss of another (the host)".

Knowing that parasite species are more numerous than non-parasitic species [127,128], it follows that coextinctions are more numerous than extinctions [129]. Dobson et al. (2008) [130] estimated that 3–5% of helminths are threatened by extinction in the next 50–100 years. However, Dunn (2009) [129] mentioned that there is no well documented case of the coextinction of a vertebrate parasite. Rózsa (1992) [131] pointed out that even a decrease in numbers within a host population, without the danger of extinction, could jeopardize the survival of certain parasite species. Koh et al. (2004) [126] calculated a risk of extinction of 593 species of monogeneans associated with 746 endangered species of fish. Justine (2007) [132] remarked that such a prediction underestimated the number of parasites in fish in rich ecosystems such as coral reefs. For Moir et al (2010) [133], this discrepancy highlights how biogeographic variation and knowledge gaps in dependant species biodiversity may lead to biased estimates of co-extinction risk.

Should we be concerned by the extinction of parasites? This is hard to defend to the general public, because "parasites tend to lack charisma" [129] and many blood-sucking parasites transmit diseases [134]. However, parasites play a major role in the balance of populations and the evolution of their hosts [114,129,130,135], and, as such, are an important and irreplaceable part of biodiversity and ecosystems.

The numerical evaluation of parasitic biodiversity in coral reef fish provided in the present study allows a more precise prediction of the risk of coextinction if a coral reef fish species becomes extinct, or simply has its population decreased [133]. Coral reefs are threatened across the planet [136-140] and special threats exist in New Caledonia [141,142]. Our results suggest that extinction of a coral reef fish species of average size would eventually result in the co-extinction of at least ten species of parasites.

Conclusions

As surprising as it might seem for studies mainly based on lists of parasites and morphological identifications, the present paper and our two previous similar papers [7,8] are pioneering works in the field of biodiversity of parasites of tropical coral reef fishes. Our main discovery of a parasitic biodiversity at least ten times higher than fish biodiversity has potential implications in the evaluation of loss of biodiversity when a coral reef fish species is threatened or becomes extinct.

Methods

Methods used in this paper are essentially the same as for the two previous papers of this series [7,8] and for brevity are not repeated here. For parasite collection, we used two methods targeting two sets of organs, designated as "gills" and "abdominal organs". We generally used the "gut wash" method [143] but in some circumstances we used a simplified method [144]. Full details and possible methodological flaws were discussed previously [7,8]. In addition, we soaked a few fish in saline in order to collect surface monogeneans. A high number (239) of fork-tailed threadfin breams, *Nemipterus furcosus*, were examined, including specimens examined for research and specimens examined by students during practical courses at the University of New Caledonia.

Parasite specimens, generally collected by J.-L. Justine and his team, and sometimes by visiting colleagues, were forwarded to respective specialists: I. Beveridge (trypanorhynch cestodes), G. A. Boxshall (copepods), R. A. Bray and T. L. Miller (digeneans), F. Moravec (nematodes), J.-P. Trilles (isopods), I. D. Whittington (capsalid monogeneans), and J.-L. Justine (other monogeneans). Hirudineans were examined by E. Burreson (College of William and Mary, Gloucester Point, Virginia, USA); a few monogeneans were examined by L. Euzet (Sète, France); some anisakids were identified at the generic level by S. Shamsi (Charles Sturt University, Wagga Wagga, NSW, Australia). Gills of several lutjanids, prepared with hot water and formalin, were examined by D. C. Kritsky (Idaho State University, Pocatello, Idaho, USA) and monogeneans were described [45]. Sometimes external isopods were brought by fishermen and provided "anecdotal" collections.

The names of cestode orders and suborders follow Khalil et al. [145] updated by recent work [98,146,147]. Polyopisthocotylean and monopisthocotylean monogeneans are treated as two independent groups, because monophyly of the monogeneans is not established [148-152]. However, since polyopisthocotyleans were rare, results for both groups were often pooled. Monogeneans sometimes included in the Dactylogyridae are here considered as members of the Ancyrocephalidae [153,154]. Many specimens have been deposited in recognized collections (Appendix 3); other specimens under study are still in the collections of the various authors but will be eventually deposited in the collection of the Muséum National d'Histoire Naturelle, Paris, France (MNHN) and/or in other recognized, curated collections.

Research carried out on animals (fish) was performed in accordance with the ethical requirements of the IRD (Institut de Recherche pour le Développement, France) and University of Adelaide Animal Ethics Approval S-020-2008 for work by I.D. Whittington.

Appendix 1. Host-parasite list

New, unpublished records indicated by [0]; other records: [6,24,25,45,47,58,59,71,78,79,81,87,88,107,155-166].

8 fish species with low sample size * were included in Table 1 but not kept in final calculations of parasite numbers (Table 3).

Caesionidae
Caesio cuning (Bloch, 1791)

Dige: Hemi: *Lecithochirium* sp. (digestive tract) [0]
Dige: Leci: *Aponurus* sp. (digestive tract) [0]
Both: Unid: unidentified larvae (digestive tract) [0]
Tetr: Unid: unidentified larvae (digestive tract) [0]
Nema: Cucu: *Cucullanus* sp. (digestive tract) [0]

Remarks: 8 specimens examined (4 for gills, 8 for abdominal organs)
HPCs: 5; SLIP-HPCs: 0.

Lutjanidae, reef-associated
Aprion virescens Valenciennes, 1830

Mono: Caps: *Pseudonitzschia uku* (gills) [58]
Dige: Acan: *Pleorchis uku*, immature (intestine) [81]
Dige: Acan: *Stephanostomum uku* (intestine) [81]

Remarks: 3 specimens examined (3 for gills, 3 for abdominal organs)
HPCs: 3; SLIP-HPCs: 3.

Lutjanus adetii (Castelnau, 1873)

Dige: Opec: *Hamacreadium mutabile* (digestive tract) [0]
Dige: Scle: *Prosogonotrema bilabiatum* immature (digestive tract) [0] (NGR)
Both: Unid: unidentified larvae (digestive tract) [0]
Nema: Anis: unidentified larvae (abdominal cavity) [0]

Remarks: 5 specimens examined (0 for gills, 5 for abdominal organs)
HPCs: 4; SLIP-HPCs: 2.

Lutjanus argentimaculatus (Forsskål, 1775)

Isop: Gnat: Praniza larvae (gills) [0]
Cope: Hats: *Hatschekia* n. sp. 20 (gills) [0]
Cope: Lerp: *Parabrachiella lutiani* (gills) [0] (NHR)
Mono: Ancy: *Euryhaliotrema* sp. (gills) [45]
Mono: Ancy: *Haliotrematoides novaecaledoniae* (gills) [45]
Mono: Ancy: *Haliotrematoides potens* (gills) [45]
Mono: Caps: *Metabenedeniella* sp. (pectoral fins) [59]

Mono: Caps: *Trilobiodiscus lutiani* (gills) [59] (NGR)
Dige: Cryp: *Retrovarium manteri* (digestive tract) [0] (NGR)
Dige: Cryp: *Retrovarium saccatum* (digestive tract) [0] (NHR)
Dige: Cryp: *Siphoderina hirastricta* (digestive tract) [0] (NGR)
Dige: Opec: *Hamacreadium mutabile* (digestive tract) [0] (NHR)
Tetr: Unid: unidentified larvae (digestive tract) [0]

Remarks: 4 specimens examined (3 for gills, 3 for abdominal organs)
HPCs: 13; SLIP-HPCs: 8.

Lutjanus fulviflamma (Forsskål, 1775)

Cope: Cali: Chalimus larvae (gills) [0]
Cope: Hats: *Hatschekia tanysoma* (gills) [0] (NGR)
Mono: Ancy: *Euryhaliotrema* sp.(gills) [45]
Mono: Ancy: *Haliotrematoides patellacirrus* (gills) [45]
Mono: Ancy: *Haliotrematoides tainophallus* (gills) [45]
Poly: Unid (gills) [0]
Dige: Cryp: *Euryakaina marina* (intestine) [78]
Dige: Opec: *Hamacreadium mutabile* (digestive tract) [0]
Nema: Anis: unidentified larvae (abdominal cavity) [0]

Remarks: 17 specimens examined (11 for gills, 11 for abdominal organs)
HPCs: 9; SLIP-HPCs: 5.

Lutjanus fulvus (Forster, 1801) *

Mono: Ancy: *Euryhaliotrema* sp. (gills) [45]
Mono: Ancy: *Haliotrematoides longitubocirrus* (gills) [45]
Mono: Ancy: *Haliotrematoides patellacirrus* (gills) [45]
Mono: Ancy: unidentified (gills) [45]

Remarks: 2 specimens examined (1 for gills, 0 for abdominal organs, additional gills examined by D. C. Kritsky). * Fish species not kept for final parasite counts (Table 3) because of low sample size.
HPCs: 4; SLIP-HPCs: 2.

Lutjanus gibbus (Forsskål, 1775) *

Cope: Hats *Hatschekia clava* (gills) [0] (NHR)
Mono: Ancy: unidentified (gills) [0]

Remarks: 2 specimens examined (2 for gills, 1 for abdominal organs). * Fish species not kept for final parasite counts (Table 3) because of low sample size.
HPCs: 2; SLIP-HPCs: 1.

Lutjanus kasmira (Forsskål, 1775)

Cope: Hats: *Hatschekia* n. sp. 19 (gills) [0]
Mono: Dipl: *Diplectanum* cf. *fusiforme* (gills) [0]
Mono: Dipl: *Diplectanum* cf. *spirale* (gills) [0]
Mono: Ancy: unidentified sp. 1 (gills) [0] [45]
Mono: Ancy: unidentified sp. 2 (gills) [0]
Mono: Ancy: unidentified sp. 3 (gills) [0]
Mono: Ancy: unidentified sp. 4 (gills) [0]
Dige: Cryp: *Siphoderina* sp. (digestive tract) [0]
Dige: Didy: unidentified larvae (digestive tract) [0]
Dige: Hemi: *Lecithochirium* sp. (digestive tract) [0]
Dige: Opec: *Hamacreadium mutabile* (digestive tract) [163]
Dige: Tran: *Transversotrema borboleta* (under scales) [0] (NGR)
Both: Unid: unidentified larvae (digestive tract) [0]
Tetr: Unid: unidentified larvae (digestive tract) [0]
Nema: Anis: unidentified larvae (abdominal cavity) [0]

Remarks: 16 specimens examined (12 for gills, 12 for abdominal organs, 2 soaked bodies for skin parasites, additional gills examined by D. C. Kritsky)
HPCs: 15; SLIP-HPCs: 1.

Lutjanus monostigma (Cuvier, 1828) *

Isop: Cora: *Argathona macronema* (body) [0] (NHR)

Remarks: 0 specimen examined (external isopod collected from one fish otherwise not examined). * Fish species not kept for final parasite counts (Table 3) because of low sample size.
HPCs: 1; SLIP-HPCs: 1.

Lutjanus quinquelineatus (Bloch, 1790)

Mono: Ancy: *Euryhaliotrema* sp. (gills) [45]
Mono: Ancy: *Haliotrematoides lanx* (gills) [45]
Mono: Ancy: *Haliotrematoides longitubocirrus* (gills) [45]
Mono: Ancy: *Haliotrematoides patellacirrus* (gills) [45]
Mono: Ancy: unidentified (gills) [45]
Dige: Cryp: *Varialvus charadrus* (intestine) [79]
Dige: Opec: *Hamacreadium mutabile* (intestine) [0]
Tetr: Unid: unidentified larvae (digestive tract) [0]
Nema: Anis: unidentified larvae (abdominal cavity) [0]

Remarks: 12 specimens examined (0 for gills, 6 for abdominal organs, additional gills examined by D. C. Kritsky)
HPCs: 9; SLIP-HPCs: 5.

Lutjanus rivulatus (Cuvier, 1828) *

Mono: Ancy: unidentified sp. (gills) [0]
Nema: Anis: unidentified larvae (abdominal cavity) [0]
Hiru: Pisc: *Trachelobdella* sp. immature (gills) [0]

Remarks: 2 specimens examined (2 for gills, 2 for abdominal organs; including 1 juvenile). * Fish species not kept for final parasite counts (Table 3) because of low sample size.
HPCs: 3; SLIP-HPCs: 0.

Lutjanus russellii (Bleeker, 1849)

Isop: Gnat: Praniza larvae (gills) [0]
Mono: Ancy: *Euryhaliotrema* sp. (gills) [45]
Mono: Ancy: *Haliotrematoides longitubocirrus* (gills) [45]
Mono: Ancy: *Haliotrematoides patellacirrus* (gills) [45]
Mono: Caps: Capsalidae sp. 6 (body washing) [59]
Poly: Unid: unidentified immature (gills) [0]
Dige: Didy: unidentified larvae (intestine) [0]
Dige: Opec: *Hamacreadium mutabile* (intestine) [0]

Remarks: 6 specimens examined (0 for gills, 1 for abdominal organs, additional gills examined by D. C. Kritsky)
HPCs: 8; SLIP-HPCs: 3.

Lutjanus vitta (Quoy & Gaimard, 1824)

Cope: Cali: chalimus larvae (gills) [0]
Mono: Ancy: *Euryhaliotrema* sp. (gills) [45]
Mono: Ancy: *Haliotrematoides longitubocirrus* (gills) [45]
Mono: Ancy: *Haliotrematoides patellacirrus* (gills) [45]
Mono: Ancy: unidentified (gills) [45]
Mono: Caps: Capsalidae sp. 7 (branchiostegal membranes) [59]
Mono: Dipl: *Diplectanum* cf. *fusiforme* (gills) [0]
Dige: Cryp: *Euryakaina manilensis* (intestine) [78]
Dige: Cryp: *Varialvus charadrus* (intestine) [79]
Dige: Didy: unidentified larvae (intestine) [0]
Dige: Hemi: *Lecithochirium* sp. (intestine) [0]
Dige: Opec: *Hamacreadium mutabile* (intestine) [0] (NHR)
Tetr: Unid: unidentified (intestine) [0]
Tryp: Laci: *Callitetrarhynchus gracilis* (abdominal cavity) [0]
Tryp: Laci: *Pseudolacistorhynchus heroniensis* (abdominal cavity) [0] (NHR)
Nema: Anis: *Raphidascaris (Ichthyascaris)* sp. (intestine) [161]

Nema: Anis: *Terranova* sp. larvae (abdominal cavity) [0]
Nema: Cama: unidentified larvae (intestine) [0]
Nema: Phil: *Philometra brevicollis* (gonads) [160]; as "sp" [165]
Nema: Phil: *Philometra mira* (gonads) [160]; as "sp" [165]

Remarks: 42 specimens examined (19 for gills, 31 for abdominal organs, 5 unproductive soaked bodies, additional gills examined by D. C. Kritsky)
HPCs: 20; SLIP-HPCs: 9.

Macolor niger (Forsskål, 1775) *

Cope: Cali: *Caligus* n. sp. 2 (gills) [0]
Cope: Diss: *Dissonus excavatus* (gills) [155]
Cope: Hats: *Hatschekia* n. sp. 18 (gills) [0]
Mono: Caps: unidentified (gills) [0]
Mono: Gyro: unidentified (gills) [0]
Dige: Hemi: *Lecithochirium* sp. (intestine) [0]

Remarks: 2 specimens examined (2 for gills, 1 for abdominal organs). * Fish species not kept for final parasite counts (Table 3) because of low sample size.
HPCs: 6; SLIP-HPCs: 1.

Lutjanidae, deep water
Etelis carbunculus Cuvier, 1828

Isop: Cymo: *Anilocra gigantea* (body) [0]
Cope: Cali: *Caligus brevis* (body) [0] (NHR)
Cope: Cali: chalimus larvae (gills) [0]
Cope: Hats: *Hatschekia* n. sp. 21 (gills) [0]
Cope: Lerp: *Parabrachiella* sp. 2 (body) [0]
Mono: Caps: *Benedenia elongata* (gills) [0] (NHR) (NGR)
Dige: Cryp: *Siphoderina ulaula* (digestive tract) [0] (NGR)
Dige: Didy: unidentified adults (under scales) [0]
Dige: Opec: *Neolebouria blatta* (digestive tract) [87]
Tetr: Unid: unidentified larvae (digestive tract) [0]
Tryp: Unid: unidentified larvae, sterile cysts (abdominal cavity) [0]
Nema: Anis: *Raphidascaris (Ichthyascaris)* sp. (digestive tract) [0]
Nema: Cucu: *Dichelyne etelidis* (digestive tract) [159]
Acan: Unid: unidentified (digestive tract) [0]

Remarks: 16 specimens examined (5 for gills, 3 for abdominal organs, occasional collect of skin isopods or didymozoids)
HPCs: 14; SLIP-HPCs: 6.

Etelis coruscans Valenciennes, 1862

Isop: Cymo: *Anilocra gigantea* (body) [0] (NHR)
Cope: Cali: *Caligus brevis* (body) [0] (NHR)
Cope: Cali: chalimus larvae (body) [0]
Cope: Penn: chalimus larvae and premetamorphic adults (body) [0]
Cope: Hats: *Hatschekia* n. sp. 21 (gills) [0]
Mono: Caps: *Benedenia elongata* (gills) [0] (NHR) (NGR)
Mono: Caps: *Lagenivaginopseudobenedenia etelis* (gills) [0] (NHR)(NGR)
Poly: Micr: *Allomicrocotyla* sp. (gills) Euzet det.
Dige: Cryp: *Siphoderina* cf. *onaga* (digestive tract) [0]
Dige: Cryp: *Siphoderina* n. sp. (digestive tract) [0]
Dige: Fell: *Tergestia magna* (digestive tract) [0] (NGR) (NHR)
Tryp: Unid: unidentified larvae, sterile cysts (abdominal cavity) [0]
Nema: Anis: *Raphidascaris (Ichthyascaris) etelidis* (digestive tract) [161]
Nema: Cucu: *Dichelyne etelidis* (digestive tract) [159]

Remarks: 18 specimens examined (11 for gills, 5 for abdominal organs, occasional collect of skin isopods or didymozoids); The taxon listed as *Lagenivaginopseudobenedenia* sp. and sequenced in [59] is actually likely to be *Benedenia elongata* (I.D. Whittington, unpublished)
HPCs: 14; SLIP-HPCs: 7.

Pristipomoides argyrogrammicus (Valenciennes, 1832)

Isop: Cymo: *Anilocra longicauda* (body) [0] (NHR) (NGR)
Cope: Lerp: *Clavellotis* sp. (pectoral fins) [6,162]
Mono: Caps: *Benedenia elongata* (gills) [0] (NHR) (NGR)
Mono: Caps: Capsalidae sp. 17 (head) [59]
Mono: Dipl: *Diplectanum* cf. *curvivagina* (gills) [0]
Dige: Cryp: *Metadena rooseveltiae* (digestive tract) [0] (NHR)(NGR)
Dige: Cryp: *Siphoderina* n. sp. (digestive tract) [0]
Dige: Cryp: *Siphoderina ulaula* (digestive tract) [0] (NHR)(NGR)
Dige: Opec: *Neolebouria blatta* (digestive tract) [87]
Tetr: Unid: unidentified (intestine) [0]
Tryp: Unid: unidentified larvae, sterile cysts (abdominal cavity) [0]
Nema: Anis: unidentified larvae (abdominal cavity) [0]
Nema: Capi: *Pseudocapillaria novaecaledoniensis* (digestive tract) [158]

Remarks: 20 specimens examined (14 for gills, 18 for abdominal organs, 1 soaked body). The taxon listed as

Capsalidae sp. 17 and sequenced in [166] (no mounted voucher specimen available) is actually likely to be *Benedenia elongata* (I.D. Whittington, unpublished)
HPCs: 13; SLIP-HPCs: 6.

Pristipomoides auricilla (Jordan, Evermann & Tanaka, 1927)

Isop: Gnat: Praniza larvae (gills) [0]
Mono: Dipl: *Diplectanum* cf. *opakapaka* (gills) [0]
Dige: Lepo: *Lepidapedoides kalikali* (stomach) [0] (NGR)
Tetr: Unid: unidentified larvae (intestine) [0]
Tryp: Unid: unidentified larvae, sterile cysts (abdominal cavity) [0]
Nema: Anis: unidentified larvae (abdominal cavity) [0]
Nema: Cama: unidentified adults (digestive tract) [0]
Nema: Cucu: *Cucullanus bourdini* (digestive tract) [159]

Remarks: 2 specimens examined (2 for gills, 2 for abdominal organs)
HPCs: 8; SLIP-HPCs: 2.

Pristipomoides filamentosus (Valenciennes, 1830) *

Isop: Aegi: *Aega musorstom* (body) [0] (NHR)
Isop: Cymo: *Anilocra gigantea* (body) [0] (NHR)
Cope: Penn: *Lernaeolophus sultanus* (body) [0] (NHR)
Mono: Dipl: *Diplectanum* sp. (gills) [0]
Dige: Didy: unidentified adults (digestive tract) [0]
Dige: Didy: unidentified larvae (digestive tract) [0]
Nema: Anis: *Raphidascaris (Ichthyascaris) etelidis* (digestive tract) [161]
Nema: Cama: *Camallanus* sp. (digestive tract) [0]
Nema: Cucu: *Cucullanus bourdini* (digestive tract) [159]

Remarks: 7 specimens examined (2 for gills, 2 for abdominal organs, occasional collect of external isopods or copepods). * Fish species not kept for final parasite counts (Table 3) because of low sample size.
HPCs: 9; SLIP-HPCs: 5.

Nemipteridae
Nemipterus furcosus (Valenciennes, 1830)

Isop: Gnat: Praniza larvae (gills) [0]
Mono: Caps: unidentified (gills) ("*Benedenia* sp." [156])
Mono: Caps: Capsalidae sp. 13 (pelvic and anal fins, gills, branchiostegal membranes) [0] (branchiostegal membranes) [59]
Poly: Micr: unidentified immature (gills) [156]
Dige: Cryp: *Adlardia novaecaledoniae* (intestine) [71,164]

Dige: Didy: unidentified larvae (digestive tract) [0]
Dige: Hemi: *Ectenurus trachuri* (digestive tract) [0]
(NGR)(NHR)
Dige: Opec: *Macvicaria jagannathi* (digestive tract) [87]
Dige: Opec: *Neolebouria lineatus* (digestive tract) [87]
Both: Unid: unidentified larvae (abdominal cavity and intestine) [0]
Tetr: Unid: unidentified larvae (intestine) [0]
Tryp: Laci: *Callitetrarhynchus gracilis* larvae (abdominal cavity) [0] (NHR)
Tryp: Laci: *Floriceps minacanthus* larvae (abdominal cavity) [0]
Tryp: Otob: *Otobothrium mugilis* larvae (intestine) [0]
Tryp: Tent: *Nybelinia goreensis* larvae (intestine) [0]
Tryp: Tent: *Nybelinia indica* larvae (intestine) [0] (NHR)(NGR)
Tryp: Tent: *Nybelinia queenslandensis* larvae (intestine) [0] (NHR)(NGR)
Nema: Anis: *Anisakis* sp. larvae (abdominal cavity) [0]
Nema: Anis: *Hysterothylacium* sp. larvae (abdominal cavity) [0]
Nema: Anis: *Raphidascaris (Ichthyascaris) nemipteri* adults (intestine) [157]
Nema: Cama: *Camallanus carangis* adults (intestine) [166]
Nema: Gnto: unidentified larvae (intestine) [0]
Nema: Phys: unidentified larvae (intestine) [0]
Nema: Tric: *Huffmanela branchialis* eggs (gills) [24]
Turb: Unid: unidentified adults (gills) [107]

Remarks: 239 specimens examined (111 for gills, 160 for abdominal organs, branchiostegal membranes examined)
HPCs: 25; SLIP-HPCs: 13.

Pentapodus aureofasciatus Russell, 2001

Isop: Gnat: Praniza larvae (gills) [156]
Cope: Hats: *Hatschekia* n. sp. 17 (gills) [0]; as "sp." [47]
Mono: Ancy: unidentified (gills) [0]
Mono: Dipl: *Calydiscoides limae* (gills) [156]
Poly: Micr: unidentified immature (gills) [156]
Dige: Hemi: *Lecithochirium* sp. (digestive tract) [0]
Dige: Hemi: *Lecithocladium* sp. (digestive tract) [0]
Dige: Opec: *Neochoanostoma* sp. (digestive tract) [0]
Dige: Opec: *Neolebouria* sp. (digestive tract) [0]
Tetr: Unid: unidentified larvae (intestine) [0]
Nema: Anis: unidentified larvae (abdominal cavity) [0]
Nema: Tric: *Huffmanela* sp. eggs (gills) [24] [25]

Remarks: 23 specimens examined (19 for gills, 12 for abdominal organs)
HPCs: 12; SLIP-HPCs: 1.

Pentapodus nagasakiensis (Tanaka, 1915) *

Dige: Hemi: *Lecithochirium* sp. (digestive tract) [0]
Dige: Opec: *Macvicaria* sp. (digestive tract) [0]
Tetr: Unid: unidentified larvae (intestine) [0]
Nema: Anis: unidentified larvae (abdominal cavity) [0]

Remarks: 2 specimens examined (2 for gills, 2 for abdominal organs). * Fish species not kept for final parasite counts (Table 3) because of low sample size.
HPCs: 4; SLIP-HPCs: 0.

Scolopsis bilineata (Bloch, 1793)

Isop: Gnat: Praniza larvae (gills) [0]
Dige: Monr: *Allobacciger macrorchis* (intestine) [0] (NGR)(NHR)
Dige: Opec: *Allopodocotyle* sp. (digestive tract) [0]
Nema: Cama: unidentified larvae (intestine) [0]
Hiru: Pisc: *Trachelobdella* sp. immature (gills) [0]

Remarks: 12 specimens examined (8 for gills, 9 for abdominal organs)
HPCs: 5; SLIP-HPCs: 1.

Scolopsis taenioptera (Cuvier, 1830) *

Nema: Anis: unidentified larvae (abdominal cavity) [0]

Remarks: 3 specimens examined (2 for gills, 3 for abdominal organs). * Fish species not kept for final parasite counts (Table 3) because of low sample size.
HPCs: 1; SLIP-HPCs: 0.

Appendix 2. Parasite-host list
8 fish species with low sample size * were included in Table 1 but not kept in final calculations of parasite numbers (Table 3).

Isopoda

Minimized number of taxa: 5
Number of SLIPs: total 4; Lutjanidae: reef: 1-0*, deep-sea: 3-3*, all: 4-3*; Nemipteridae: 0.
Number of non-SLIP taxa: 0
Undistinguishable larval taxa: 1 (gnathiids)

Aegi: *Aega musorstom*
    Pristipomoides filamentosus (NHR)
Cora: *Argathona macronema*
    Lutjanus monostigma * (NHR)
Cymo: *Anilocra gigantea*
    Etelis carbunculus
    Etelis coruscans (NHR)

Pristipomoides filamentosus (NHR)
Cymo: *Anilocra longicauda*
   Pristipomoides argyrogrammicus (NHR)(NGR)
Gnat: Praniza larvae
   Lutjanus argentimaculatus
   Lutjanus russellii
   Nemipterus furcosus
   Pentapodus aureofasciatus
   Pristipomoides auricilla *
   Scolopsis bilineata

Copepoda

Minimized number of taxa: 14
Number of SLIPs: total 6; Lutjanidae: reef: 4-2*; deep-sea: 2-2*, all: 6-4*; Nemipteridae: 0-0*.
Undistinguishable larval taxa: 2
Note: for minimizing number of taxa, we considered that the caligid and pennellid larvae could correspond to their adult counterparts found on same or similar fish.

Cali: *Caligus brevis*
   Etelis carbunculus (NHR)
   Etelis coruscans (NHR)
Cali: *Caligus* n. sp. 2
   Macolor niger *
Cali: chalimus larvae
   Etelis carbunculus
   Etelis coruscans
   Lutjanus fulviflamma
   Lutjanus vitta
Diss: *Dissonus excavatus*
   Macolor niger *
Hats: *Hatschekia clava*
   Lutjanus gibbus * (NHR)
Hats: *Hatschekia* n. sp. 17
   Pentapodus aureofasciatus
Hats: *Hatschekia* n. sp. 18
   Macolor niger *
Hats: *Hatschekia* n. sp. 19
   Lutjanus kasmira
Hats: *Hatschekia* n. sp. 20
   Lutjanus argentimaculatus
Hats: *Hatschekia* n. sp. 21
   Etelis carbunculus
   Etelis coruscans
Hats: *Hatschekia tanysoma*
   Lutjanus fulviflamma (NGR)
Lerp: *Clavellotis* sp.
   Pristipomoides argyrogrammicus
Lerp: *Parabrachiella lutiani*
   Lutjanus argentimaculatus (NHR)
Lerp: *Parabrachiella* sp. 2
   Etelis carbunculus

Penn: chalimus larvae and premetamorphic adults
   Etelis coruscans
Penn: *Lernaeolophus sultanus*
   Pristipomoides filamentosus (NHR)

Monopisthocotylea

Minimized number of taxa: 23
Number of SLIPs: total 11; Lutjanidae: reef: 8-8*, deep-sea: 2-2*, all: 10-10*; Nemipteridae: 1-1*.
Number of non-SLIP taxa: 10
Undistinguishable larval taxa: 0
Note: Note: for minimizing number of taxa, *Euryhaliotrema* spp. and Ancyrocephalids Gen. spp. were each counted as one species (but this is certainly an underestimate), and the two unidentified capsalids and *Diplectanum* sp. were not counted (because they could, respectively, correspond to their better identified counterparts).

Ancy: *Euryhaliotrema* sp. (several spp.)
   Lutjanus argentimaculatus
   Lutjanus fulviflamma
   Lutjanus fulvus *
   Lutjanus quinquelineatus
   Lutjanus russellii
   Lutjanus vitta
Ancy: *Haliotrematoides lanx*
   Lutjanus quinquelineatus
Ancy: *Haliotrematoides longitubocirrus*
   Lutjanus fulvus *
   Lutjanus quinquelineatus
   Lutjanus russellii
   Lutjanus vitta
Ancy: *Haliotrematoides novaecaledoniae*
   Lutjanus argentimaculatus
Ancy: *Haliotrematoides patellacirrus*
   Lutjanus fulviflamma
   Lutjanus fulvus *
   Lutjanus quinquelineatus
   Lutjanus russellii
   Lutjanus vitta
Ancy: *Haliotrematoides potens*
   Lutjanus argentimaculatus
Ancy: *Haliotrematoides tainophallus*
   Lutjanus fulviflamma
Ancy: unidentified sp. (often several sp.)
   Lutjanus fulvus *
   Lutjanus gibbus *
   Lutjanus kasmira
   Lutjanus quinquelineatus
   Lutjanus rivulatus *
   Lutjanus vitta
   Pentapodus aureofasciatus

Caps: *Benedenia elongata*
  Etelis carbunculus (NHR)(NGR)
  Etelis coruscans (NHR)(NGR)
  Pristipomoides argyrogrammicus (NHR)(NGR)
Caps: Capsalidae sp. 6
  Lutjanus russellii
Caps: Capsalidae sp. 7
  Lutjanus vitta
Caps: Capsalidae sp. 13
  Nemipterus furcosus
Caps: Capsalidae sp. 17
  Pristipomoides argyrogrammicus
Caps: *Lagenivaginopseudobenedenia etelis*
  Etelis coruscans (NHR)(NGR)
Caps: *Metabenedeniella* sp.
  Lutjanus argentimaculatus
Caps: *Pseudonitzschia uku*
  Aprion virescens
Caps: *Trilobiodiscus lutiani*
  Lutjanus argentimaculatus (NGR)
Caps: unidentified species
  Macolor niger *
  Nemipterus furcosus
Dipl: *Calydiscoides limae*
  Pentapodus aureofasciatus
Dipl: *Diplectanum* cf. *curvivagina*
  Pristipomoides argyrogrammicus
Dipl: *Diplectanum* cf. *fusiforme*
  Lutjanus kasmira
  Lutjanus vitta
Dipl: *Diplectanum* cf. *opakapaka*
  Pristipomoides auricilla *
Dipl: *Diplectanum* cf. *spirale*
  Lutjanus kasmira
Dipl: *Diplectanum* sp.
  Pristipomoides filamentosus
Gyro: unidentified species
  Macolor niger *

Polyopisthocotylea

Minimized number of taxa: 2
Number of SLIPs: 0
Number of non-SLIP taxa: 1
Undistinguishable larval taxa: 0
Note: for minimizing number of taxa, we considered all unidentified records as a single species.

Micr: *Allomicrocotyla* sp.
  Etelis coruscans
Micr: unidentified species
  Nemipterus furcosus
  Pentapodus aureofasciatus
unidentified family: unidentified species

Lutjanus fulviflamma
Lutjanus russellii

Digenea

Minimized number of taxa: 33
Number of SLIPs: total 20; Lutjanidae: reef: 10-10*, deep-sea: 5-4*, all 15-14*; Nemipteridae: 5-5*.
Number of non-SLIP taxa: 2
Undistinguishable larval taxa: 1 (didymozoid juveniles)
Note: for minimizing number of taxa, we considered that the single non-SLIP taxon was *Siphoderina* cf. *onaga*; the 2 *Siphoderina* sp. from deep-sea and coral lutjanids were counted as 2 taxa; unidentified adult didymozoids and unidentified juveniles were counted each as 1 taxon.

Acan: *Pleorchis uku*
  Aprion virescens
Acan: *Stephanostomum uku*
  Aprion virescens
Cryp: *Adlardia novaecaledoniae*
  Nemipterus furcosus
Cryp: *Euryakaina manilensis*
  Lutjanus vitta
Cryp: *Euryakaina marina*
  Lutjanus fulviflamma
Cryp: *Metadena rooseveltiae*
  Pristipomoides argyrogrammicus (NHR)(NGR)
Cryp: *Retrovarium manteri*
  Lutjanus argentimaculatus (NGR)
Cryp: *Retrovarium saccatum*
  Lutjanus argentimaculatus (NHR)
Cryp: *Siphoderina* cf. *onaga*
  Etelis coruscans
Cryp: *Siphoderina hirastricta*
  Lutjanus argentimaculatus
Cryp: *Siphoderina* n. sp.
  Etelis coruscans
  Pristipomoides argyrogrammicus
Cryp: *Siphoderina* sp.
  Lutjanus kasmira
Cryp: *Siphoderina ulaula*
  Etelis carbunculus(NGR)
  Pristipomoides argyrogrammicus (NHR)(NGR)
Cryp: *Varialvus charadrus*
  Lutjanus quinquelineatus
  Lutjanus vitta
Didy: unidentified adults
  Etelis carbunculus
  Pristipomoides filamentosus
Didy: unidentified juveniles
  Lutjanus kasmira
  Lutjanus russellii
  Lutjanus vitta

Nemipterus furcosus
Pristipomoides filamentosus
Fell: *Tergestia magna*
   Etelis coruscans (NHR)(NGR)
Hemi: *Ectenurus trachuri*
   Nemipterus furcosus (NHR)(NGR)
Hemi: *Lecithochirium* sp.
   Caesio cuning
   Lutjanus kasmira
   Lutjanus vitta
   Macolor niger *
   Pentapodus nagasakiensis *
   Pentapodus aureofasciatus
Hemi: *Lecithocladium* sp.
   Pentapodus aureofasciatus
Leci: *Aponurus* sp.
   Caesio cuning
Lepo: *Lepidapedoides kalikali*
   Pristipomoides auricilla * (NGR)
Monr: *Allobacciger macrorchis*
   Scolopsis bilineata
(NHR)(NGR)
Opec: *Allopodocotyle* sp.
   Scolopsis bilineata
Opec: *Hamacreadium mutabile*
   Lutjanus adetii
   Lutjanus argentimaculatus (NHR)
   Lutjanus fulviflamma
   Lutjanus kasmira
   Lutjanus quinquelineatus
   Lutjanus russellii
   Lutjanus vitta (NHR)
Opec: *Macvicaria jagannathi*
   Nemipterus furcosus
Opec: *Macvicaria* sp.
   Pentapodus nagasakiensis *
Opec: *Neochoanostoma* sp.
   Pentapodus aureofasciatus
Opec: *Neolebouria blatta*
   Etelis carbunculus
   Pristipomoides argyrogrammicus
Opec: *Neolebouria lineatus*
   Nemipterus furcosus
Opec: *Neolebouria* sp.
   Pentapodus aureofasciatus
Scle: *Prosogonotrema bilabiatum*
   Lutjanus adetii (NGR)
Tran: *Transversotrema borboleta*
   Lutjanus kasmira (NGR)

Bothriocephalidea

Minimized number of taxa: 1
Number of SLIPs: 0

Number of non-SLIP taxa: 0
Undistinguishable larval taxa: 1

unidentified family: unidentified species, larvae
   Caesio cuning
   Lutjanus adetii
   Lutjanus kasmira
   Nemipterus furcosus

Tetraphyllidea

Minimized number of taxa: 1
Number of SLIPs: 0
Number of non-SLIP taxa: 0
Undistinguishable larval taxa: 1

unidentified family: unidentified species, larvae
   Caesio cuning
   Etelis carbunculus
   Lutjanus argentimaculatus
   Lutjanus kasmira
   Lutjanus quinquelineatus
   Lutjanus vitta
   Nemipterus furcosus
   Pentapodus aureofasciatus
   Pentapodus nagasakiensis *
   Pristipomoides argyrogrammicus
   Pristipomoides auricilla *

Trypanorhyncha

Minimized number of taxa: 8
Number of SLIPs: 7
Number of SLIPs: total 7; Lutjanidae: reef: 2-2*, deep-sea: 0-0*, all 2-2*; Nemipteridae: 6-6*.
Number of non-SLIP taxa: 0
Undistinguishable larval taxa: 1
Note: for minimizing number of taxa, we considered that all unproductive cysts from deep-sea lutjanids corresponded to 1 taxon and was distinct from the four other cyst-producing species.

Laci: *Callitetrarhynchus gracilis*, larvae
   Lutjanus vitta
   Nemipterus furcosus (NHR)
Laci: *Floriceps minacanthus*, larvae
   Nemipterus furcosus
Laci: *Pseudolacistorhynchus heroniensis*, larvae
   Lutjanus vitta (NHR)
Otob: *Otobothrium mugilis*, larvae
   Nemipterus furcosus (NHR)(NGR)
Tent: *Nybelinia goreensis*, larvae
   Nemipterus furcosus

Tentaculariidae: *Nybelinia indica*, larvae
  Nemipterus furcosus (NHR)(NGR)
Tentaculariidae: *Nybelinia queenslandensis*, larvae
  Nemipterus furcosus (NHR)(NGR)
unidentified family: unidentified species, larvae
  Etelis carbunculus
  Etelis coruscans
  Pristipomoides argyrogrammicus
  Pristipomoides auricilla *

Nematoda

Minimized number of taxa: 17
Number of SLIPs: total 9; Lutjanidae: reef: 2-2*, deep-sea: 4-4*, all 6-6*; Nemipteridae: 3-3*.
Number of non-SLIP taxa: 6
Undistinguishable larval taxa: 1 (anisakids, gnathostomatids)
Note: for minimizing number of taxa, we considered that unidentified anisakids corresponded to one of the 3 identified larval anisakid genera; *Camallanus* sp. and unidentified camallanids were counted as a single species; *Huffmanela* sp. was distinguished as different; the *Raphidascaris (Ichthyascaris)* sp. from reef lutjanids was different from that from deep-sea.

Anis: *Anisakis* sp., larvae
  Nemipterus furcosus
Anis: *Hysterothylacium* sp., larvae
  Nemipterus furcosus
Anis: *Raphidascaris (Ichthyascaris) etelidis*
  Etelis coruscans
  Pristipomoides filamentosus
Anis: *Raphidascaris (Ichthyascaris) nemipteri*
  Nemipterus furcosus
Anis: *Raphidascaris (Ichthyascaris)* sp.
  Etelis carbunculus
  Lutjanus vitta
Anis: *Terranova* sp., larvae
  Lutjanus vitta
Anis: unidentified species, larvae
  Lutjanus adetii
  Lutjanus fulviflamma
  Lutjanus kasmira
  Lutjanus quinquelineatus
  Lutjanus rivulatus *
  Pentapodus aureofasciatus
  Pentapodus nagasakiensis *
  Pristipomoides argyrogrammicus
  Pristipomoides auricilla *
  Scolopsis taenioptera *
Cama: *Camallanus carangis*
  Nemipterus furcosus

Cama: *Camallanus* sp.
  Pristipomoides filamentosus
Cama: unidentified species
  Lutjanus vitta
  Pristipomoides auricilla *
  Scolopsis bilineata
Capi: *Pseudocapillaria novaecaledoniensis*
  Pristipomoides argyrogrammicus
Cucu: *Cucullanus bourdini*
  Pristipomoides auricilla *
  Pristipomoides filamentosus
Cucu: *Dichelyne etelidis*
  Etelis carbunculus
  Etelis coruscans
Gnto: unidentified species
  Nemipterus furcosus
Phil: *Philometra brevicollis*
  Lutjanus vitta
Phil: *Philometra mira*
  Lutjanus vitta
Phys: unidentified species
  Nemipterus furcosus
Tric: *Huffmanela branchialis*
  Nemipterus furcosus
Tric: *Huffmanela* sp.
  Pentapodus aureofasciatus

Hirudinea

Minimized number of taxa: 1
Number of SLIPs: 0
Number of non-SLIP taxa: 1
Undistinguishable larval taxa: 0

Pisc: *Trachelobdella* sp., juvenile
  Lutjanus rivulatus *
  Scolopsis bilineata

Turbellaria

Minimized number of taxa: 1
Number of SLIPs: 0
Number of non-SLIP taxa: 1
Undistinguishable larval taxa: 0

unidentified family: unidentified species
  Nemipterus furcosus

Acanthocephala

Minimized number of taxa: 1
Number of SLIPs: 0
Number of non-SLIP taxa: 1
Undistinguishable larval taxa: 0

unidentified family: unidentified species
  Etelis carbunculus

Appendix 3. Material deposited
Pisces

Nemipterus furcosus, MNHN 2005–0768, 2006–1330.
  Pentapodus aureofasciatus, MNHN 2004–1168, 2004–1169, 2004–2164, 2004–2172.

Isopoda

Aegi: *Aega musorstom* ex *Pr. filamentosus*, MNHN Is6295, Is6296, Is6297, Is6298.
Cora: *Argathona macronema* ex *Lu. monostigma*, MNHN Is6299.
Cymo: *Anilocra gigantea* ex *Et. carbunculus*, MNHN-IU-2009-1710, IU-2009-1712; ex *Et. coruscans*, MNHN Is6293, MNHN-IU-2009-1711; ex *Pr. filamentosus*, MNHN Is6292.
Cymo: *Anilocra longicauda* ex *Pr. argyrogrammicus*, MNHN Is6294.

Copepoda

Cali: *Caligus brevis*, ex *Et. coruscans*, MNHN Cp8059, BMNH 2010.767–769; ex *Et. carbunculus*, BMNH 2010.770–771.
Diss: *Dissonus excavatus*, ex *Ma. niger*, MNHN Cp2436, BMNH 2007.316–325.
Hats: *Hatschekia clava*, ex *Lu. gibbus*, MNHN Cp8067.
Hats: *Hatschekia tanysoma*, ex *Lu. fulviflamma*, MNHN Cp8068–8069.
Penn: *Lernaeolophus sultanus*, ex *Pr. filamentosus*, BMNH 2010.750.
Lerp: *Parabrachiella lutiani*, ex *Lu. argentimaculatus*, MNHN Cp8060, BMNH 2010.786–791.

Monogenea

Ancy: *Haliotrematoides lanx* ex *Lu. quinquelineatus*, slides MNHN JNC1590, JNC2145, JNC2140, USNPC 101344–5, BMNH 2008.11.19.36–37.
Ancy: *Haliotrematoides longitubocirrus* ex *Lu. vitta*, slides MNHN JNC2306, USNPC 101349, 101350–352, BMNH 2008.11.19.38–39; ex *Lu. russellii*, MNHN JNC1584, USNPC 101347; ex. *Lu. fulvus*, USNPC 101346; ex. *Lu. quinquelineatus*, MNHN JNC1588, USNPC 101348.
Ancy: *Haliotrematoides novaecaledoniae* ex *Lu. argentimaculatus*, slides MNHN JNC2332, USNPC 101337, BMNH 2008.11.19.24–27.
Ancy: *Haliotrematoides patellacirrus* ex *Lu. russellii*, slides MNHN JNC1582, JNC1583, JNC1584, JNC1585,

USNPC 101338, BMNH 2008.11.19.28–29; ex *Lu. fulviflamma*, slide MNHN JNC2531; ex *Lu. fulvus*, slides MNHN JNC1591, JNC1592, USNPC 101341, BMNH 2008.11.19.32–33; ex *Lu. quinquelineatus*, slides MNHN JNC2146, JNC2147, JNC2142, JNC2141, JNC2144, USNPC 101342, 101343, BMNH 2008.11.19.34–35; ex. *Lu. vitta*, MNHN 2470.
Ancy: *Haliotrematoides potens* ex *Lu. argentimaculatus*, slides MNHN JNC2332, USNPC 101336, BMNH 2008.11.19.23.
Ancy: *Haliotrematoides tainophallus* ex *Lu. fulviflamma*, slides MNHN JNC2531.
Caps: *Benedenia elongata* ex gills *Et. carbunculus* (fish JNC2459) SAMA AHC 35066 (6 slides).
Caps: *Benedenia elongata* ex gills *Et. coruscans* (fish JNC2448) MNHN JNC2448 A1 (1 slide).
Caps: *Benedenia elongata* ex gills *Pr. argyrogrammicus* (from fish JNC2449) MNHN JNC2449 B1 (1 slide), (from fish JNC2603) MNHN JNC2603 A3 (1 slide), SAMA AHC 35067 (2 slides), (from fish JNC2604) MNHN JNC2604 A1 (1 slide), (from fish JNC2729) AHC 35068 (1 slide of a single specimen that was 'slivered'; sliver fixed in 95% ethanol; possibly conspecific with Capsalidae sp. 17 of Perkins 2010), (from fish JNC2730) SAMA AHC 35069 (1 slide).
Caps: Capsalidae sp. 7 (see Perkins 2010) ex branchiostegal membranes *Lu. vitta* (from fish JNC2686) SAMA AHC 29706 (1 slide).
Caps: *Lagenivaginopseudobenedenia etelis* ex gills *Et. coruscans* (from fish JNC2616) MNHN JNC2616 A1 (1 slide), JN111, JN115, JN119.
Caps: *Metabenedeniella* sp. (see Perkins 2010) ex pectoral fins *Lu. argentimaculatus* (from fish JNC2735) SAMA AHC 29714 (3 slides).
Caps: *Pseudonitzschia uku* (see Perkins et al. 2009) ex gills *Aprion virescens* (from fish JNC1557) MNHN JNC1557 A1 (1 slide), SAMA AHC 35070 (2 slides).
Caps: *Trilobiodiscus lutiani* (see Perkins 2010) ex gills *Lu. argentimaculatus* (from JNC2332), USNPC 101526 (4 slides), (from JNC2735) SAMA AHC 29713 (1 slide).
Caps: Capsalidae sp. 13 (see Perkins 2010) ex branchiostegal membranes *Ne. furcosus* (from fish JNC2692) SAMA AHC 29707 (2 slides), AHC 35073 (4 slides).
Caps: Capsalidae sp. 13 of Perkins (2010) ex gills *Ne. furcosus* (from fish JNC971) MNHN JNC971A6 (1 slide), SAMA AHC 35071 (2 slides), (from fish JNC2692) SAMA AHC 35072 (5 slides), (from fish JNC2693) SAMA AHC 35075 (4 slides), (from fish JNC2694) SAMA AHC 35076 (6 slides), (from fish JNC2695) SAMA AHC 35077 (2 slides).
Caps: Capsalidae sp. 13 of Perkins (2010) ex pelvic and anal fins *Ne. furcosus* (from fish JNC2694), SAMA AHC 35074 (3 slides).

Dipl: *Diplectanum* cf. *curvivagina* ex *Pr. argyrogrammicus*, slides MHNH JNC2426, JNC2449, JNC2456, JNC2729, JNC2996.

Dipl: *Diplectanum* cf. *opakapaka* ex *Pr. auricilla*, slides MHNH JNC2457.

Dipl: *Diplectanum* sp. ex *Pr. filamentosus*, slides MHNH JNC2452, JNC2460.

Polyopisthocotylea

Micr: *Allomicrocotyla* sp. ex. *Et. coruscans*, slides JN114.

Unidentified polyopisthocotylean ex *Lu. russellii*, slide JNC1582.

Digenea

Acan: *Pleorchis uku* ex *Aprion virescens*, MNHN JNC2568.

Acan: *Stephanostomum uku* ex *Ap. virescens*, MNHN JNC1557C.

Cryp: *Adlardia novaecaledoniae* ex *Ne. furcosus*, MNHN JNC2291–1, 16, MNHN JNC2289-1–5, MNHN JNC2291-2–4, MNHN JNC2331B-1–4, MNHN JNC2288-1–3, MNHN JNC2398-1, 10; BMNH 2008.12.30.1–3.

Cryp: *Euryakaina manilensis* ex *Lu. vitta*, MNHN JNC2285, MNHN JNC2286, MNHN JNC2306, MNHN JNC2470, MNHN JNC2686, MNHN JNC2862, MNHN JNC2887, MNHN JNC2897, MNHN JNC2898, MNHN JNC2900, MNHN JNC2902; BMHN 2010.4.23.1–12.

Cryp: *Euryakaina marina* ex *Lu. fulviflamma*, MNHN JNC1269B.

Cryp: *Retrovarium manteri* ex *Lu. argentimaculatus*, MNHN JNC 2735–2, BMNH 2012.2.15.7

Cryp: *Retrovarium saccatum* ex *Lu. argentimaculatus*, MNHN JNC 2735–3

Cryp: *Siphoderina hirastricta* ex *Lu. argentimaculatus*, MNHN JNC 2735–1

Cryp: *Varialvus charadrus* ex *Lu. quinquelineatus*, MNHN JNC2143; ex *Lu. vitta* MNHN JNC2285, MNHN JNC2688, MNHN JNC2689.

Fell: *Tergestia magna* ex *Et. coruscans*, MNHN JNC 2616B-1, JNC2617B-1; ex *Pr. argyrogrammicus*, MNHN JNC2820B-1.

Hemi: *Ectenurus trachuri* ex *Ne. furcosus*, MNHN JNC2586-1.

Lepo: *Lepidapedoides kalikali* ex *Pr. auricilla*, MNHN JNC2457-1, JNC2468-1, BMNH 2012.2.15.6.

Monr: *Allobacciger macrorchis* ex *Sc. bilineata*, MNHN JNC2522-1, BMNH 2012.2.15.16.

Opec: *Hamacreadium mutabile* ex *Lu. adetii*, MNHN JNC3021-1; ex *Lu. argentimaculatus*, MNHN

JNC2735-2, BMNH 2012.2.15.15; ex *Lu. fulviflamma*, MNHN JNC2531-1, BMNH 2012.2.15.13; ex *Lu. kasmira*, MNHN JNC2708-1-2, BMNH 2012.2.15.14; ex *Lu. quinquelineatus*, MNHN JNC 2142–1; ex *Lu. russellii*, MNHN JNC2191-1; ex *Lu. vitta*, MNHN JNC 2285–1, MNHN 2306–1, MNHN JNC2470-1, MNHN JNC2686-1, MNHN JNC2900-1, MNHN JNC2896-1, MNHN JNC2898-1, MNHN JNC2899-1, BMNH 2012.2.15.9–12.

Opec: *Macvicaria jagannathi* ex *Ne. furcosus*, MNHN JNC2331A-1–3, JNC2366A-1., BMNH 2009.2.12.11-14.

Opec: *Neolebouria blatta* ex *Et. carbunculus*, MNHN JNC2427; ex *Pr. argyrogrammicus*, MNHN JNC2464-1, MNHN JNC2426, MNHN JNC2456, MNHN JNC2461, MNHN JNC2464-66, MNHN JNC2604-05, MNHN JNC 2729, MNHN JNC2995-1, MNHN JNC2996A-1; BMNH 2009.4.7.1–7, 2012.2.15.8.

Opec: *Neolebouria lineatus* ex *Ne. furcosus*, MNHN JNC2398-1–2; BMNH 2009.2.12.15.

Scle: *Prosogonotrema bilabiatum* ex *Lu. adetii*, MNHN JNC3022-1.

Tran: *Transversotrema borboleta* ex *Lu. kasmira*, MNHN JNC2708.

Trypanorhyncha

Laci: *Callitetrarhynchus gracilis* ex *Ne. furcosus*, slide MNHN JNC2596.

Laci: *Floriceps minacanthus* ex *Ne. furcosus*, slide MNHN JNC3019.

Otob: *Otobothrium mugilis* ex *Ne. furcosus*, slides MNHN JNC2610, JNC2586.

Tent: *Nybelinia goreensis* ex *Ne. furcosus*, slide MNHN JNC2612.

Tent: *Nybelinia indica* ex *Ne. furcosus*, slides MNHN JNC2288, JNC2611, JNC3016.

Nematoda

Anis: *Raphidascaris (Ichthyascaris) etelidis* ex *Et. coruscans*, MNHN JNC2616, JNC2617, JNC2623, HCIP N-969; ex *Pr. filamentosus*, MNHN JNC2460.

Anis: *Raphidascaris (Ichthyascaris) nemipteri* ex *Ne. furcosus*, MNHN JNC218, JNC330, JNC214, JNC217, JNC278, JNC311, HCIP N-836.

Cama: *Camallanus carangis* ex *Ne. furcosus*, MNHN JNC276, JNC280, JNC465, JNC467, JNC1261, HCIP N-859.

Capi: *Pseudocapillaria novaecaledoniensis* ex *Pr. argyrogrammicus*, MNHN JNC2604, HCIP N-930.

Cucu: *Cucullanus bourdini* ex *Pr. auricilla* MNHN JNC2457, ex *Pr. filamentosus* JNC2460, HCIP N-949.

Cucu: *Dichelyne etelidis* ex *Et. carbunculus*, MNHN JNC2459, HCIP N-948; ex *Et. coruscans*, MNHN JNC2617.

Phil: *Philometra brevicollis* ex *Lu. vitta*, MNHN JNC2901, JNC2038, HCIP N-967.

Phil: *Philometra mira* ex *Lu. vitta*, MNHN JNC2901, HCIP N-968.

Tric: *Huffmanela branchialis* ex *Ne. furcosus*, NHN JNC216, HCIP N-816, BMNH 2004.2.18.1, SAMA AHC 32856.

Tric: *Huffmanela* sp. ex *Pe. aureofasciatus*, MNHN JNC1040.

Abbreviations

NGR: New geographic record; NHR: New host record; HPC: Host-parasite combination; SLIP: Species-level identified parasite; SLIP-HPC: Species-level identified parasite – host-parasite combination.

Institutional abbreviations

BMNH: Natural History Museum, London, United Kingdom; HCIP: Helminthological Collection, Institute of Parasitology, Biology Centre, Academy of Sciences of the Czech Republic, České Budějovice, Czech Republic; MNHN: Muséum National d'Histoire Naturelle, Paris, France; SAMA AHC: South Australian Museum Adelaide, Australian Helminthological Collection, Adelaide, Australia; USNPC: United States National Parasite Collection, Beltsville, USA.

Abbreviations for higher parasite taxa

The following abbreviations are used in Tables and Appendices.: For all: Unid: Unidentified family; Isop: Isopoda; Families: Aegi: Aegidae; Cora: Corallanidae; Cymo: Cymothoidae; Gnat: Gnathiidae; Cope: Copepoda; Families: Cali: Caligidae; Diss: Dissonidae; Hats: Hatschekiidae; Lerp: Lernaeopodidae; Penn: Pennellidae; Mono: Monogenea; Monop: Monopisthocotylea; Poly: Polyopisthocotylea; Families: Ancy: Ancyrocephalidae; Caps: Capsalidae; Dipl: Diplectanidae; Gyro: Gyrodactylidae; Micr: Microcotylidae; Dige: Digenea; Families: Acan: Acanthocolpidae; Cryp: Cryptogonimidae; Didy: Didymozoidae; Fell: Fellodistomatidae; Hemi: Hemiuridae; Leci: Lecithasteridae; Lepo: Lepocreadiidae; Monr: Monorchiidae; Opec: Opecoelidae; Scler: Sclerodistomidae; Tran: Transversotrematidae; Tryp: Cestoda Trypanorhyncha; Families: Laci: Lacistorhynchidae; Otob: Otobothriidae; Tent: Tentaculariidae; Both: Cestoda Bothriocephalidea (no family identified); Tetr: Cestoda Tetraphyllidea (no family identified); Nema: Nematoda; Families: Anis: Anisakidae; Cama: Camallanidae; Capi: Capillariidae; Cucu: Cucullanidae; Gnto: Gnathostomatidae; Phil: Philometridae; Phys: Physalopteridae; Tric: Trichosomoididae; Hiru: Hirudinea (no family identified); Family: Pisc: Piscicolidae; Turb: 'Turbellaria'; Acantho: Acanthocephala (no family identified).

Competing interests

The authors declare that they have no competing interests.

Authors' contributions

JLJ collected fish and parasites and compiled and compared results. JLJ IB GAB RAB TLM FM JPT IDW identified parasites. GAB JLJ made figures. JLJ IB GAB RAB TLM FM JPT IDW wrote the paper. Authors are in alphabetical order, except JLJ. All authors read and approved the final manuscript.

Authors' information

JLJ: specialist of phylogeny and taxonomy of parasitic worms, mainly monogeneans; professor and curator of the collections of parasitic worms of MNHN, Paris, France, has spent more than eight years (2003–2011) in Nouméa, New Caledonia, South Pacific, collecting and studying parasites from fish; UMR 7138 Systématique, Adaptation, Évolution, Muséum National d'Histoire Naturelle, Case postale 51, 55, rue Buffon, 75231 Paris cedex 05, France.

IB: specialist of phylogeny and taxonomy of parasitic worms, including trypanorhynch cestodes; Department of Veterinary Science, University of Melbourne, Veterinary Clinical Centre, Werribee 3030, Victoria, Australia.

GAB: specialist of all aspects of biology and taxonomy of copepods, including parasitic groups; Department of Zoology, Natural History Museum, Cromwell Road, London SW7 5BD, UK.

RAB: specialist of phylogeny and taxonomy of parasitic worms, mainly digeneans and cestodes; Department of Zoology, Natural History Museum, Cromwell Road, London SW7 5BD, UK.

TLM: specialist of phylogeny and taxonomy of digeneans, especially cryptogonimids; Biodiversity Program, Queensland Museum, PO Box 3300, South Brisbane, Queensland, 4101 Australia

FM: specialist of biology and taxonomy of parasitic nematodes; Institute of Parasitology, Biology Centre, Academy of Sciences of the Czech Republic, Branišovská 31, 370 05 České Budějovice, Czech Republic.

JPT: specialist of taxonomy and biogeography of parasitic isopods; Équipe Adaptation écophysiologique et Ontogenèse, UMR 5119 (CNRS-UM2-IRD-UM1-IFREMER), Université Montpellier 2, Place Eugène Bataillon, 34095 Montpellier cedex 05, France.

IDW: specialist of biology, taxonomy and phylogeny of monogeneans, including capsalids; Monogenean Research Laboratory, The South Australian Museum, Adelaide 5000, & Marine Parasitology Laboratory, & Australian Centre for Evolutionary Biology and Biodiversity, The University of Adelaide, North Terrace, Adelaide 5005, South Australia, Australia.

Acknowledgements

Numerous students helped for the parasitological survey. The identification of certain hosts was kindly confirmed (from photographs) by John E. Randall (Bishop Museum, Hawaii) or Ronald Fricke (Staatliches Museum für Naturkunde, Stuttgart, Germany). Parasite identifications were kindly provided by E. Burreson, L. Euzet, D. C. Kritsky and S. Shamsi. Partly supported (FM) by the Czech Science Foundation (grant no. P505/12/G112), and research project of the Institute of Parasitology BC AS CR (Z60220518). Partly supported (IDW) by Australian Research Council grant no. DP0556780 (2005–07) awarded to IDW and Steve Donnellan (South Australian Museum) which permitted IDW to visit Nouméa for 3 weeks in October-November 2008 ably assisted by Drs Lizzie Perkins and Vanessa Glennon who were supported by a J.D. Smyth Travel Award from the Australian Society for Parasitology Inc. awarded to Dr Lizzie Perkins. Partly supported (JLJ) by ATM barcode (MNHN) and ATM Biodiversité actuelle et fossile (MNHN).

Author details

[1]UMR 7138 Systématique, Adaptation, Évolution, Muséum National d'Histoire Naturelle, Case postale 51, 55, rue Buffon, 75231 Paris cedex 05, France. [2]Department of Veterinary Science, University of Melbourne, Veterinary Clinical Centre, Werribee 3030, Victoria, Australia. [3]Department of Zoology, Natural History Museum, Cromwell Road, London SW7 5BD, UK. [4]Biodiversity Program, Queensland Museum, PO Box 3300, South Brisbane, Queensland 4101, Australia. [5]Institute of Parasitology, Biology Centre, Academy of Sciences of the Czech Republic, Branišovská 31 370 05, České Budějovice, Czech Republic. [6]Équipe Adaptation écophysiologique et Ontogenèse, UMR 5119 (CNRS-UM2-IRD-UM1-IFREMER), Université Montpellier 2, Place Eugène Bataillon, 34095 Montpellier cedex 05, France. [7]Monogenean Research Laboratory, The South Australian Museum, Adelaide 5000, & Marine Parasitology Laboratory, & Australian Centre for Evolutionary Biology and Biodiversity, The University of Adelaide, North Terrace, Adelaide 5005, South Australia, Australia.

References

1. Reaka-Kudla ML: **The global biodiversity of coral reefs: a comparison with rain forests**. In *Biodiversity II: Understanding and protecting our biological resources*. Edited by Reaka-Kudla ML, Wilson DE. E.O. W. Washington: Joseph Henry Press; 1997:83–108.
2. Rohde K: **Marine parasitology in Australia**. *Search (Sydney)* 1976, **7**:477–482.
3. Rohde K: **Species diversity of parasites on the Great Barrier Reef**. *Z ParasitenKde* 1976, **50**:93–94.
4. Cribb TH, Bray RA, Barker SC, Adlard RD, Anderson GR: **Ecology and diversity of digenean trematodes of reef and inshore fishes of Queensland**. *Int J Parasitol* 1994, **24**:851–860.

5. Whittington ID: Diversity "down under": monogeneans in the Antipodes (Australia) with a prediction of monogenean biodiversity worldwide. *Int J Parasitol* 1998, **28**:1481–1493.

6. Justine J-L: Parasites of coral reef fish: how much do we know? With a bibliography of fish parasites in New Caledonia. *Belg J Zool* 2010, **140**:155–190. Suppl.

7. Justine J-L, Beveridge I, Boxshall GA, Bray RA, Moravec F, Trilles J-P, Whittington ID: An annotated list of parasites (Isopoda, Copepoda, Monogenea, Digenea, Cestoda and Nematoda) collected in groupers (Serranidae, Epinephelinae) in New Caledonia emphasizes parasite biodiversity in coral reef fish. *Folia Parasitol* 2010, **57**:237–262.

8. Justine J-L, Beveridge I, Boxshall GA, Bray RA, Moravec F, Whittington ID: An annotated list of fish parasites (Copepoda, Monogenea, Digenea, Cestoda and Nematoda) collected from Emperors and Emperor Bream (Lethrinidae) in New Caledonia further highlights parasite biodiversity estimates on coral reef fish. *Zootaxa* 2010, **2691**:1–40.

9. Allen GR: *FAO species catalogue. Vol. 6. Snappers of the world. An annotated and illustrated catalogue of lutjanid species known to date. FAO Fisheries Synopsis 125.* Rome: FAO; 1985.

10. Carpenter KE: *FAO species catalogue. Vol 8. Fusilier fishes of the world. An annotated and illustrated catalogue of caesionid species known to date. FAO Fisheries Synopsis 125.* Rome: FAO; 1988.

11. Russell BC: *FAO Species catalogue. Vol. 12. Nemipterid Fishes of the World. (Threadfin breams, whiptail breams, monocle breams, dwarf monocle breams, and coral breams). Family Nemipteridae. An annotated and illustrated catalogue of nemipterid species known to date. FAO Fisheries Synopsis 125.* Rome: FAO; 1990.

12. Miller TL, Cribb TH: Phylogenetic relationships of some common Indo-Pacific snappers (Perciformes: Lutjanidae) based on mitochondrial DNA sequences, with comments on the taxonomic position of the Caesioninae. *Mol Phylogenet Evol* 2007, **44**:450–460.

13. Gold JR, Voelker G, Renshaw MA: Phylogenetic relationships of tropical western Atlantic snappers in subfamily Lutjaninae (Lutjanidae: Perciformes) inferred from mitochondrial DNA sequences. *Biol J Linn Soc* 2011, **102**:915–929.

14. Guo Y, Wang Z, Liu C, Liu L, Liu Y: Phylogenetic relationships of South China Sea snappers (Genus *Lutjanus*; Family Lutjanidae) based on mitochondrial DNA sequences. *Mar Biotechnol* 2007, **9**:682–688.

15. Guo Y, Wong Z, Liu C, Liu Y: Sequencing and analysis of the complete mitochondrial DNA of Russell's snapper (*L. russellii*). *Progr Nat Sci* 2008, **18**:1233–1238.

16. Eschmeyer WN, Fong JD: Pisces. In *Animal biodiversity: An outline of higher-level classification and survey of taxonomic richness, Zootaxa, Volume 3148.* Edited by Zhang Z-Q.; 2011:27–38.

17. Fricke R, Kulbicki M, Wantiez L: Checklist of the fishes of New Caledonia, and their distribution in the Southwest Pacific Ocean (Pisces). *Stuttg Beitr Natkd Ser A (Biol)* 2011, **4**:341–463.

18. Kulbicki M, Bozec Y-M, Labrosse P, Letourneur Y, Mou-Tham G, Wantiez L: Diet composition of carnivorous fishes from coral reef lagoons of New Caledonia. *Aquat Living Resour* 2005, **18**:231–250.

19. Fourmanoir P: *Pêche profonde en Nouvelle-Calédonie et aux Loyauté.* Nouméa: Commission du Pacifique Sud / South Pacific Commission (CPS/SPC); 1977:8. www.spc.int.

20. Marie AD, Justine J-L: Monocotylids (Monogenea: Monopisthocotylea) from *Aetobatus* cf. *narinari* off New Caledonia, with a description of *Decacotyle elpora* n. sp. *Syst Parasitol* 2005, **60**:175–185.

21. Justine J-L, Dupoux C, Cribb TH: Resolution of the discrepant host-specificity of *Pseudorhabdosynochus* species (Monogenea, Diplectanidae) from serranid fishes in the tropical Indo-Pacific. *Acta Parasitol* 2009, **54**:119–130.

22. Justine J-L, Briand MJ: Three new species, *Lamellodiscus tubulicornis* n. sp., *L. magnicornis* n. sp. and *L. parvicornis* n. sp. (Monogenea: Diplectanidae) from *Gymnocranius* spp. (Lethrinidae: Monotaxinae) off New Caledonia, with proposal of the new morphological group 'tubulicornis' within *Lamellodiscus* Johnston & Tiegs, 1922. *Syst Parasitol* 2010, **75**:159–179.

23. Schoelinck C, Cruaud C, Justine J-L: Are all species of *Pseudorhabdosynochus* strictly host specific? – a molecular study. *Parasitol Int* 2012, **61**:356–359.

24. Justine J-L: Three new species of *Huffmanela* Moravec, 1987 (Nematoda: Trichosomoididae) from the gills of marine fish off New Caledonia. *Syst Parasitol* 2004, **59**:29–37.

25. Justine J-L: *Huffmanela* spp. (Nematoda, Trichosomoididae) parasites in coral reef fishes off New Caledonia, with descriptions of *H. balista* n. sp. and *H. longa* n. sp. *Zootaxa* 2007, **1628**:23–41.

26. Trilles JP: Sur quatre Isopodes Cymothoïdés du Pacifique (Nouvelle Calédonie). *Cahiers ORSTOM, série Océan* 1972, **10**:3–17.

27. Bruce NL, Harrison-Nelson EB: New records of fish parasitic marine isopod crustaceans (Cymothoidae, subfamily Anilocrinae) from the Indo-west Pacific. *Proc Biol Soc Wash* 1988, **101**:585–602.

28. Herklots J: Deux nouveaux genres de Crustacés vivant en parasites sur des poissons, *Epichthyes* et *Ichthyoxenos*. *Arch Néerl Sci Ex Nat* 1870, **5**:120–137.

29. Schiœdte JC, Meinert FW: Symbolae ad Monographiam Cymothoarum Crustaceorum Isopodum Familiae 2. Anilocridae. *Naturhistorikan Tidsskrift* 1881, **13**(3):1–166.

30. Trilles J-P: Les Cymothoidae (Crustacea, Isopoda) du Monde (Prodrome pour une Faune). *Stud Mar* 1994, **21/22**:5–288.

31. Bruce NL: Australian *Pleopodias* Richardson, 1910, and *Anilocra* Leach, 1818 (Isopoda: Cymothoidae), crustacean parasites of marine fishes. *Rec Aust Mus* 1987, **39**:85–130.

32. Bruce NL: Reassessment of the isopod crustacean *Aega deshaysiana* (Milne Edwards, 1840) (Cymothoida: Aegidae): a world-wide complex of 21 species. *Zool J Linn Soc* 2004, **142**:135–222.

33. Delaney M: Phylogeny and Biogeography of the Marine Isopod Family Corallanidae (Crustacea, Isopoda, Flabellifera). *Contr Sci* 1989, **409**:1–75.

34. Smit NJ, Basson L: *Gnathia pantherina* sp. n. (Crustacea: Isopoda: Gnathiidae), a temporary ectoparasite of some elasmobranch species from southern Africa. *Folia Parasitol* 2002, **49**:137–151.

35. Smit NJ, Basson L, Van As JG: Life cycle of the temporary fish parasite, *Gnathia africana* (Crustacea: Isopoda: Gnathiidae). *Folia Parasitol* 2003, **50**:135–142.

36. J-s H, Kim I-H: New species of *Hatschekia* Poche, 1902 (Copepoda: Hatschekiidae) parasitic on marine fishes of Kuwait. *Syst Parasitol* 2001, **49**:73–79.

37. Kabata Z: Copepoda parasitic on Australian fishes, XIII - Family Hatschekiidae. *J Nat Hist* 1991, **25**:91–121.

38. Boxshall GA, Halsey SH: *An introduction to copepod diversity.* London: Ray Society; 2004.

39. Raibaut A, Combes C, Benoit F: Analysis of the parasitic copepod species richness among Mediterranean fish. *J Mar Syst* 1998, **15**:185–206.

40. Shiino SM: Note on a new parasitic copepod *Caligus brevis* n. sp. *Bull Jpn Soc Sci Fish* 1954, **20**:178–183.

41. Hewitt GC: Some New Zealand parasitic Copepoda of the Family Caligidae. *Trans R Soc New Zeal Zool* 1963, **4**:61–115.

42. Ho J-S, Lin C-L: *Sea Lice of Taiwan (Copepoda: Siphonostomatoida: Caligidae).* Keelung: The Sueichan Press; 2004.

43. Kritsky DC, Bullard SA, Bakenhaster MD: First report of gastrocotylinean post-oncomiracidia (Platyhelminthes: Monogenoidea: Heteronchoinea) on gills of flyingfish (Exocoetidae), snapper (Lutjanidae), dolphinfish (Coryphaenidae), and amberjack (Carangidae) from the Gulf of Mexico: Decoy hosts and the dilution effect. *Parasitol Int* 2011, **60**:274–282.

44. Lester RJG, Sewell KB: Checklist of parasites from Heron Island, Great Barrier Reef. *Aust J Zool* 1989, **37**:101–128.

45. Kritsky DC, Yang T, Sun Y: Dactylogyrids (Monogenoidea, Polyonchoinea) parasitizing the gills of snappers (Perciformes, Lutjanidae): Proposal of *Haliotrematoides* n. gen. and descriptions of new and previously described species from marine fishes of the Red Sea, the eastern and Indo-west Pacific Ocean, Gulf of Mexico and Caribbean Sea. *Zootaxa* 1970, **2009**:1–51.

46. Fajer-Avila EJ, Velasquez-Medina SP, Betancourt-Lozano M: Effectiveness of treatments against eggs, and adults of *Haliotrema* sp and *Euryhaliotrema* sp (Monogenea: Ancyrocephalinae) infecting red snapper, *Lutjanus guttatus*. *Aquaculture* 2007, **264**:66–72.

47. Justine J-L: Species of *Calydiscoides* Young, 1969 (Monogenea: Diplectanidae) from lethrinid fishes, with the redescription of all of the type-specimens and the description of *C. euzeti* n. sp. from *Lethrinus rubrioperculatus* and *L. xanthochilus* off New Caledonia. *Syst Parasitol* 2007, **67**:187–209.

48. Lim LHS: Species of *Calydiscoides* Young, 1969 (Monogenea; Diplectanidae Bychowsky, 1957; Lamellodiscinae Oliver, 1969) from nemipterid fishes off Peninsular Malaysia. *Syst Parasitol* 2003, **55**:115–126.

49. Yamaguti S: *Monogenetic Trematodes of Hawaiian Fishes*. Honolulu: University of Hawaii Press; 1968.

50. Kent ML, Heidel JR, Marie A, Moriwake A, Moriwake V, Alexander B, Watral V, Kelley CD: **Diseases of Opakapaka** *Pristipomoides filamentosus*. Fish Health Section. In *Diseases in Asian Aquaculture V*. Edited by Walker P, Lester R, Bondad-Reantaso MG. Manila: Asian Fisheries Society; 2005:183–195.

51. Oliver G, Paperna I: **Diplectanidae Bychowsky, 1957 (Monogenea, Monopisthocotylea), parasites de Perciformes de Méditerranée orientale, de la mer Rouge et de l'océan Indien.** *Bull Mus Natl Hist Nat, Paris* 1984, **4**(6, section A):49–65.

52. Vignon M, Morat F, Galzin R, Sasal P: **Evidence for spatial limitation of the bluestripe snapper** *Lutjanus kasmira* in French Polynesia from parasite and otolith shape analysis. *J Fish Biol* 2008, **73**:2305–2320.

53. Vignon M, Sasal P, Rigby MC, Galzin R: **Multiple parasite introduction and host management plan: case study of lutjanid fish in the Hawaiian Archipelago.** *Dis Aquat Org* 2009, **85**:133–145.

54. Nagibina LF: **New species of the genus** *Diplectanum* (Monogenoidea, Diplectanidae). In: Fauna, systematics and phylogeny of Monogenoidea. *Tr Biologo-Pochvennogo Inst, Nov Ser* 1976, **35**:89–96.

55. Deveney MR, Whittington ID: **Three new species of** *Benedenia* Diesing, 1858 from the Great Barrier Reef, Australia with a key to species of the genus. *Zootaxa* 2010, **2348**:1–22.

56. Whittington ID: **Revision of** *Benedeniella* Johnston, 1929 (Monogenea: Capsalidae), its assignment to Entobdellinae Bychowsky, 1957 and comments on subfamilial composition. *Zootaxa* 2010, **2519**:1–30.

57. Whittington ID, Deveney MR: **New** *Benedenia* species (Monogenea: Capsalidae) from *Diagramma labiosum* (Perciformes: Haemulidae) on the Great Barrier Reef, Australia, with oncomiracidial descriptions and a report of egg attachment to the host. *J Parasitol* 2011, **97**:1026–1034.

58. Perkins EM, Donnellan SC, Bertozzi T, Chisholm LA, Whittington ID: **Looks can deceive: Molecular phylogeny of a family of flatworm ectoparasites (Monogenea: Capsalidae) does not reflect current morphological classification.** *Mol Phylogenet Evol* 2009, **52**:705–714.

59. Perkins EM: *Family ties: molecular phylogenetics, evolution and radiation of flatworm parasites (Monogenea: Capsalidae)*. PhD, The University of Adelaide; 2010.

60. Whittington ID: **The Capsalidae (Monogenea: Monopisthocotylea): a review of diversity, classification and phylogeny with a note about species complexes.** *Folia Parasitol* 2004, **51**:109–122.

61. Whittington ID: **Benedeniine capsalid monogeneans from Australian fishes: Pathogenic species, site-specificity and camouflage.** *J Helminthol* 1996, **70**:177–184.

62. Whittington ID, Deveney MR, Wyborn SJ: **A revision of** *Benedenia* Diesing, 1858 including a redescription of *B. sciaenae* (van Beneden, 1856) Odhner, 1905 and recognition of *Menziesia* Gibson, 1976 (Monogenea: Capsalidae). *J Nat Hist* 1856, **2001**(35):663–777.

63. Whittington ID, Ernst I: **Migration, site-specificity and development of** *Benedenia lutjani* (Monogenea: Capsalidae) on the surface of its host, *Lutjanus carponotatus* (Pisces: Lutjanidae). *Parasitology* 2002, **124**:423–434.

64. Horton MA, Whittington ID: **A new species of** *Metabenedeniella* (Monogenea, Capsalidae) from the dorsal fin of *Diagramma pictum* (Perciformes, Haemulidae) from the Great Barrier Reef, Australia with a revision of the genus. *J Parasitol* 1994, **80**:998–1007.

65. In *FishBase. World Wide Web electronic publication*. Edited by Froese R, Pauly D.; 2012. www.fishbase.org.

66. Yamaguti S: **New monogenetic trematodes from Hawaiian fishes, II.** *Pac Sci* 1966, **20**:419–434.

67. Bychowsky BE, Nagibina L: **New Capsalidae (Monogenoidea) from Pacific fishes.** *Parazitologiya* 1967, **1**:521–528.

68. Grutter AS: **Spatial and temporal variations of the ectoparasites of seven reef fish species from Lizard Island and Heron Island, Australia.** *Mar Ecol Prog Ser* 1994, **115**:21–30.

69. Cribb TH, Bray RA, Olson PD, Littlewood DTJ: **Life cycle evolution in the Digenea: a new perspective from phylogeny.** *Adv Parasitol* 2003, **54**:197–254.

70. Gibson DI, Bray RA: **The Hemiuridae (Digenea) of fishes from the north-east Atlantic.** *Bull Br Mus (Nat Hist) (Zool Ser)* 1986, **51**:1–125.

71. Miller TL, Bray RA, Goiran C, Justine J-L, Cribb TH: *Adlardia novaecaledoniae* n.g., n. sp. (Digenea: Cryptogonimidae) from the fork-tailed threadfin bream *Nemipterus furcosus* (Val.) (Perciformes: Nemipteridae) off New Caledonia. *Syst Parasitol* 2009, **73**:151–160.

72. Velasquez CC: **Cryptogonimidae (Digenea: Trematoda) from Philippine food fishes.** *J Parasitol* 1961, **47**:914–918.

73. Manter HW: **Studies in the digenetic trematodes of fishes of Fiji. III. Families Acanthocolpidae, Fellodistomatidae, and Cryptogonimidae.** *J Parasitol* 1963, **49**:443–450.

74. Yamaguti S: *Digenetic trematodes of Hawaiian fishes*. Tokyo: Keigaku; 1970.

75. Gu C-D, Shen J-W: **Digenetic trematodes of fishes from the Xisha Islands, Guangdong Province, China. I.** *Stud Mar Sin* 1983, **20**:157–184. In Chinese.

76. Miller TL, Cribb TH: **Coevolution of** *Retrovarium* n. gen. (Digenea: Cryptogonimidae) in Lutjanidae and Haemulidae (Perciformes) in the Indo-West Pacific. *Int J Parasitol* 2007, **37**:1023–1045.

77. Hafeezullah M, Siddiqi AH: **Digenetic trematodes of marine fishes of India. Part I. Bucephalidae and Cryptogonimidae.** *Indian J Helminthol* 1970, **22**:1–22.

78. Miller TL, Adlard RD, Bray RA, Justine J-L, Cribb TH: **Cryptic species of** *Euryakaina* n. g. (Digenea: Cryptogonimidae) from sympatric lutjanids in the Indo-West Pacific. *Syst Parasitol* 2010, **77**:185–204.

79. Miller TL, Bray RA, Justine J-L, Cribb TH: *Varialvus* gen. nov. (Digenea, Cryptogonimidae), from species of Lutjanidae (Perciformes) off the Great Barrier Reef, New Caledonia and the Maldives. *Acta Parasitol* 2010, **55**:327–339.

80. Bray RA, Webster BL, Bartoli P, Littlewood DTJ: **Relationships within the Acanthocolpidae Lühe, 1906 and their place among the Digenea.** *Acta Parasitol* 2005, **50**:281–291.

81. Bray RA, Justine J-L: **Acanthocolpidae (Digenea) of marine fishes off New Caledonia, with the descriptions of two new species.** *Folia Parasitol* 2011, **58**:35–47.

82. Bray RA, Cribb TH: **Digeneans of the family Opecoelidae Ozaki, 1925 from the southern Great Barrier Reef, including a new genus and three new species.** *J Nat Hist* 1989, **23**:429–473.

83. McCoy OR: **The life-history of a marine trematode,** *Hamacreadium mutabile* Linton, 1910. *Parasitology* 1929, **21**:220–225.

84. Gupta PC, Singh RB: **Four new digenetic trematodes from marine fishes off Puri coast, Bay of Bengal.** *Indian J Parasitol* 1985, **9**:25–34.

85. Gupta PC, Gupta VC: **On three opecoelid trematodes (Digenea) from marine fishes, Bay of Bengal, India.** *Proc Parasitol* 1988, **6**:59–68.

86. Aken'ova TOL, Cribb TH: **Two new species of** *Neolebouria* Gibson, 1976 (Digenea: Opecoelidae) from temperate marine fishes of Australia. *Syst Parasitol* 2001, **49**:65–71.

87. Bray RA, Justine J-L: *Neolebouria blatta* n. sp. (Digenea: Opecoelidae) from *Pristipomoides argyrogrammicus* (Valenciennes) and *Etelis carbunculus* Cuvier (Perciformes: Lutjanidae) off New Caledonia. *Syst Parasitol* 2009, **74**:161–167.

88. Bray RA, Justine J-L: **Opecoelids (Platyhelminthes: Digenea) from the fork-tailed threadfin bream** *Nemipterus furcosus* (Valenciennes, 1830) (Perciformes: Nemipteridae), with preliminary keys to the problematic genera *Macvicaria* Gibson & Bray, 1982 and *Neolebouria* Gibson, 1976. *Acta Parasitol* 2009, **54**:218–229.

89. Hunter JA, Cribb TH: **A cryptic complex of species related to** *Transversotrema licinum* Manter, 1970 from fishes of the Indo-West Pacific, including descriptions of ten new species of *Transversotrema* Witenberg, 1944 (Digenea: Transversotrematidae). *Zootaxa* 2012, **3176**:1–44.

90. Cribb TH, Pichelin S, Dufour V, Bray RA, Chauvet C, Faliex E, Galzin R, Lo CM, Lo-Yat A, Morand S, *et al*: **Parasites of recruiting coral reef fish larvae in New Caledonia.** *Int J Parasitol* 2000, **30**:1445–1451.

91. Pozdnyakov SE, Gibson DI: **Family Didymozoidae Monticelli, 1888.** In *Keys to the Trematoda*, Volume 3. Edited by Bray RA, Gibson DI, Jones A. London: CAB International and The Natural History Museum; 2008:631–734.

92. Bray RA: **Hemiuridae (Digenea) from marine fishes of the southern Indian Ocean: Dinurinae, Elytrophallinae, Glomericirrinae and Plerurinae.** *Syst Parasitol* 1990, **17**:183–217.

93. Nasir P: **Monotypic status of** *Prosogonotrema* Perez Vigueras, 1940 (Trematoda: Digenea). *Riv Parassitol* 1973, **34**:271–276.

94. Nahhas FM, Sey O, Nishimoto R: **Digenetic trematodes of marine fishes from the Kuwaiti Coast of the Arabian Gulf: Families Pleorchiidae, Fellodistomidae, and Cryptogonimidae, with a description of two new species,** *Neoparacryptogonimus sphericus* and *Paracryptogonimus ramadani*. *J Helminthol Soc Wash* 1998, **65**:129–140.

95. Machida M: **Monorchiidae (Trematoda, Digenea) from fishes of Japanese and adjacent waters.** *Bull Natl Sci Mus, Tokyo Ser A Zool* 2005, **31**:123–136.

96. Kuramochi T: Digenean trematodes of fishes caught in Sagami Bay, off Izu Islands and off Ogasawara Islands. *Mem Natl Mus Nat Sci, Tokyo* 2011, **47**:51–63.

97. Korotaeva VD: New species of Trematoda of the genus *Tergestia* Stossich, 1899 (Trematoda, Fellodistomatidae). *Izv Tikhookean Nauchno-Issled Inst Rybn Khoz Okeanogr* 1972, **81**:263–266.

98. Olson PD, Caira JN, Jensen K, Overstreet RM, Palm HW, Beveridge I: Evolution of the trypanorhynch tapeworms: Parasite phylogeny supports independent lineages of sharks and rays. *Int J Parasitol* 2010, **40**:223–242.

99. Palm HW: *The Trypanorhyncha Diesing, 1863*. Bogor: PKSPL-IPB Press; 2004.

100. Andrews AH, Kalish JM, Newman SJ, Johnston JM: Bomb radiocarbon dating of three important reef-fish species using Indo-Pacific Δ^{14}C chronologies. *Mar Freshw Res* 2011, **62**:1259–1269.

101. Moravec F: *Dracunculoid and anguillicoloid nematodes parasitic in vertebrates*. Praha: Academia; 2006.

102. Santos Cavalcanti ET, Takemoto RM, Alves LC, Chellappa S: First record of endoparasite *Philometra* sp. (Nematoda: Philometridae) in lane snapper *Lutjanus synagris* from the coast of Rio Grande do Norte, Brazil. *Mar Biodiv Rec* 2011, **3**:e93.

103. Justine J-L: *Huffmanela plectropomi* n. sp. (Nematoda: Trichosomoididae: Huffmanelinae) from the coralgrouper *Plectropomus leopardus* (Lacépède) off New Caledonia. *Syst Parasitol* 2011, **79**:139–143.

104. Bosques Rodríguez LJ: *Metazoan parasites of snappers, Lutjanidae (Pisces) from Puerto Rico*. PhD thesis, University of Puerto Rico; 2004.

105. Bullard SA, Barse AM, Curran SS, Morris JA: First record of a digenean from invasive Lionfish, *Pterois* cf. *volitans*, (Scorpaeniformes: Scorpaenidae) in the Northwestern Atlantic Ocean. *J Parasitol* 2011, **97**:833–837.

106. Cannon LRG, Lester RJG: Two turbellarians parasitic in fish. *Dis Aquat Org* 1988, **5**:15–22.

107. Justine J-L, Leblanc P, Keller F, Lester RJG: Turbellarian black spot disease in bluespine unicornfish *Naso unicornis* in New Caledonia caused by the parasitic turbellarian *Piscinquilinus* sp. *Dis Aquat Org* 2009, **85**:245–249.

108. Foata J, Quilichini Y, Justine J-L, Bray RA, Marchand B: Ultrastructural study of spermiogenesis and the spermatozoon of *Cavisoma magnum* (Southwell, 1927) (Acanthocephala, Palaeacanthocephala, Cavisomidae), from *Siganus lineatus* (Pisces, Teleostei, Siganidae) (Valenciennes, 1835) in New Caledonia. *Micron* 2012, **43**:141–149.

109. Rohde K: Gill Monogenea of deepwater and surface fish in southeastern Australia. *Hydrobiologia* 1988, **160**:271–283.

110. Boxshall GA: Host specificity in copepod parasites of deep-sea fishes. *J Mar Syst* 1998, **15**:215–223.

111. Bray RA, Littlewood DTJ, Herniou EA, Williams B, Henderson RE: Digenean parasites of deep-sea teleosts: a review and case studies of intrageneric phylogenies. *Parasitology* 1999, **119**:S125–S144.

112. Klimpel S, Busch MW, Kellermans E, Kleinertz S, Palm HW: Metazoan deep-sea fish parasites. *Acta Biol Benrodis Supplbd* 2009, **11**:1–384.

113. Kulbicki M, Guillemot N, Amand M: A general approach to length-weight relationships for New Caledonian lagoon fishes. *Cybium* 2005, **29**:235–253.

114. Poulin R, Morand S: *Parasite Biodiversity*. Washington: Smithsonian Books; 2004.

115. Luque JL, Poulin R: Metazoan parasite species richness in Neotropical fishes: hotspots and the geography of biodiversity. *Parasitology* 2007, **134**:865–878.

116. McNamara MKA, Adlard RD, Bray RA, Sasal P, Cribb TH: Monorchiids (Platyhelminthes: Digenea) of chaetodontid fishes (Perciformes): Biogeographical patterns in the tropical Indo-West Pacific. *Parasitol Int* 2012, **61**:288–306.

117. McNamara MKA, Cribb TH: Taxonomy, host specificity and dietary implications of *Hurleytrematoides* (Digenea: Monorchiidae) from chaetodontid fishes on the Great Barrier Reef. *Parasitol Int* 2011, **60**:255–269.

118. Gunter NL, Adlard RD: Seven new species of *Ceratomyxa* Thélohan, 1892 (Myxozoa) from the gall-bladders of serranid fishes from the Grear Barrier Reef, Australia. *Syst Parasitol* 2009, **73**:1–11.

119. Gunter NL, Whipps CM, Adlard RD: *Ceratomyxa* (Myxozoa: Bivalvulida): Robust taxon or genus of convenience? *Int J Parasitol* 2009, **39**:1395–1405.

120. Beveridge I, Justine J-L: Two new species of *Prochristianella* Dollfus, 1946 (Platyhelminthes, Cestoda) from the blue-spotted stingray, *Neotrygon kuhlii* (Müller & Henle, 1841) off New Caledonia. *Zoosystema* 2010, **32**:643–652.

121. Bray RA, Cribb TH, Justine J-L: *Multitestis* Manter 1931 (Digenea: Lepocreadiidae) in ephippid and chaetodontid fishes (Perciformes) in the south-western Pacific Ocean and the Indian Ocean off Western Australia. *Zootaxa* 2010, **2427**:36–46.

122. Bray RA, Cribb TH, Justine J-L: *Diploproctodaeum* spp. (Digenea: Lepocreadiidae) in Australian and New Caledonian waters including two new species from Tetraodontiformes and new records of related species. *Acta Parasitol* 2010, **51**:313–326.

123. Bray RA, Justine J-L: Bucephaline digeneans (Bucephalidae) in *Sphyraena putnamae* Jordan & Seale (Sphyraenidae) from the lagoon off New Caledonia. *Syst Parasitol* 2011, **79**:123–138.

124. Moravec F, Justine J-L: Two new genera and species of cystidicolids (Nematoda, Cystidicolidae) from marine fishes off New Caledonia. *Parasitol Int* 2010, **59**:198–205.

125. Stork NE, Lyal CHC: Extinction or "co-extinction" rates. *Nature* 1993, **366**:307–307.

126. Koh LP, Dunn RR, Sodhi NS, Colwell RK, Proctor HC, Smith VS: Species coextinctions and the biodiversity crisis. *Science* 2004, **305**:1632–1634.

127. Windsor DA: Equal rights for parasites. *Conserv Biol* 1995, **9**:1–2.

128. Windsor DA: Most of species on Earth are parasites. *Int J Parasitol* 1998, **28**:1939–1941.

129. Dunn RR, Harris NC, Colwell RK, Koh LP, Sodhi NS: The sixth mass coextinction: are most endangered species parasites and mutualists? *Proc Roy Soc B Biol Sci* 2009, **276**:3037–3045.

130. Dobson A, Lafferty KD, Kuris AM, Hechinger RF, Jetz W: Homage to Linnaeus: How many parasites? How many hosts? *Proc Natl Acad Sci U S A* 2008, **105**:11482–11489.

131. Rózsa L: Endangered parasite species. *Int J Parasitol* 1992, **22**:265–266.

132. Justine J-L: Parasite biodiversity in a coral reef fish: twelve species of monogeneans on the gills of the grouper *Epinephelus maculatus* (Perciformes: Serranidae) off New Caledonia, with a description of eight new species of *Pseudorhabdosynochus* (Monogenea: Diplectanidae). *Syst Parasitol* 2007, **66**:81–129.

133. Moir ML, Vesk PA, Brennan KEC, Keith DA, Hughes L, McCarthy MA: Current constraints and future directions in estimating coextinction. *Conserv Biol* 2010, **24**:682–690.

134. Mihalca A, Gherman C, Cozma V: Coendangered hard-ticks: threatened or threatening? *Parasites Vectors* 2011, **4**:71.

135. Cardoso P, Erwin TL, Borges PAV, New TR: The seven impediments in invertebrate conservation and how to overcome them. *Biol Conserv* 2011, **144**:2647–2655.

136. Hughes TP: Catastrophes, phase shifts, and large-scale degradation of a Caribbean coral reef. *Science* 1994, **265**:1547–1551.

137. Roberts CM, McClean CJ, Veron JEN, Hawkins JP, Allen GR, McAllister DE, Mittermeier CG, Schueler FW, Spalding M, Wells F, *et al*: Marine biodiversity hotspots and conservation priorities for tropical reefs. *Science* 2002, **295**:1280–1284.

138. Pandolfi JM, Bradbury RH, Sala E, Hughes TP, Bjorndal KA, Cooke RG, McArdle D, McClenachan L, Newman MJH, Paredes G, *et al*: Global trajectories of the long-term decline of coral reef ecosystems. *Science* 2003, **301**:955–958.

139. Jones GP, McCormick MI, Srinivasan M, Eagle JV: Coral decline threatens fish biodiversity in marine reserves. *Proc Natl Acad Sci U S A* 2004, **101**:8251–8253.

140. Bruno JF, Selig ER: Regional decline of coral cover in the Indo-Pacific: Timing, extent, and subregional comparisons. *PLoS ONE* 2007, **2**:e711.

141. Pascal M, Richer de Forges B, Le Guyader H, Simberloff D: Mining and other threats to the New Caledonia biodiversity hotspot. *Conserv Biol* 2008, **22**:498–499.

142. Richer de Forges B, Pascal M: La Nouvelle-Calédonie, un "point chaud" de la biodiversité mondiale gravement menacé par l'exploitation minière. *J Soc Oceanistes* 2008, **126–127**:95–112.

143. Cribb TH, Bray RA: Gut wash, body soak, blender, and heat-fixation: approaches to the effective collection, fixation and preservation of trematodes of fishes. *Syst Parasitol* 2010, **76**:1–7.

144. Justine J-L, Briand MJ, Bray RA: A quick and simple method, usable in the field, for collecting parasites in suitable condition for both morphological and molecular studies. *Parasitol Res* 2012, **111**:341–351.

145. Khalil LF, Jones A, Bray RA: *Keys to the Cestode Parasites of Vertebrates*. Wallingford: CAB International; 1994.

146. Kuchta R, Scholz T, Brabec J, Bray RA: **Suppression of the tapeworm order Pseudophyllidea (Platyhelminthes: Eucestoda) and the proposal of two new orders, Bothriocephalidea and Diphyllobothriidea.** *Int J Parasitol* 2008, **38**:49–55.

147. Healy CJ, Caira JN, Jensen K, Webster BL, Littlewood DTJ: **Proposal for a new tapeworm order, Rhinebothriidea.** *Int J Parasitol* 2009, **39**:497–511.

148. Jovelin R, Justine J-L: **Phylogenetic relationships within the polyopisthocotylean monogeneans (Platyhelminthes) inferred from partial 28 S rDNA sequences.** *Int J Parasitol* 2001, **31**:393–401.

149. Justine J-L: **Non-monophyly of the monogeneans?** *Int J Parasitol* 1998, **28**:1653–1657.

150. Mollaret I, Jamieson BGM, Adlard RD, Hugall A, Lecointre G, Chombard C, Justine J-L: **Phylogenetic analysis of the Monogenea and their relationships with Digenea and Eucestoda inferred from 28 S rDNA sequences.** *Mol Biochem Parasitol* 1997, **90**:433–438.

151. Mollaret I, Jamieson BGM, Justine J-L: **Phylogeny of the Monopisthocotylea and Polyopisthocotylea (Platyhelminthes) inferred from 28 S rDNA sequences.** *Int J Parasitol* 2000, **30**:171–185.

152. Perkins EM, Donnellan SC, Bertozzi T, Whittington ID: **Closing the mitochondrial circle on paraphyly of the Monogenea (Platyhelminthes) infers evolution in the diet of parasitic flatworms.** *Int J Parasitol* 2010, **40**:1237–1245.

153. Lim LHS, Timofeeva TA, Gibson DI: **Dactylogyridean monogeneans of the siluriform fishes of the Old World.** *Syst Parasitol* 2001, **50**:159–197.

154. Lim LHS, Justine J-L: **Two new species of ancyrocephalid monogeneans from *Lethrinus rubrioperculatus* Sato (Perciformes: Lethrinidae) off New Caledonia, with the proposal of *Lethrinitrema* n. g.** *Syst Parasitol* 2011, **78**:123–128.

155. Boxshall GA, Lin C-L, Ho J-S, Ohtsuka S, Venmathi Maran BA, Justine J-L: **A revision of the family Dissonidae Kurtz, 1924 (Copepoda: Siphonostomatoida).** *Syst Parasitol* 2008, **70**:81–106.

156. Justine J-L, Brena PF: ***Calydiscoides limae* sp. nov. (Monogenea, Diplectanidae) from *Pentapodus aureofasciatus* (Perciformes, Nemipteridae) off New Caledonia.** *Acta Parasitol* 2009, **54**:22–27.

157. Moravec F, Justine J-L: **Two anisakid nematodes from marine fishes off New Caledonia, including *Raphidascaris (Ichthyascaris) nemipteri* n. sp. from *Nemipterus furcosus*.** *Syst Parasitol* 2005, **62**:101–110.

158. Moravec F, Justine J-L: **Some trichinelloid nematodes from marine fishes off New Caledonia, including description of *Pseudocapillaria novaecaledoniensis* sp. nov. (Capillariidae).** *Acta Parasitol* 2010, **55**:71–80.

159. Moravec F, Justine J-L: **Cucullanid nematodes (Nematoda: Cucullanidae) from deep-sea marine fishes off New Caledonia, including *Dichelyne etelidis* n. sp.** *Syst Parasitol* 2011, **78**:95–108.

160. Moravec F, Justine J-L: **Two new gonad-infecting *Philometra* species (Nematoda: Philometridae) from the marine fish *Lutjanus vitta* (Perciformes: Lutjanidae) off New Caledonia.** *Folia Parasitol* 2011, **58**:302–310.

161. Moravec F, Justine J-L: ***Raphidascaris (Ichthyascaris) etelidis* n. sp (Nematoda, Anisakidae), a new ascaridoid nematode from lutjanid fishes off New Caledonia.** *Zoosystema* 2012, **34**:113–121.

162. Boxshall GA, Huys R: **Copepoda of New Caledonia.** In *Compendium of Marine Species from New Caledonia (Vol Documents Scientifiques et Techniques II7, Deuxième Édition, pp 259–265)*. Edited by Payri CE, de Forges BR. Nouméa, New Caledonia: Institut de Recherche pour le Développement; 2007. 435 pp + Color Plates.

163. Bray RA, Justine J-L: ***Pseudopycnadena tendu* sp. nov. (Digenea, Opecoelidae) in the yellow-spotted triggerfish *Pseudobalistes fuscus* (Perciformes, Balistidae) and additional opecoelids parasitizing fishes from the waters off New Caledonia.** *Acta Parasitol* 2007, **52**:13–17.

164. Quilichini Y, Foata J, Justine J-L, Bray RA, Marchand B: **Sperm ultrastructure of the digenean *Siphoderina elongata* (Platyhelminthes, Cryptogonimidae) intestinal parasite of *Nemipterus furcosus* (Pisces, Teleostei).** *Parasitol Res* 2009, **105**:87–95.

165. Moravec F, Justine J-L: **Some philometrid nematodes (Philometridae), including four new species of *Philometra*, from marine fishes off New Caledonia.** *Acta Parasitol* 2008, **53**:369–381.

166. Moravec F, Justine J-L, Rigby MC: **Some camallanid nematodes from marine perciform fishes off New Caledonia.** *Folia Parasitol* 2006, **53**:223–239.

167. Bleeker P: **Recherches sur les Crustacés de l'Inde Archipélagique II. Sur les Isopodes cymothoadiens de l'Archipel Indien.** *Verhand Natuurkunde Ver, Batavia* 1857, **2**:20–40.

168. Monod T: **Tanaidacea et Isopoda. Mission Robert-Ph. Dollfus en Égypte.** *Mém Inst Égypte* 1933, **21**:161–264.

169. Milne Edwards H: **Ordre des Copépodes.** In *Histoire naturelle des Crustacés, comprenant l'anatomie, la physiologie et la classification de ces animaux. Volume 3*; 1840:411–529.

170. Heller C: **Crustaceen.** *Reise der Oesterreichischen Fregatte 'Novara' um die Erde* 1865, **2**(3):1–280.

171. Pillai NK: **Description of some *Brachiella* and *Clavellopsis* with comments on *Isobranchia* Heegaard.** *Crustaceana* 1968, (Supplement 1):119–135.

172. Shiino SM: **Copepods parasitic on Japanese fishes. 5. Five species of the family Pandaridae.** *Rep Fac Fish, Pref Univ Mie* 1954, **1**:291–332.

173. Bychowsky BE, Nagibina LF: **New and little known species of the genus *Haliotrema* Johnston et Tiegs, 1922 (Monogenoidea). 2.** *Zool Zh* 1971, **50**:25–40.

174. Egorova TP: **A taxonomic review of the subfamily Benedeniinae (Monogenoidea: Capsalidae).** *Parazitologiya* 1997, 438–451.

175. Yamaguti S: **New monogenetic trematodes from Hawaiian fishes, I.** *Pac Sci* 1965, **19**:55–95.

176. Bijukumar A: **Digenetic trematode parasites of the flatfishes (Pleuronectiformes) of the Kerala coast, India.** *Acta Parasitol* 1997, **42**:149–157.

177. Linton E: **Helminth fauna of the dry Tortugas II. Trematodes.** *Pap Tortugas Lab Carnegie Inst Washing* 1910, **4**:11–98.

178. Miller TL, Cribb TH: **Family Cryptogonimidae Ward, 1917.** In *Keys to the Trematoda*, Volume 3. Edited by Bray RA, Gibson DI, Jones A. Wallingford: CAB International; 2008:51–112.

179. Vigueras IP: **Prosogonotremidae n. fam. y *Prosogonotrema bilabiatum* n. gen. n. sp. (Trematoda, Distomata) parasito de *Ocyurus chrysurus* Bloch (Pisces).** *Mem Soc Cubana Hist Nat* 1940, **14**:249–252.

180. Yamaguti S: **Studies on the helminth fauna of Japan. Part 2. Trematodes of fishes, I.** *Jap J Zool* 1934, **5**:249–541.

181. Campbell RA, Beveridge I: ***Floriceps minacanthus* sp. nov. (Cestoda: Trypanorhyncha) from Australian fishes.** *Trans R Soc S Aust* 1987, **111**:189–194.

182. Chandra KJ: ***Nybelinia indica* n. sp. (Cestoda: Trypanorhyncha) from teleost fishes off Waltair coast, Bay of Bengal.** *Riv Parassitol* 1986, **3**:199–202.

183. Dollfus RP: **Sur une collection de Tétrarhynques homéacanthes de la famille des Tentaculariidae récoltés principalement dans la région de Dakar.** *Bull Inst Fond Afr Noire, Sér A* 1960, **22**:788–852.

184. Hiscock ID: **A new species of *Otobothrium* (Cestoda, Trypanorhyncha) from Australian fish.** *Parasitology* 1954, **44**:65–70.

185. Jones MK, Beveridge I: ***Nybelinia queenslandensis* sp. n. (Cestoda: Trypanorhyncha) parasitic in *Carcharhinus melanopterus*, from Australia, with observations on the fine structure of the scolex including the rhyncheal system.** *Folia Parasitol* 1998, **45**:295–311.

186. Pintner T: **Wenigbekanntes und Unbekanntes von Rüssel-bandwürmern. II.** *Öster Akad Wissenschn Sitzunzber, Math Naturwissene Kla, Abt 1* 1931, **140**:777–820.

187. Rudolphi CA: *Entozoorum synopsis, cui accedunt mantissa duplex et indices locupletissimi*. Berolini; 1819.

188. Sakanari J: ***Grillotia heroniensis*, sp. nov., and *G. overstreeti*, sp. nov., (Cestoda: Trypanorhyncha) from Great Barrier Reef fishes.** *Aust J Zool* 1989, **37**:81–87.

189. Olsen LS: **Some nematodes parasitic in marine fishes.** *Publi Inst Mar Sci Univ Texas* 1952, **2**:173–215.

190. Petter AJ, Le Bel J: **Two new species in the genus *Cucullanus* (Nematoda - Cucullanidae) from the Australian Region.** *Mém Inst Oswaldo Cruz, Rio de Janeiro* 1992, **87**(suppl. I):201–206.

9

Detritus-based assemblage responses under salinity stress conditions in a disused aquatic artificial ecosystem

Fulvio Cerfolli[1*], Bruno Bellisario[1] and Corrado Battisti[2]

Abstract

Background: Despite the plethora of approaches, the sensitivity of the methods to measure the relationship between the abundance and biomass curves in stressed detritus-based ecosystems still remain to be refined. In this work, we report the comparison between biomass and abundance in a set of detritus-based macrozoobenthic assemblages located in six sampling pools with different salinity in an artificial aquatic ecosystem (disused Tarquinia Saltworks), using two diversity/dominance approaches (Abundance/Biomass Comparisons, or ABC, and Whittaker plots). We also evaluated the contribution of abundances and biomasses diversity (Simpson index) and nestedness, which measures the order by which macroinvertebrates colonized the detrital resource.

Results: The outputs obtained by both ABC curves and Whittaker plots highlight two different thresholds in assemblage structure: between about 44 and 50 practical salinity unit (psu) and between 50 and 87 psu, respectively. The first threshold was due to a turnover in taxon composition between assemblages, the second threshold (evidenced by Whittaker plots) was due to a change in taxon richness (lower in pools with higher salinity: i.e. > 50 psu). Moreover, a normal-shaped pattern in diversity (Simpson index) emerged, suggestive of an intermediate disturbance effect. The nested pattern did not show significant differences when considering the density and biomass of the sampled taxa, providing similar threshold of salinity in the relative contribution of macrozoobenthos on nestedness.

Conclusions: The use of detailed (ABC and Whittaker plots) and macroscopic (Simpson index and nestedness) approaches is proposed to identify thresholds in the structuring and functioning of detritus-based community of disused aquatic ecosystems: in particular, the inclusion of the parameter of biomass (scarcely utilized in community-based research) appears crucial. The responses of macrozoobenthic assemblages to the salinity stress conditions, in term of abundance and biomass, using a detritus food source (*Phragmites australis* leaves), may also highlight, by comparing macroscopic and detailed approaches, structuring and functioning patterns to consider for the management of disused artificial ecosystems.

Keywords: Leaf-detritus, Macrozoobenthos, Abundance/Biomass Comparisons, Whittaker plots, Simpson index, Nestedness, Patchy environment

Introduction

The aquatic ecosystems represent a test bench to study the structures and functioning of the heterotrophic macrobenthic assemblages under stress conditions [1,2]. In the trophic structures based on autochthonous and allochthonous plant detritus inputs, the role of heterotrophy is crucial to canalize energy and organic materials [3,4]. To understand the importance of detritus-based macrozoobenthic assemblages in terms of trophic exchanges, investigations on the donor-controlled properties of detritus-based energy channels are essential [5]. It is well known that in the aquatic ecosystems, the chemical stressors (i.e. salinity, dissolved oxygen, pH) represent a driving force for the organization of the ecological community assemblages in relative species-specific abundances and biomasses [6,7]. In community ecology, large datasets on the abundance and biomass of species

* Correspondence: fulviocerfolli@unitus.it
[1]Department of Ecological and Biological Sciences (DEB), Ichthyogenic Experimental Marine Centre (CISMAR), Tuscia University, Borgo Le Saline, 01016, Tarquinia, VT, Italy
Full list of author information is available at the end of the article

assemblages [8] may be analyzed using different approaches [9-11]. Among them, the Abundance/Biomass Comparisons (or ABC curves) and the diversity/dominance diagrams (or Whittaker plots) [12-14] have been largely utilized. All the diagrams obtained by these analyses make explicit the frequency ratio (or dominance) among species, either calculated on individual abundance (e.g., in diversity/dominance diagrams) or cumulatively, based on abundance and biomass at the same time (e.g., ABC curves). These representations provide an explicit information on the structure of species assemblages (e.g., diversity and evenness), but they also allow to assess the stress level that might functionally affect the organisms: for instance, in the Whittaker plots, more elevated abundance curves represent the less diverse and more stressed assemblages, while in the ABC curves the comparison between the biomass and abundance curves is used to make inferences on the level of disturbance affecting the taxonomic assemblages [15]. Therefore, knowing the abundance and biomass of the taxonomic assemblages, the Simpson index is profitably used to verify the presence of a normal-shaped pattern in diversity [11]. Concerning other approaches, it is useful to analyse specific mechanisms involved in the structuring processes of ecological communities, since they may exert a strong influence on both the stability and functioning of ecosystems [16], partly reflecting the extent to which interspecific competition is involved in communities composition [17]. One possible approach to investigate patterns and mechanisms involved in community composition, to be proactively coupled with the approaches set out above, is the calculation of nestedness [18], which refers to the 'linkage order' observed between elements of different sets (i.e. species/inlands [19] and plant and pollinator [18]). Macroscopically, the nestedness in habitat/resource colonization occurs when the species present in species-poor sites are proper subsets of the assemblages found in species-rich sites [20]. A perfect nested structure occurs when all species-poor sites are proper subsets of the assemblages found in richer species sites [21]. However, absence of nestedness does not always mean absence of structural pattern, as many other assemblages can be observed in the structuring process of ecological communities (e.g., gradients and compartments) [22,23]. Therefore, the correct evaluation of the mechanisms involved in such structures may provide useful information on the stability and functioning of detritus-based communities [24].

The sensitivity of the methods to weigh the relationship between the biomass and abundance curves in stressed detritus-based ecosystems remains to be explored in detail [3,25] in particular in disused aquatic ecosystems. Indeed, the ABC method and Whittaker plots have been recently applied to some vertebrate and invertebrate assemblages [26-29], but not to detritus-based macrozoobenthic

assemblages, a significant component of heterotrophic food webs [3].

Recent works emphasized that stress conditions disrupt the abundance/biomass relationships (in ABC curves) and the evenness (in Whittaker plots) in species assemblages [15] and affects the diversity and the biomass of communities of primary producers in streams [30]. Conversely, diversity indices (Shannon's and Simpson's), abundance and biomass, during breakdown have been largely adopted in detritus-based communities [31].

Moreover, useful topological properties of network assemblage (e.g. nestedness) have been used to measure the role of salinity in the structuring and functioning of artificial aquatic ecosystems [32] with emphasis on predicting the mechanisms behind the ecological patterns in macrozoobenthic assemblages. In this work, we compared biomass with abundance in a set of detritus-based macrozoobenthic assemblages sampling on *Phragmites australis* (Cav.) Trin. ex Steud., leaf detritus in six sites with different salinity, located in the disused Tarquinia Saltworks, an artificial aquatic ecosystem of central Italy. We applied ABC curves and Whittaker plots to compare the cumulative abundance and biomass data (ABC curves) and the ranking in relative abundance (Whittaker plots) obtained from these macrozoobenthic assemblages to evaluate their responses under salinity stress conditions. We extended the Abundance/Biomass Comparison method using the Simpson index, a macroscopic approach, to investigate the responses of donor-controlled communities [33] due to salinity variation in terms of diversity.

To test the sensitivity of the structuring and functioning of the macrozoobenthic assemblages, we also measured nestedness, a well-known structural characteristic of the complex networks [18,21] with a linkage to the functional attributes of the systems [32].

Methods
Study area
The study area is the aquatic ecosystem of disused Tarquinia Saltworks, a patchy environment (central Italy, 42°12' N, 11°43' E), composed by a series of about 100 pools whose connection is ensured by a surrounding drainage system. The exchange of waters is provided by a single connection with the sea located north of the area (Figure 1). Isolation and hydrological connectivity give rise to a wide salinity gradient [7], spanning from hypohaline (mean annual salinity 8.515 psu or gL^{-1}) to hyperhaline waters (mean annual salinity 115.000 psu or gL^{-1}), (Table 1).

Field and laboratory methods
The spatial characteristics of the pools within the study area (e.g. isolation and connectivity), give rise to a wide environmental gradient, showing a spatial pattern in the

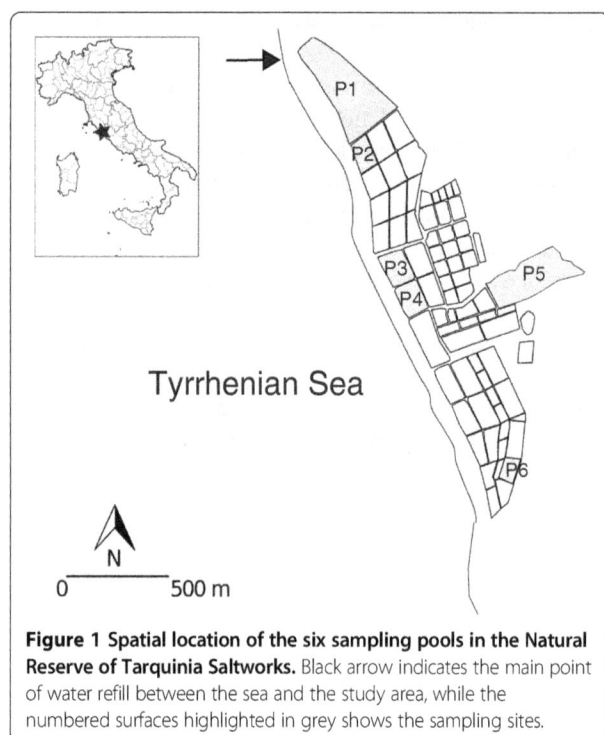

Figure 1 Spatial location of the six sampling pools in the Natural Reserve of Tarquinia Saltworks. Black arrow indicates the main point of water refill between the sea and the study area, while the numbered surfaces highlighted in grey shows the sampling sites.

as means with n = 4 replicates), after storing at 60°C for at least 72 h, and the determination of the ash free dry weight (AFDW) after ignition in a muffle furnace at 500°C for 6 h [34]. The loss upon oxidation is referred to as AFDW. AFDW, obtained by subtracting the ash content (>60% of total dry mass) was considered to provide a better comparison with other macroinvertebrate taxa than dry weight [35].

Data analysis

For each taxon in each sampling site, we obtained values of abundance (N) and biomass (B), and for both we calculated their relative (frN and frB) and cumulative frequency (Table 2). We then ranked the taxa from the most to the least important based on either cumulative abundance or biomass along the x-axis in a Cartesian space in order to obtain two curves for these parameters (ABC curves).

Whittaker plots were obtained building a taxon rank/relative frequency diagrams, and utilizing the data set (taxon/frequency) applied to values in frequency on abundance (log-transformed to improve normality and for literature comparison [11]) of the six detritus-based assemblages studied. A first-degree equation was calculated by fit analysis for each detritus-based assemblage, including all taxa. The equation is $FrA = b^{ar}$, where FrA is the relative frequency (on the abundance) of each taxon in each pool; r is the rank of each taxon in the pool assemblage; a is the angular coefficient (negative) of the regression line, indicating the mean decrease of the relative frequency of the taxon with increase of the taxon rank (slope of the line); b is a coefficient (intercept value) that reflects the trend value of the first dominant taxon of the pool assemblage represented in the regression line in equation.

For each regression line we obtained the coefficient of determination (R^2) as an estimate of the variance explained [36]. For each taxon assemblage we also calculated the Simpson diversity indexes as $D = 1 - \Sigma \, fr^2$, both on the abundance and biomass frequencies (D_N and D_B). The index provides a good estimate of diversity with a relatively small sample size, being less sensitive to taxon richness and capturing the variance of the taxon abundance (or biomass) distribution [11]. To compare the abundance and biomass frequency distributions between pools we performed the Kolmogorov-Smirnov 2-sample test. Probability level (p) was set at 0.05. We investigated taxon assemblages as whole units where relative abundance frequency of the

variability of the salinity levels, pH and dissolved oxygen concentration [32].

Six sampling sites (pools) were randomly selected, covering the maximal range of salinity variation from hypohaline (mean annual salinity 8.515 psu) to hyperhaline waters (mean annual salinity more than 115.000 psu), (Table 1), to perform a macrozoobenthic colonization experiment of *Phragmites australis* (Cav.) Trin. ex Steud. leaf detritus, naturally present in the area, under different salinity stress conditions. To measure the relationship between cumulative abundance and biomass frequencies and salinity, we placed in each pool 48 protected (mesh size: 5 × 5 mm) and 48 unprotected (mesh size: 10 × 10 mm) leaf packs. We measured, on a monthly sampling with r = 4 replicates: i) the dry weight of leaf detritus in both protected and unprotected leaf packs, after storing at 60°C for at least 72 h (leaf pack initial weight: 2.000 ± 0.004 g dry mass) [7]; ii) the number of colonizing taxa (first column in Table 2); iii) the number of individuals for each taxon (expressed as means with n = 4 replicates); iv) the dry biomass of individuals for each taxon (expressed

Table 1 Main chemical–physical parameters measured in the six sampling pools

POOL	P5	P1	P2	P3	P4	P6
Salinity (psu)	8.515 (±4.292)	44.769 (±4.746)	50.531 (±5.126)	87.000 (±10.083)	100.154 (±12.548)	115.000 (±29.958)
[O₂] mgL⁻¹	10.03 (±1.644)	5.765 (±2.543)	6.068 (±2.703)	5.505 (±1.568)	5.865 (±1.216)	5.536 (±1.779)
pH	8.953 (±0.566)	8.112 (±0.243)	8.104 (±0.252)	8.067 (±0.296)	8.134 (±0.236)	8.034 (±0.187)

Values are expressed as mean ± SD.

Table 2 Detritus-based assemblages in the six pools studied, ordered by increasing salinity

POOL	P5		P1		P2		P3		P4		P6	
Taxon	FrN	FrB	FrN	FrB	FrN	FrB	FrN	FrB	FrN	FrB	FrN	FrB
Chironomus sp. (larvae)	0.167	0.014	0.239	0.014	0.536	0.371	0.757	0.180	0.928	0.586	0.994	0.980
Gammarus aequicauda	0	0	0.268	0.023	0.034	0.008	0	0	0	0	0	0
Perinereis culltrifera	0	0	0.058	0.062	0.003	0.010	0	0	0	0	0	0
Nereis diversicolor	0	0	0.262	0.490	0.034	0.175	0	0	0	0	0	0
Hydrobia acuta (complex)	0	0	0.087	0.072	0.140	0.312	0.216	0.728	0.003	0.028	0	0
Cerastoderma glaucum	0	0	0.003	0.013	0.003	0.042	0.004	0.070	0.005	0.225	0	0
(other) Coleoptera (larvae)	0.014	0.002	0.026	0.002	0	0	0	0	0	0	0	0
Cerithium vulgatum	0	0	0.052	0.324	0	0	0	0	0	0	0	0
Gordiidae	0	0	0.003	0	0	0	0	0	0	0	0	0
Spio decorates	0	0	0	0	0.007	0.004	0	0	0	0	0	0
(other) Diptera (larvae)	0.007	0.001	0	0	0.003	0.001	0	0	0	0	0	0
Monocorophium insidiosum	0	0	0	0	0.225	0.063	0	0	0	0	0	0
Idotea balthica	0	0	0	0	0.014	0.012	0	0	0	0	0	0
Haliplus sp.	0.530	0.182	0	0	0	0	0.023	0.022	0.064	0.162	0.004	0.015
Micronecta sp	0.035	0.008	0	0	0	0	0	0	0	0	0	0
Anisoptera (nymphae)	0.220	0.705	0	0	0	0	0	0	0	0	0	0
Acilius sp (larvae)	0.021	0.086	0	0	0	0	0	0	0	0	0	0
Hydrophilus sp	0.007	0.002	0	0	0	0	0	0	0	0	0.002	0.005
Taxonomic units	8		9		10		4		4		3	
Abundance (total N ind)	287		343		293		518		643		519	
Biomass (grams, AFDW)		2.679		4.727		1.483		1.739		0.815		0.421

Abundance frequency (frN) and biomass frequency (frB) for each taxon, number of taxonomic units, total number of macrozoobenthic individuals (N) and total macrozoobenthic biomass (B) are reported.

taxon is a value aimed to obtain regression lines in the Whittaker plots. Therefore, the results did not allow a discussion on possible implications on assembly rules among taxa determined by interspecific competition, predation or disturbance at species level.

We evaluated also the nested pattern of macrozoobenthic assemblage on *P. australis* leaf detritus in the six sampling sites by taking into account the weights of the association between taxa and sites. Although nestedness originally relies on the presence/absence of a particular association in the association matrices [19], recent advances suggest the role of weighting the intensity of such association for a thorough understanding of the mechanism involved [37].

Two bipartite networks were then created to study the role of abundances and biomasses on nestedness. A bipartite network is defined by two distinct sets of nodes (in our case macroinvertebrates and resource/sites), where links may occur only between the nodes of different sets but not within nodes of the same set. To account for the relative contribution of species abundance and biomass on nestedness, we used a weighted version of nestedness based on the Manhattan distance (WINE, Weighted-Interaction

Nestedness Estimator) [38]. Although WINE has been criticized due to its tendency to overestimate nestedness for matrices with no co-occurrence among species and/or for matrices with sites of identical richness [39], some authors suggested its capability to measure the relative contribution of species to nestedness when dealing with abundance data [40]. Weighted nestedness was measured with the WINE function implemented in the "bipartite" package of R [41]. WINE takes into account the weight or intensity of each entry (e.g. the abundances and biomasses of sampled macroinvertebrates). The nestedness score of the data matrix is normalized by comparing it to the average score of equivalent random matrices and to the score of the maximal nestedness matrix to obtain the weighted-interaction nestedness estimator [37]:

$$\eta_w = (d^w - d_{rnd})/(d_{max} - d_{rnd}) \quad (1)$$

where d^w is the mean weighted distance of all its non-zero elements, d_{rnd} is the average value of 1,000 replication random matrix and d_{max} the distance of the completely packed matrix. η_w varies between 0, when the score of the original data matrix is close to the average score of the equivalent random matrices, and 1, as it gets closer to the nestedness of the maximal nestedness matrix. To assess the significance of nestedness, a Z-score measuring the difference between d^w and d_{rnd} is calculated. Z values below −1.65 or above 1.65 indicate approximate statistical significance at the 5% error level (one-tailed test). WINE also calculates a weighted-interaction distance (d_{ij}^{w}), which estimates nestedness taking into account the number of events in the links, in our case the abundance and biomass of macrozoobenthos sampled on leaf detritus. A distance-based permutational multivariate analysis of variance was then performed [42] to test the influence of salinity on the contribution of abundance and biomass on nestedness. The 'adonis' function in package 'vegan', implemented in the R software environment [43], was used for partitioning distance matrices among sources of variation. Although similar to the classic PERMANOVA, the function 'adonis' is more robust, as it can accept both categorical and continuous variables. We used average salinity as fixed factor, to test for its influence on the relative contribution of abundance and biomass in different assemblages. The Bray-Curtis resemblance matrices were constructed, and significance was tested by performing 999 permutations of both abundances and estimated biomasses within each group, which were defined following the salinity gradient.

We finally compared the ranking of sites derived from nestedness (measured on both the abundance and densities) with the ranking yielded by the angular coefficient given by the best fitting model of the Whittaker plots, to look for a correspondence between nestedness and pattern of species distribution.

Results

Fluctuations in the level of pH and dissolved oxygen concentration slightly affected the environmental conditions within the pools, as showed by the PCA ordination that explained 83.41% of total variance, mainly attributable to the variation of salinity within the pools ($r = 0.96$).

A total of 2,603 individuals (11.864 AFDW grams) were collected, belonging to S = 18 macrozoobenthic taxa in P = 6 pools.

In the Abundance/Biomass Comparison method, we detected a change in relative patterns of cumulative abundance and biomass. In particular, between about 44 e 50 psu, we observed a change in the relative position of abundance and biomass curves. Until 44 psu, the biomass curves cumulate before the abundance curves, and then the abundance curves cumulate before the biomass curves (Figure 2). The frequency distribution for abundance and biomass was significantly different in all the pools, except in the pool with elevated salinity level (Table 3).

In Whittaker plots, we observed two sets of regression lines well fitting the assemblage values (i.e. all with a high coefficient of determination > 0.80). A first set of regression lines, characterized by a low slope (ranging between −0.60 and −0.66), including the three detritus-based assemblages living in pools with a salinity lower than 50 psu; a second set, characterized by a higher slope (ranging between −1.81 and −3.12), including the three assemblages living in pools with a salinity higher than 50 psu (Figure 2abc).

Comparing the Simpson indexes calculated on abundance (N) and biomass (B) frequencies, two patterns apparently normal-shaped but shifting among them were observed: D_N peaks in pool n. 1 (psu = 44.769) and D_B peaks in pool n. 2 (psu = 50.531) (Figure 3).

The results obtained by both ABC curves and Whittaker plots highlight two different thresholds: between 44–50 and 50–87 psu, respectively. The first threshold (evidenced by the ABC curves) seems due to a turnover in taxon composition between assemblages (from taxon with higher biomass and less abundance to taxon with lower biomass and higher abundance); the second threshold (evidenced by Whittaker plots) is due to a change in taxon richness (lower in pools with higher salinity: i.e. > 50 psu).

A significant nested structure was found for the association between macrozoobenthos and the detrital resource (measure of the availability of *P. australis* detritus following the variation patterns of decomposition [6]) in different pools either when considering the abundance ($\eta_w = 0.265$, $Z = 2.142$, $p = 0.02$) or the biomass of sampled taxa ($\eta_w = 0.412$, $Z = 3.414$, $p < 0.001$). The contribution of macrozoobenthos to nestedness varied between pools following the salinity gradient ($F > 75$ and $p = 0.01$ for both abundances and biomasses), highlighting a threshold in the salinity values able to affect the nested assemblage of macroinvertebrates on leaf detritus.

The ranking comparison between nestedness and pattern of species distribution was also highly significant (Spearman's rank coefficient test: $r_s = 0.889$, $p < 0.01$).

Discussion

Following the Abundance/Biomass Comparison model, a change in relative location of the cumulative biomass vs. abundance curves implies a change in the level of disturbance affecting specific assemblages. The theory on the ABC curves assumes that when the abundance curves cumulate before (i.e. are higher), an assemblage may be stressed by a disturbance [11]. In our detritus-based assemblages, this disturbance could be due to an increasing level of salinity, starting from values higher than 44 psu. As for the influence of salinity, the adoption of ordination analyses is useful in providing information on the relative importance of chemical factors on the macrozoobenthic communities. The ABC curves emphasize the different ecological role that the abundance and biomass parameters play at the taxon assemblage level. In particular, abundance curves indicate a relative distribution of the spatial niche and dominance of the taxonomic unities, while biomass curves

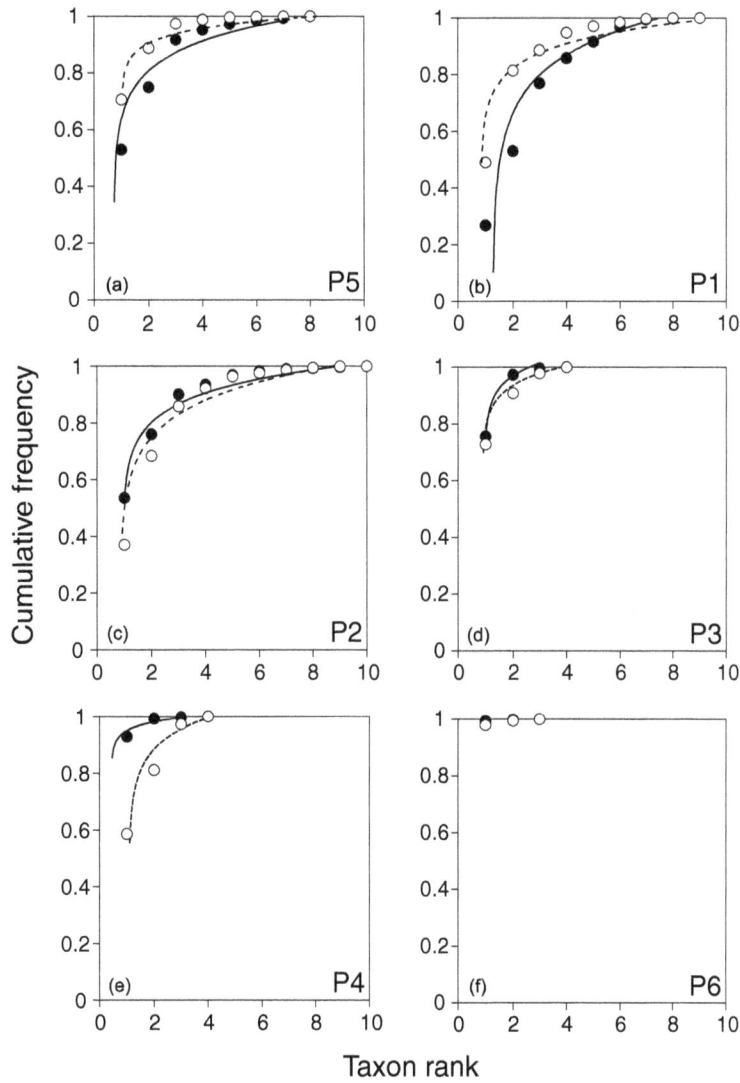

Figure 2 Abundance/Biomass Comparisons (ABC curves) for the six sampling pools. Values of cumulative frequency on abundance (black circles; continuous line) and biomass (white circles; dashed line) along a taxon ranking (x-axis) are reported. Figures 2**abc** refer to pools with salinity < 50 psu, while Figures 2**def** to pools with values of salinity > 50 psu.

indicate the relative distribution of the energy flow in the assemblage, according to the trophic resources used by taxa [10]. The Abundance/Biomass Comparisons are based on the assumption that in disturbed habitats, small-sized and generalist taxa (i.e., with low body weight, low trophic level and/or r-selected) tend to increase in their abundance. As a consequence, in these ecosystems the abundance curves approach an asymptote before the biomass curves. On the contrary, in undisturbed habitats the opposite pattern could be observed, with the biomass curves cumulating before

Table 3 Comparison between abundance and biomass frequencies by Kolmogorov-Smirnov 2-sample test (Z with p) for macrozoobenthic assemblages in the six sampling pools

POOL	P5	P1	P2	P3	P4	P6
Z (p)	2 (<0.01)	1.741 (<0.01)	2.415 (< 0.01)	1.414 (< 0.05)	1.414 (<0.05)	1.225 (= 0.1)
FrA	$235.8e^{-0.66x}$	$334e^{-0.61x}$	$194.28e^{-0.60x}$	$2933.5e^{-1.81x}$	$2703e^{-1.97x}$	$5214.5e^{-3.12x}$
R^2	0.95	0.89	0.95	0.99	0.93	0.83
a	122	181.4	107	480	377	230.26

The relative equations (FrA), the coefficients of determination (R^2) and the angular coefficients (a) are also listed for each pool.

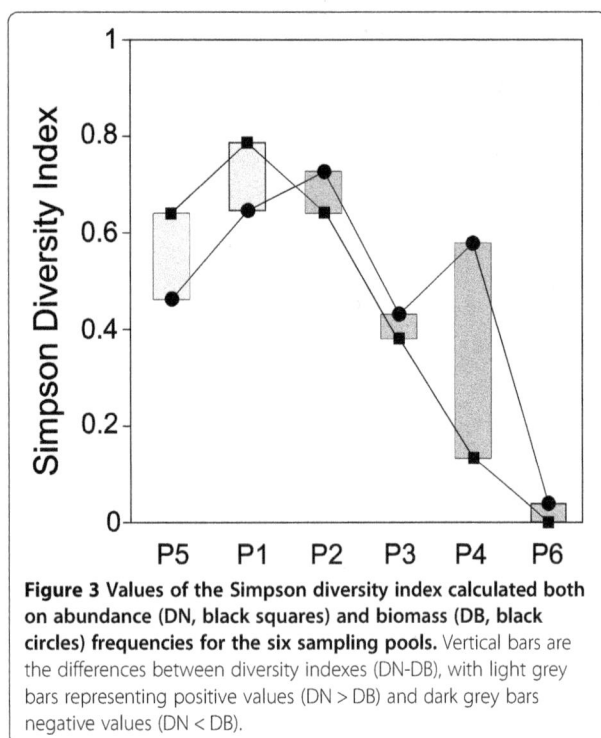

Figure 3 Values of the Simpson diversity index calculated both on abundance (DN, black squares) and biomass (DB, black circles) frequencies for the six sampling pools. Vertical bars are the differences between diversity indexes (DN-DB), with light grey bars representing positive values (DN > DB) and dark grey bars negative values (DN < DB).

the abundance curves. This may suggest that a higher number of large-sized taxa of high trophic level occur in a more complex and diverse assemblage. Following this approach, early cumulating abundance curves may indicate that the resources are used by few dominant (i.e., more abundant) taxa with a broad spatial niche (i.e., generalist), while early cumulating biomass curves may indicate that the individuals with high body weight (and taxa with a high biomass) largely occur in the undisturbed assemblage typical of stable ecosystems [11]. Therefore, when abundance and biomass curves are compared, we may obtain information on the level of anthropogenic or natural disturbance that affects a taxon assemblage by alternating relative dominance patterns among large and specialized taxa *versus* small and generalist ones.

Following this general model, the observed pattern suggests that the detritus-based macrozoobenthic assemblages are stressed by a different level of disturbance starting from a salinity level of 50 psu. Between about 44 and 50 psu, we observed a pattern where the relationship between abundance and biomass of macrozoobenthic assemblages allows the presence of taxa with many individuals but low biomass.

Differently, Whittaker plots emphasized a threshold between 50 and 87 psu of salinity, especially due to a strong reduction in taxon richness and an over-dominance of the remnant generalist taxa with a high relative abundance. Indeed, when salinity increases at very high level, only a low number of taxa (i.e. larvae of *Chironomus* sp.) survive, so demographically increasing their dominance [7,32].

Therefore, the information on change in both taxon composition and taxon richness is intercepted from two diversity/dominance approaches (ABC curves and Whittaker plots). When analyzing assemblages along gradients, the use of different approaches at assemblage level may help to identify different thresholds: in particular, the parameter of biomass (scarcely utilized in community-based research) appears crucial to describe the relationship between abundance and biomass [35].

Finally, we observed a normal-shaped pattern in diversity (Simpson index), suggesting the presence of an intermediate disturbance effect [44,45]. This effect predicts that diversity will be the highest in assemblages with intermediate levels of disturbances. In this case, taxon richness is maximized because of the coexistence between salinity-tolerant generalist taxa and other more sensitive taxa. These assemblages differ in term of abundance and biomass, thereby inducing a shift (between 44 and 50 psu) in the higher values of two diversity indexes.

Concerning the metrics of network analysis, the contribution of macrozoobenthic assemblages to nestedness varied among pools following the salinity gradient ($F > 75$ and $p = 0.01$ for both abundances and biomasses), highlighting the presence of a threshold in the salinity values able to influence the nested assemblages on leaf detritus.

The disused Tarquinia Saltworks showed, in the early stages, an intense process of macrozoobenthic colonization [46]. The single pools, characterized by different residual salinity, thus become, macroscopically, a patchy environment with distinct trophic structures [32]. Along with the structuring of the green trophic structures, due to the colonization of aquatic submersed vegetation (i.e. *Ruppia cirrhosa* (Petag.) Grande) and microalgae, other trophic dynamics are established upon the input of allochthonous plant debris such as that resulting from riparian plants (i.e. *Phragmites australis)*. The detritus-based macrozoobenthic assemblages constitute the primary consumers in the brown trophic food webs of Mediterranean coastal lagoons [47]. However, the long-term disuse of large aquatic artificial systems leads to a loss of structural and functional heterogeneity resulting in a homogenization of trophic structures and a decrease of the values of biological diversity.

Our work suggests that the heterogeneity of the macrozoobenthic assemblages of patchy aquatic ecosystems into disuse, a result due to the processes in the short to medium term ecological colonization, is maintained over the long term through targeted contrast homogenization and abundance and biomass monitoring activities on macrozoobenthic assemblages, independently from the scale [48]. The highest values of taxon diversity fall between 44 and 50 psu, at intermediate values in salinity. As a first approximation, the analysis of shifts in ABC curves enables to understand when, in these artificial patchy environments, the action is needed to maintain in confined areas high levels of taxon diversity and diversification of trophic structures [49], with taxon of higher biomass. Above 44 psu in salinity, the taxa tends to decrease their biomass and a taxon turn-over occurs qualitatively changing the assemblage composition. Above 50 psu in salinity, these systems tend to abruptly decrease their taxon richness due to extreme conditions favouring only widely diffused taxa.

The results of the analysis of both macroscopic (diversity indexes and network metrics) and detailed patterns (ABC and Whittaker plots), based on abundance and biomass analysis, suggest that the sampling of abundances and bio-masses of the macrozoobenthic assemblages is useful to increase the predictive capacity to test the sensitivity of the responses of detritus-based communities under salinity stress.

The exploitation of more ecological techniques to un-ravel the abundance and biomass relationships is necessary to refine the management criteria of aquatic environments with high heterogeneity, especially disused artificial ecosystems as they can represent a good model for patchy ecosystem management [50].

These results highlight also the relevance to adopt more kinds of community-based approaches to probe the patterns of macrozoobenthic assemblages, before adopting a patchy ecosystem management criterion.

Competing interests
The authors declare that they have no competing interests.

Authors' contributions
FC and CB contributed to the conceptual development of the work. FC wrote the ms and carried out the statistical analyses with CB and BB. All the authors read and approved the final version of the manuscript.

Acknowledgements
We wish to thank the Italian State Forestry Department for access to the sampling sites within the Natural Reserve of Tarquinia Saltworks. We are grateful to Giuseppe Nascetti, Claudio Carere and two anonymous reviewers, whose constructive comments and suggestions greatly improved the manuscript.

Author details
[1]Department of Ecological and Biological Sciences (DEB), Ichthyogenic Experimental Marine Centre (CISMAR), Tuscia University, Borgo Le Saline, 01016, Tarquinia, VT, Italy. [2]'Torre Flavia' LTER (Long Term Environmental Research) Station, Environmental Service, Provincia di Roma, via Tiburtina, 691, 00159, Rome, Italy.

References

1. Williams WD: **Salinity as a determinant of the structure of biological communities in salt lakes.** Hydrobiologia 1998, **381:**191–201.
2. Dobson M, Frid C: Ecology of aquatic systems. Oxford: Oxford University Press; 2009.
3. Moore JC, Berlow EL, Coleman DC, de Ruiter PC, Dong Q, Hastings A, Johnson NC, McCann KS, Melville K, Morin PJ, Nadelhoffer K, Rosemond AD, Post DM, Sabo JL, Scow KM, Vanni MJ, Wall DH: **Detritus, trophic dynamics and biodiversity.** Ecol Lett 2004, **7:**584–600.
4. Belgrano A, Scharler UM, Dunne J, Ulanowicz RE: Aquatic food webs: an ecosystem approach. Oxford: Oxford University Press; 2005.
5. Moore JC, de Ruiter PC: Energetic food webs: an analysis of real and model ecosystems. Oxford: Oxford University Press; 2012.
6. Velasco J, Millán A, Hernández J, Gutiérrez-Cánovas C, Abellan P, Sanchez-Fernandez D, Ruiz M: **Response of biotic communities to salinity changes in a Mediterranean hypersaline stream.** Saline Systems 2006, **2:**12.
7. Bellisario B, Cerfolli F, Nascetti G: **Spatial network structure and robustness of detritus-based communities in a patchy environment.** Ecol Res 2010, **25:**813–821.
8. Fauth JE, Bernardo J, Camara M, Rasetarits WJ, Van Buskirk J, McCollum SA: **Simplifying the jargon of community ecology: a conceptual approach.** Am Nat 1996, **147:**282–286.
9. Wiens JA: The ecology of bird communities, Vol 2, processes and variations. Cambridge: Cambridge University Press; 1989.
10. Krebs CJ: Ecology: the experimental analysis of distribution and abundance. 5th edition. London: Benjamin-Cummings Pub Co; 2001.
11. Magurran AE: Measuring biological diversity. New York: Blackwell; 2004.
12. Lambshead PJD, Platt HM, Shaw KM: **The detection of differences among assemblages of marine benthic species based on an assessment of dominance and diversity.** J Nat Hist 1983, **17:**859–874.
13. Warwick RM: **A new method for detecting pollution effects on marine macro-benthic communities.** Mar Biol 1986, **92:**557–562.
14. Warwick RM, Clarke KR: **Relearning the ABC-taxonomic changes and abundance biomass relationships in disturbed benthic communities.** Mar Biol 1994, **118:**739–744.
15. Dornelas M, Soykan CU, Ugland KI: **Biodiversity and disturbance.** In Biological diversity: frontiers in measurements and assessments. Edited by Magurran A, McGill BJ. Oxford, New York: Oxford University Press; 2011:237–251.
16. Ives AR, Carpenter SR: **Stability and diversity of ecosystems.** Science 2007, **317:**58–62.
17. Diamond JM: **Assembly of species communities.** In Ecology and evolution of communities. Edited by Cody ML, Diamond JM. Cambridge, Massachusetts, USA: Harvard Univ. Press; 1975:342–444.
18. Bascompte J, Jordano P, Melián CJ, Olesen JM: **The nested assembly of plant animal mutualistic networks.** Proc Natl Acad Sci U S A 2003, **100:**9383–9387.
19. Atmar W, Patterson BD: **The measure of order and disorder in the distribution of species in fragmented habitat.** Oecologia 1993, **96:**373–382.
20. Patterson BD, Atmar W: **Nested subsets and the structure of insular mammalian faunas and archipelagos.** Biol J Linnean Soc 1986, **28:**65–82.
21. Almeida-Neto M, Guimarães PR, Lewinsohn TM: **On nestedness analyses: rethinking matrix temperature and anti-nestedness.** Oikos 2007, **116:**716–722.
22. Leibold MA, Mikkelson GM: **Coherence, species turnover, and boundary clumping: elements of meta-community structure.** Oikos 2002, **97:**237–250.
23. Lewinsohn TM, Prado PI, Jordano P, Bascompte J, Olesen JM: **Structure in plant–animal interaction assemblages.** Oikos 2006, **113:**174–184.
24. De Angelis DL: Dynamics of nutrient cycling and food webs. New York: Chapman and Hall; 1992.
25. Meire PM, Dereu J: **Use of the abundance/biomass comparison method for detecting environmental stress: some considerations based on intertidal macrozoobenthos and bird communities.** J Appl Ecol 1990, **27:**210–223.
26. Penczak T, Kruk A: **Applicability of the abundance/biomass comparison method for detecting human impact on fish populations in the Pilica river.** Pol Fish Res 1999, **39:**229–240.
27. Magurran AE, Phillip SAT: **Implications of species loss in freshwater fish assemblages.** Ecography 2001, **24:**645–650.
28. Benassi G, Battisti C, Luiselli L: **Applying abundance/biomass comparisons in breeding bird assemblages of a set of remnant wetlands in central Italy.** J Mediterr Ecol 2009, **10:**13–18.
29. Prete S, Battisti C, Marini F, Ciucci P: **Applying abundance/biomass comparisons on a small mammal assemblage from Barn owl (Tyto alba) pellets (Mount Soratte, central Italy): a cautionary note.** Rend Fis Acc Lincei 2012, **23:**349–354.
30. Niyogi DK, Lewis WM Jr, McKnight DM: **Effects of stress from mine drainage on diversity, biomass, and function of primary producers in mountain streams.** Ecosystems 2002, **5:**554–567.
31. Kominosky J, Pringle C: **Resource–consumer diversity: testing the effects of leaf litter species diversity on stream macroinvertebrate communities.** Freshw Biol 2009, **54:**1461–1473.
32. Bellisario B, Cerfolli F, Nascetti G: **The interplay between network structure and functioning of detritus-based communities in patchy aquatic environment.** Aquat Ecol 2012, **46:**431–441.

33. Pimm SL: *Food webs.* London (UK): Chapman and Hall; 1982.

34. Payne WJ: **Energy yields and growth of heterotrophs.** *Annu Rev Microbiol* 1970, **24**:17–52.

35. Vignes F, Fedele M, Pinna M, Mancinelli G, Basset A: **Variability of lekanesphaera monodi metabolic rates with habitat trophic status.** *Acta Oecol* 2012, **41**:58–64.

36. Dytham C: *Choosing and using statistic.* Wiley-Blackwell, UK: A Biologist's guide; 2011.

37. Selva N, Fortuna MA: **The nested structure of a scavenger community.** *Proc R Soc B* 2007, **274**:1101–1108.

38. Galeano J, Pastor JM, Iriondo JM: **Weighted-interaction nestedness estimator (WINE): a new estimator to calculate over frequency matrices.** *Environ Model Softw* 2009, **24**:1342–1346.

39. Corso G, de Araujo AIL, de Almeida AM: **A new nestedness estimator in community networks.** *arXiv* 2008. 0803.0007v1.

40. Ulrich W, Almeida-Neto M, Gotelli NJ: **A consumer's guide to nestedness analysis.** *Oikos* 2009, **118**:3–17.

41. R Development Core Team: *R: a language and environment for statistical computing.* Vienna, Austria: R Foundation for Statistical Computing; 2009.

42. Anderson MJ: **A new method for non-parametric multivariate analysis of variance.** *Austral Ecol* 2001, **26**:32–46.

43. R Development Core Team: *R: a language and environment for statistical computing.* Vienna, Austria: R Foundation for Statistical Computing; 2011.

44. Connell JH: **Diversity in tropical rain forests and coral reefs.** *Science* 1978, **199**:1302–1310.

45. Collins SL, Glenn SM: **Intermediate disturbance and its relationship to within and between-patch dynamics.** *New Zeal J Ecol* 1997, **21**:103–110.

46. Bellisario B, Novelli C, Angeletti D, Cerfolli F, Cimmaruta R, Nascetti G: **The ecological restoration of Tarquinia saltern drives temporal changes in benthic community structure.** *Transitional Waters Bulletin* 2010, **4**:105–114.

47. Alemanno S, Mancinelli G, Basset A: **Detritus processing in Tri-trophic food chains: a modelling approach.** *Internat Rev Hydrobiol* 2007, **92**:103–116.

48. Williams WD, Boulton AJ, Taaffe RG: **Salinity as a determinant of salt lake fauna: a question of scale.** *Hydrobiologia* 1990, **197**:257–266.

49. Magurran AE: **Biodiversity in the context of ecosystem function.** In *Marine biodiversity & ecosystem functioning - frameworks, methodologies and integration.* Edited by Solan M, Aspden RJ, Paterson DM. Oxford: Oxford University Press; 2012:16–23.

50. Bellisario B, Carere C, Cerfolli F, Angeletti D, Nascetti G, Cimmarura R: **Infaunal macrobenthic community dynamics in a manipulated hyperhaline ecosystem: a long term study.** *Aquat Biosyst* 2013, **9**:20.

10

Effects of flow restoration on mussel growth in a Wild and Scenic North American River

Brandon J Sansom[1,4*], Daniel J Hornbach[2,3], Mark C Hove[3] and Jason S Kilgore[1]

Abstract

Background: Freshwater mussels remain among the most imperiled species in North America due primarily to habitat loss or degradation. Understanding how mussels respond to habitat changes can improve conservation efforts. Mussels deposit rings in their shell in which age and growth information can be read, and thus used to evaluate how mussels respond to changes in habitat. However, discrepancies between methodological approaches to obtain life history information from growth rings has led to considerable uncertainty regarding the life history characteristics of many mussel species. In this study we compared two processing methods, internal and external ring examination, to obtain age and growth information of two populations of mussels in the St. Croix River, MN, and evaluated how mussel growth responded to changes in the operation of a hydroelectric dam.

Results: External ring counts consistently underestimated internal ring counts by 4 years. Despite this difference, internal and external growth patterns were consistent. In 2000, the hydroelectric dam switched from operating on a peaking schedule to run-of-the-river/partial peaking. Growth patterns between an upstream and downstream site of the dam were similar both before and after the change in operation. At the downstream site, however, older mussels had higher growth rates after the change in operation than the same sized mussels collected before the change.

Conclusions: Because growth patterns between internal and external processing methods were consistent, we suggest that external processing is an effective method to obtain growth information despite providing inaccurate age information. External processing is advantageous over internal processing due to its non-destructive nature. Applying this information to analyze the influence of the operation change in the hydroelectric dam, we suggest that changing to run-of-the-river/partial peaking operation has benefited the growth of older mussels below the dam.

Keywords: Unionidae, Annuli, Growth processing methods, Hydroelectric dam, Flow regulation

Background

Freshwater mussels are vital components to freshwater ecosystems and are considered ecosystem engineers because of their influence on surrounding habitat [1-3]. However, freshwater mussels are one of the most imperiled taxa in North America. Nearly 300 species are present in the United States, but 72% of these are endangered, threatened, or of special concern [4,5]. Much of this decline has been linked to habitat loss or degradation [6,7], a leading cause of which has been the impoundment of streams and rivers [7,8]. Numerous studies have documented the negative effect of dams on downstream mussel communities [9-12] with dozens of mussel species being extirpated [7]. Non-point source pollution and invasive species (e.g. zebra and quagga mussels) are additional factors known to negatively affect freshwater mussels [5].

Since changes in habitat impacts mussel life history traits knowing how mussels respond to these changes are crucial for proper management [13]. During periods of growth, mussels deposit calcium carbonate in the form of aragonite or calcite, resulting in shell construction [14]. The extent of calcium carbonate deposition is dependent on a number of environmental factors

* Correspondence: brandon.sansom@ou.edu
[1]Department of Biology, Washington and Jefferson College, Washington, PA, USA
[4]Current address: Oklahoma Biological Survey and Department of Biology, University of Oklahoma, 111 East Chesapeake St, Norman, OK, 73019, USA
Full list of author information is available at the end of the article

including water temperature [15-17], food availability and quality [3,18-21], and discharge [22]. Additionally, habitat features such as substrate type [23], stream length [16], water depth [24], riparian vegetation [25], and density of the surrounding mussel community [26,27] have also been known to influence variation in growth. Mussels grow in a similar manner to trees, in which rings are deposited when favorable growth conditions cease. During cessation of growth in mussels, an insoluble organic residue forms at the interface of the mantle and shell, leaving behind a conspicuous dark growth ring, which is observable on both the internal and external surfaces [14]. Proper examination and analysis of these rings in accordance with subsequent environmental conditions are useful in providing age and growth estimates.

While ring deposition has long been known in freshwater mussels [28], there is a lack of age and growth information for many mussel species, partly due to methodological issues associated with obtaining accurate age estimates and debate over the rate of ring deposition [29]. Recent work has focused on validating whether or not ring production occurs on an annual basis, a fundamental prerequisite to growth studies [30]. Various studies have examined this issue by using mark-recapture (see [31-33]) or dendrochronology applications [22,34-36]. Mark-recapture can be effective but is a time intensive method, and the possibility of low return of marked individuals may render results inconclusive [32,37]. More recent approaches utilize cross dating, a fundamental tool in dendrochronology. Cross dating uses high-frequency patterns in ring width variation to align each individual specimen's series of growth increments to specific calendar years [38,39]. If growth within a population is synchronous, individual time series are averaged to create a master chronology (see [22,34-36]). Furthermore, false or missing ring(s) within a time series can be detected by comparing each individual to the master chronology. Finally, the growth rings of an individual mussel can be considered validated if its time series is significantly correlated to the master chronology [39]. A much less time-intensive method than mark-recapture, cross dating offers a unique way to validate ring deposition and has been successful for many species [22,35,36].

Traditional methods to view and interpret growth rings involve creating shell thin sections to view the internal growth rings. Viewing internal rings is useful for validating ring deposition rate with dendrochronology approaches, estimating age and growth rate, and identifying disturbance or false rings. Disturbance rings can be differentiated from true growth rings because annual growth rings are continuous from the shell exterior to the umbone and are broader and lighter in color than false rings [32]. However, generating thin sections requires destructive sampling, which is especially problematic for rare or endangered species. Furthermore, creating and examining thin sections is a tedious and time-consuming process.

Growth rings deposited by mussels are also prominent on the external surface of the shell and may provide an alternative method to interpret age and growth. External rings can be easily distinguished during the early years of growth but become crowded and difficult to differentiate towards the shell margin [32,33]. Additionally, disturbance rings can also appear on the external shell surface [17], and, although disturbance rings are usually thinner than true annuli, differentiating external disturbance from true annuli is difficult [32]. Furthermore, only a few studies have attempted to validate annual external ring production, and most do not observe annual production [17,32,40]. In fact, only one study, Ghent et al. [31], has conclusively demonstrated that external rings are formed annually.

Although external ring analysis may have limitations if only directly viewing and counting rings, cross dating potentially offers a valid approach to analyze general growth patterns in external rings. Because cross dating identifies frequency patterns of growth, a species of mussels with synchronous external growth patterns would suggest a regular rate of ring formation. Further work, such as confirmation through mark-recapture, would be needed to validate if external rings formed on an annual basis, but general growth information can be obtained by the periodicity of external ring formation. Therefore, if general growth patterns can be obtained from external rings, mussels will not have to be destructively sampled to analyze growth dynamics, an obvious benefit especially when working with threatened species.

In this study, we examined internal and external growth patterns of a common mussel species at two sites separated by a hydroelectric dam in the St. Croix River, MN. The first objective of the study was to examine the consistency of deriving age and growth rates between internal and external annuli for a Minnesota threatened mussel species, the mucket (*Actinonaias ligamentina*). We then used estimated growth rates from external annuli to evaluate the growth-effects of changing dam operations by comparing growth between sites upstream and downstream of the dam in 1994 (peaking) and 2010 (run-of-the-river/partial peaking) and by evaluating the long-term variation of growth within the downstream site.

Results
Growth pattern analysis
For all populations collected from Interstate State Park and Wild River State Park (hereafter, Interstate and Wild

River, respectively; Figure 1), individual growth was highly synchronous for the entire population, resulting in strong interseries correlations indicative of a high degree of non-age-related growth variation among populations (Table 1). The internal ring master chronologies extended 19 years for Wild River (collected in 1993) and 18 years for Interstate (collected in 1994) (Figure 2A). The external ring master chronology extended 11 years for Wild River and 12 years for Interstate (both collected in 2010) (Figure 2B). Alternating periods of high and low growth were apparent in all populations, and for both collection dates (i.e. 1993/94 and 2010) there was a high degree of synchrony between the two locations (Figure 2A/B).

Comparison of internal and external processing

Testing the consistency between internal and external ring processing methods on *A. ligamentina* collected from Interstate in 1994 yielded mixed results. Shell

length and processing technique significantly influenced age estimation, but the interaction of length and technique was not significant (Figure 3; Length: $F = 65.70$, df = 1, 191, $p < 0.0001$; Age: $F = 200.46$, df = 1, 191, $p < 0.0001$). Relative to internal aging, external ring counts underestimated age by 4 years. Furthermore, because the interaction between length and technique was not significant, the difference in age was consistent for all sizes of mussels. Percent growth (arcsine-square root transformed) was not significantly influenced by technique or the interaction of technique and age (Figure 4). Age of mussel, however, did significantly influence percent growth (Figure 4; Log(age): $F = 512.90$, df = 1, 211, $p < 0.0001$).

Influence of hydroelectric dam operation

Mean annual discharge was negatively correlated to internal growth at both Interstate and Wild River. This relationship was significant at Interstate ($R^2 = 0.32$,

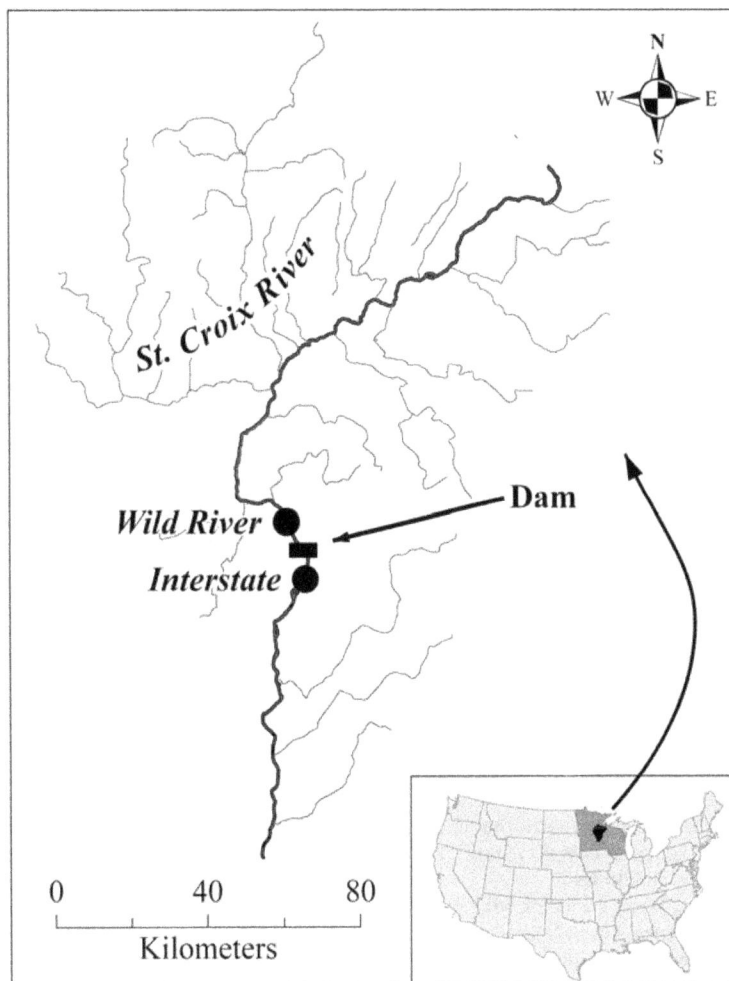

Figure 1 Wild River and Interstate State Parks (Minnesota) study sites are located in the St. Croix River drainage basin. The dam refers to the St. Croix Falls Hydro Generating Station.

Aquatic Ecosystem Management

Table 1 Cross dating statistics for _Actinonaias ligamentina_ in the St. Croix River, MN

Location	Processing method	Time series	Optimal spline flexibility	n	Percent of individuals validated		Mean interseries R (Pearson's)	
					Before quality control	After quality control	Before quality control	After quality control
Interstate	External	1998-2009	34	33	42	42*	0.567	0.567**
	Internal	1975-1994	30	49	88	100	0.261	0.570
Wild river	External	1999-2009	24	35	43	43*	0.685	0.685**
	Internal	1974-1993	30	29	81	100	0.339	0.537

Note: _n_, the number of individuals validated for each population, that is all correlations are positive and significant; Optimal Spline Flexibility, the value resulting in the highest interseries correlation for each species; * Because mussels were returned to the river after processing, no specimens could be re-evaluated after using COFECHA; ** Only those individuals that initially were validated were used in the final data set for that population.

$F = 9.5$, df $= 18$, $p = 0.007$). At Wild River, though not statistically significant, there was still a negative relationship ($R^2 = 0.14$, $F = 4.02$, df $= 19$, $p = 0.06$). Water temperature data was only available from 2000 to 2010 and thus, only the growth chronologies obtained from the 2010 collection were correlated to growing degree-days. There was no significant relationship between growing degree-days and yearly growth for either Interstate or Wild River.

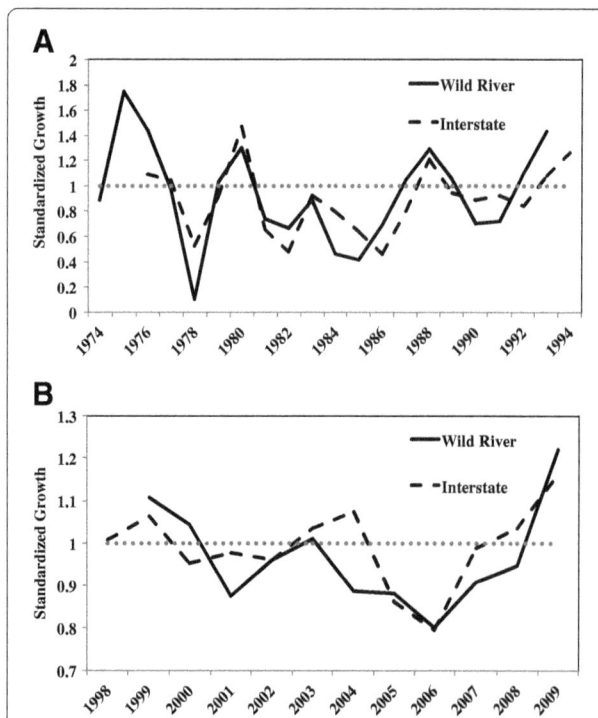

Figure 2 Master chronologies depicting the yearly-standardized growth for (A) _Actinonaias ligamentina_ collected in 1994 from Interstate State Park (_n_ = 19) and in 1993 from Wild Rive State Park (_n_ = 20) using internal processing, and (B) for _A. ligamentina_ collected in 2010 depicting yearly-standardized growth from Wild River (_n_ =11) and Interstate (_n_ =12) State Parks using external processing methods. Normal, expected growth is plotted on both representing the standardized growth index of 1. Values greater than 1 represent above average growth, whereas values less than 1 represent below average growth.

Growth between Wild River and Interstate was similar for both collection dates (Figure 2A/B), as neither the location, year, nor their interaction had a significant influence on yearly growth. Despite similarities between Wild River and Interstate, there was a long-term variation in growth at Interstate (Figure 5). Using a Ford-Walford plot we found the length of the mussel, collection date, and their interaction had significant influences on growth (Year: $F = 39.9$, df $= 1$, 1539, $p < 0.0001$; Length$_{(t)}$: $F = 51079$, df $= 1$, 1539, $p < 0.0001$; Interaction: $F = 54.8$, df $= 1$, 1539, $p < 0.0001$). Growth was generally similar between collection dates for smaller mussels. However, larger mussels (>45 mm) collected in 2010 had higher growth than the same size of mussels collected in 1994. Furthermore, mussels collected in 2010 had a higher range of growth than mussels collected in 1994.

Discussion

Our results suggest that useful growth information can be obtained from external growth rings of _Actinonaias ligamentina_. We found that percent growth was similar between different processing techniques, although age estimates were consistently different between internal and external aging methods. Using this information, we used growth patterns of external rings to examine the

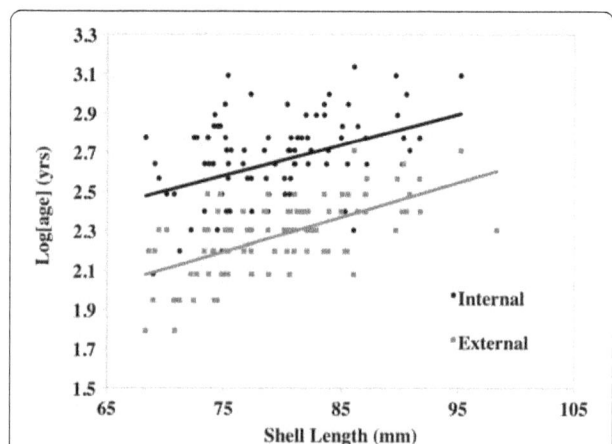

Figure 3 Comparison of internal and external aging (natural log transformed) methods for _Actinonaias ligamentina_ shells collected in 1994 from Interstate State Park.

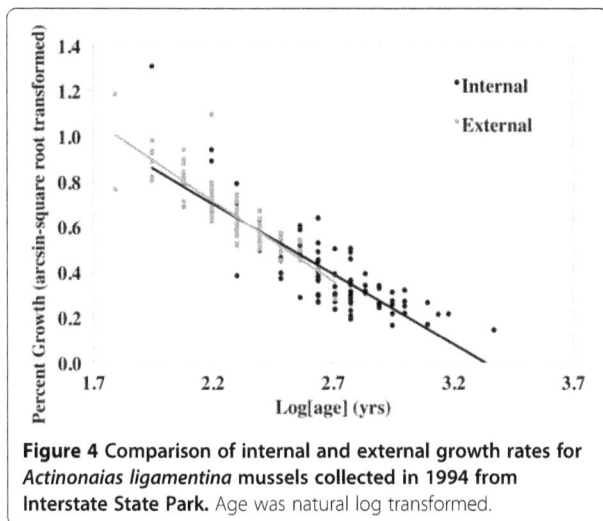

Figure 4 Comparison of internal and external growth rates for *Actinonaias ligamentina* mussels collected in 1994 from **Interstate State Park.** Age was natural log transformed.

response of *A. ligamentina* to a change in dam operation from a peaking schedule to a run-of-the-river/partial peaking flow regime. The change to run-of-the-river/ partial peaking appears to have benefited growth of older mussels below the dam.

The internal and external growth patterns obtained from this study meet the prerequisites of cross dating: *i)* growth shows non age-related patterns, and *ii)* individuals within a population exhibit synchronous growth [41]. Furthermore, identifying the regularity of ring deposition is essential to any growth study using organisms that form growth rings [30]. In cross dating studies for all taxa exhibiting ring deposition, a high degree of synchrony among individuals within a population indicates regular ring formation [22,39,41]. In this study, cross dating was an effective method to recognize synchronous growth patterns on both the internal and external surface of the shell. In addition, the strong interseries correlation for the master chronologies in this

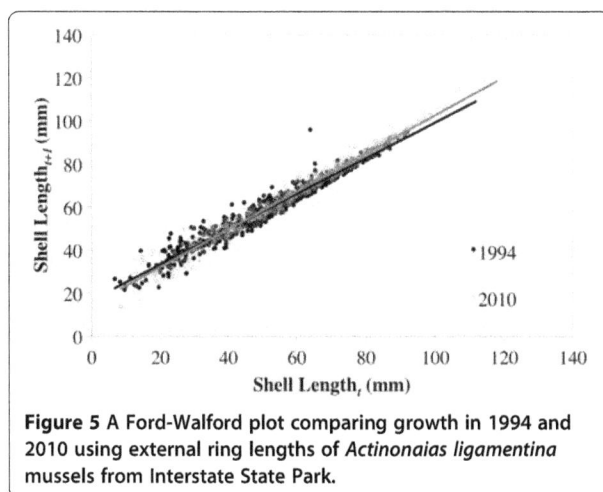

Figure 5 A Ford-Walford plot comparing growth in 1994 and 2010 using external ring lengths of *Actinonaias ligamentina* mussels from Interstate State Park.

study suggests that the assumption of regular ring formation can be validated for both internal and external rings.

The rate at which rings form has important implications for management purposes. For instance, specific calendar years of growth could be aligned to corresponding climate data to determine how populations have responded to changes in the environment. Knowing if rings form annually further strengthens the ability to compare growth to specific environmental conditions. Ring formation in many mussel species, including *A. ligamentina*, is speculated to occur annually [32,42,43]. Using mark-recapture, Moles and Layzer [44] validated internal, annual ring formation of *A. ligamentina* in the Green River, KY. Moreover, cross dating has recently become an acceptable tool to validate annual internal ring formation for many mussel species [22,34-36]. Ideally, cross dating is used to verify annual ring production by either correlating growth to environmental variables or showing a correlation between two different chronologies of the same species, one in which annual ring formation has been validated using mark-capture and the other non-validated. Rypel et al. [22] and Black et al. [34] demonstrated a negative relationship between annual growth in freshwater mussels and mean annual discharge; thus mean annual flow is a reasonable variable to use to validate annual ring formation.

Mean annual discharge was negatively correlated to both internal chronologies from Interstate and Wild River in our study. Although the gauge used to obtain discharge is located below the dam (USGS gauge 05340500) and is not an ideal measurement for the population at Wild River, the strong negative relationship still suggests a pattern associated with internal ring formation. Furthermore, the interseries correlation for the internal chronology at Wild River was similar to that of the internal chronology at Interstate. Considering the negative correlation between mean annual discharge and both internal chronologies in this study, the prior validation of annual ring formation using mark-recapture in a separate population of *A. ligamentina* [44], and the consistency of interseries correlation among other species of mussels with validated annual ring formation [22,34-36] strongly suggests that the internal growth rings for *A. ligamentina* in this study are annual rings.

Whether or not external rings are produced annually cannot be determined from this study. There were no strong relationships between any of the external chronologies and mean annual discharge. Furthermore, no prior studies have validated external ring production using mark-recapture or cross dating studies with *A. ligamentina*. Only one mark-recapture study supports annual production of external rings; Ghent et al. [31] showed clear, conspicuous annual external ring formation for *Anodonta grandis*. Other studies, however, failed to examine external ring production to support this hypothesis [17,32,40]. Although we did observe high interseries correlation of our external chronologies, without having a direct relation to discharge or conclusive mark-recapture support, the significant interseries correlations can only confirm regular patterns of ring formation. Therefore, additional studies, ideally mark-recapture, would need to validate the rate of external ring formation with *A. ligamentina* in the St. Croix River.

This study supports recent studies [22,34,35] showing that cross dating is a powerful tool for identifying growth patterns within freshwater mussel populations. Although cross dating can be a lengthy and tedious process, it is still more efficient than traditional mark-recapture studies, and can provide larger sample sizes which yields stronger growth estimations for an entire population. Also, as with any living organism, mussels are susceptible to disturbance (e.g. flooding, predation, microhabitat changes, handling while processing) that can alter normal growth. When mussels experience such stress, growth may stop, and a disturbance ring is deposited [14]. Cross dating techniques can detect these false rings by comparing each individual series to the master chronology for that population, and thereby allowing for growth increments to be correctly aligned to specific calendar years.

The inconsistencies associated with age estimation between internal and external processing methods we found are similar to the findings of Neves and Moyer [33]. In our study, we estimated that external ages were consistently 4 years less than internal ages. Such differences in aging between processing techniques are likely due to difficulties in detecting external rings. In many of our specimens, the outer prismatic layer was eroded around the umbone, likely masking early juvenile growth rings. Additionally, older growth rings become crowded near the shell margin, making differentiation between subsequent rings difficult. Finally, external ring analysis occurs in a field setting likely introducing sampling error. Thus, failure of ring detection reduces the reliability for absolute age estimation using external rings.

Although internal aging is likely more accurate, we show that important growth information can still be obtained from external growth rings. To compare internal and external growth patterns, we first standardized growth measurements to account for differences in measurement techniques. External growth was measured along the anterior to posterior axis, while internal growth was measured dorsal to ventral, perpendicular to growth lines. Our methods of standardizing growth between internal and external processing allowed

us to control for three factors influencing variation in growth. First, shell length and shell height were positively and significantly correlated ($R^2 = 0.73$, $F = 250$, df = 95, $p < 0.0001$), suggesting that growth measurements were proportional across the shell. Because length and height are proportional, using these to standardize growth as a percent of size allowed us to control for measuring different axes. Second, by using the sum of the last 5 full years of consecutive, measurable growth, we were able to remove variation associated with age-related growth. This also helped control for the third factor, which was the difference associated with age between internal and external ring counts. Here, the increment between the rings, rather than the ring count, was important. Using the last 5 full years of growth, we were able to keep the same number of growth increments constant for both methods.

The consistent pattern of growth between internal and external processing methods (Figure 4) suggests two applications. First, despite a difference in ring counts, cross dating is still efficient at identifying patterns in growth for external rings. Because the rings that likely caused discrepancies in aging estimates were either at the beginning or end of an individual chronology, the difference in ring counts should not alter the ability to identify patterns in growth. This is supported by a high interseries correlation using external rings, which is also similar to both the chronology of the internal rings in this study and other studies using cross dating [22,34-36]. Our sampling methods did, however, limit our ability to conduct quality control with external processing. Once the external measurements were recorded, we immediately returned the mussels to the river. Doing so restricted us from re-examining specimens flagged for having potential false or missing rings on the external surface. Thus, we were only able to include those specimens that were initially positive and significantly correlated to the master chronology in order to reduce the possibility of including a false/missing ring. An easy remedy to improve quality control measures would be to mark mussels and keep them in an aerated tank until cross dating is complete in case any specimens need re-examined, then return the mussels to the river. Second, and perhaps most important, obtaining growth estimates from external rings offers a non-destructive sampling method.

Because of consistent growth patterns between internal and external processing, we were able to compare the impacts of the change in dam operation on the growth of mussels at Interstate using external ring data from 1994 and 2010. Numerous studies have documented negative effects on mussel communities downstream of dams [9,11,12]. Traditional dam operation, especially hydroelectric dams, is based on a peaking schedule that greatly disrupts the natural flow regime of rivers, resulting in disruptions of the species assemblage. Implementing a run-of-the-river operation schedule theoretically re-establishes a natural flow and should have positive impacts on mussel communities. Our results partially support this concept; there were similar growth patterns between both sites in 1994 and 2010 (Figure 2), but there was a long-term variation in growth below the dam before and after the implementation of a run-of-the-river/partial peaking schedule (Figure 5).

The similarities we found in growth patterns in both 1994 and 2010 between Wild River and Interstate are particularly interesting due to the presence of the hydroelectric dam between them. Recent mussel population surveys in the St. Croix River below the dam have documented declines of another common species (*Truncilla truncata*), but no declines in the *Actinonaias ligamentina* population have been observed [45]. This suggests that *A. ligamentina* are more tolerant to higher ranges of discharge or temperature. Another possibility for the similarities in growth between the Wild River and Interstate sites in 1994 could have been a result of mussels acclimating to the peaking schedule. The hydroelectric dam was installed in 1907, which would have given the mussels nearly a century to adjust to the altered flow regimes. There are a number of ways in which organisms are known to respond to changes in hydrological regime [46]. For example, life histories of the imperiled Pacific salmon vary depending on hydrologic regime [47]. While freshwater mussels are known to have changes in shell shape that respond to hydrologic regime [48], little is known regarding life history adaptations to changing flow. Though not statistically significant, growth did differ between Wild River and Interstate in 2010, especially during the first few years after the dam changed operation in 2000 (Figure 2B). This suggests that the mussels below the dam could be responding to the change in flow conditions due to run-of-the-river/partial peaking.

The long-term difference in growth (Figure 5) suggests that the change in dam operation has benefited the growth of larger mussels, while smaller mussels do not show much difference in growth between the two collection dates. Because we do not know the exact mechanisms controlling growth of *A. ligamentina* in the St. Croix River, we cannot infer why growth was higher for adults in 2010. It is possible that older mussels generally partition the majority of their energy resources to reproduction, but under more favorable conditions they have sufficient resources available for growth and reproduction. Layzer *et al.* [49] and Galbraith and Vaughn [9] both documented a reduction in fitness at flow-regulated study sites. In our study, streamflow conditions from the peaking schedule experienced by

the mussels collected in 1994 could have resulted in lower food availability and higher fluctuations in water temperature and discharge [50]. In contrast, the natural flow regime mimicked by the run-of-the-river/partial peaking operations in 2010 could have provided a more consistent food supply and less variable, water temperature, and discharge, thus resulting in conditions more suitable for mussel reproduction and growth. Regardless of the precise mechanisms, our study indicates that the recent change in dam operation appears to benefit the growth of *A. ligamentina* in downstream reaches, particularly for older individuals.

Conclusions
As habitat for freshwater mussels continues to be impaired, providing the best management strategies is crucial. Currently, knowledge of the predominant factors influencing mussel populations responses to changes in the environment are poorly understood. Annual rings deposited by freshwater mussels record temporal changes associated with their environment, and linking patterns in growth rings to environmental changes has important conservation implications. This study supports recent applications of cross dating as an effective tool for validating annual internal ring formation and obtaining more reliable estimates of mussel growth, at the individual or population level. Our study demonstrates that cross dating can be further applied to analyze patterns in external growth rings. Our methods of cross dating resulted in consistent growth patterns between internal and external processing methods. We suggest that similar cross dating methods can be applied to other mussel species, provided growth patterns within a population are highly synchronous. Cross dating also can be conducted using non-destructive sampling methods, which is especially beneficial for management and conservation activities related to threatened or endangered freshwater mussels species.

Methods
Study site
The St. Croix River, a National Wild and Scenic Riverway, drains approximately 20,000 km^2 in east-central Minnesota and northwestern Wisconsin [51] and supports 41 species of freshwater mussels [52,53]. We collected *Actinonaias ligamentina* from Interstate and Wild River to reconstruct age and growth dynamics. *Actinonaias ligmanetina* is a common species throughout the United States with a life expectancy of around 50 years [44]. Interstate is located approximately 3.5 km downstream of a hydroelectric dam at St. Croix Falls, WI, while Wild River is about 10 km upstream the dam (Figure 1). The dam changed operation regimes from a partial hold-and-release schedule to run-of-the-river/

partial peaking in 2000. The partial hold-and-released operated on a daily peaking schedule where water was stored and only released when sufficient amount was available to produce maximum energy output, whereas the run-of-the-river/partial peaking mimics a more natural, continuous flow, while still storing and releasing water for energy production. From 1910 to 2000, average discharge was 123 m^3 s^{-1}, with a range of 50 to 243 m^3 s^{-1}. Following operational changes, the average discharge was 126 m^3 s^{-1} and ranged from 79 to 173 m^3 s^{-1}. Thus, although average discharge has remained relatively constant, the change to run-of-the-river/partial peaking has reduced the range of discharge considerably.

Collection and measurement of shell rings

Mussels were collected from Wild River and Interstate in 1993 and 1994, respectively, as part of a physiological study. During that time, thin sections were prepared for the mussels from Wild River in contrast to only the shells being preserved from Interstate. These populations were re-sampled in 2010, but only external measurements were taken on live mussels and the mussels were returned to the river immediately after processing. Thus, methodology applied to measure shell rings (i.e. internal or external processing) was based on the availability of specimens, which varied among study sites and time periods: Wild River (1993: internal; 2010: external); Interstate (1994: internal and external; 2010: external), see also Table 1.

External processing consisted of measuring the length, height, and width of the total shell, the eroded area on the umbone, and the anterior to posterior length of each external annulus for each shell (Figure 6). The same half of the shell (i.e. valve) that was used for external measurements was used for internal processing (applicable only to shells collected in 1994 from Interstate). Internal processing consisted of generating thin sections using standard methods for bivalves [33,54]. All thin sections were viewed and interpreted using a dissecting microscope (StemiDV4, Carl Zeiss, Gottingen, Germany). Internal annuli were identified, and the dorso-ventral growth increment along the prismatic and nacreous layers was measured using a linear encoder (ENC 150, Acu-Rite, Jamestown, NY) with a digital readout (QC-1000, Metronics, Bedford, NH). These measurements were linear in nature, not following the curve of the shell. Measurements began at the most recent complete growth year and continued towards the umbone. Disturbance (false) rings were identified as discontinuous from shell margin to the umbone and were not measured. Mussels < 5 years old were excluded from the sample because growth in early years is largely age-related and thus useful environmentally-influenced growth information is hard to obtain [35].

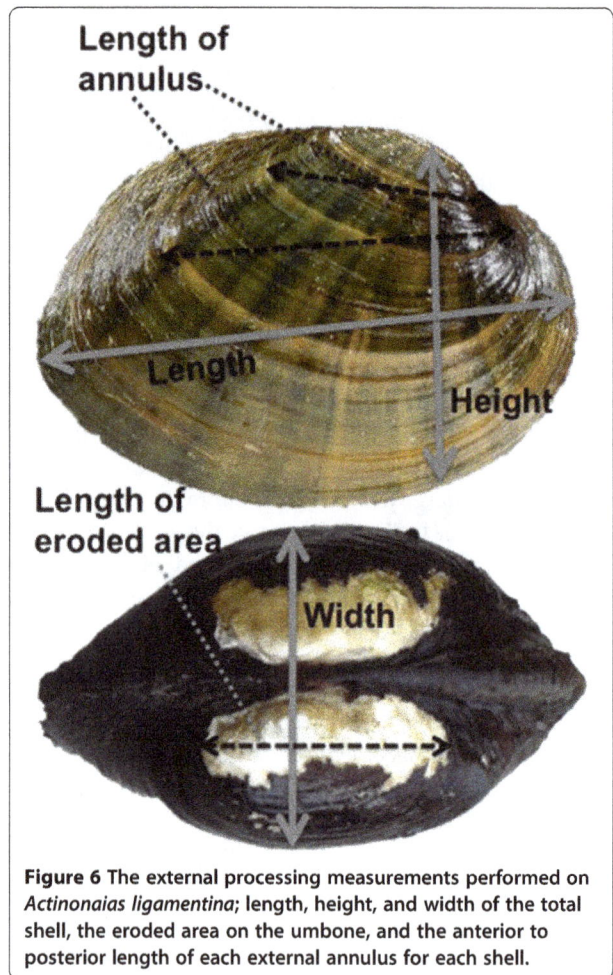

Figure 6 The external processing measurements performed on *Actinonaias ligamentina*; length, height, and width of the total shell, the eroded area on the umbone, and the anterior to posterior length of each external annulus for each shell.

Growth pattern analysis

Deposition rates of growth rings were validated using cross dating procedures following on Rypel *et al.* [22]. First, growth series for each individual were compared to the common signal (i.e. the average detrended value of shell ring increments for the population) to detect dating errors and to assess relative signal strength using COFECHA V1.26 (Dendrochronology Program Library; [39]). COFECHA removes age-related growth variation and generates a standardized index series by fitting a cubic spline to each series [39]. The optimum cubic spline depends on the average time length of the series. For each population, we determined the optimum cubic spline by choosing the value that yielded the highest interseries correlation for that specific population [22,41]; Table 1. From the standardized index, COFECHA generated a master chronology by taking the average growth index for each year. COFECHA then compares each individual time series to the master chronology, identifies potential problems with each series (e.g. false or missing annuli), and lags each time series both forward and backward to test if the series fits better at a different time interval. For internal growth chronologies, series flagged with potential problems were re-examined, and, if measurement errors were detected, the appropriate growth increments were re-measured [39]. After this quality control measure, COFECHA was then re-run with the corrected series(s). Quality control was limited for the external chronologies from 2010 due to sampling methods, as measurements were taken from live mussels that were immediately returned to the river once measurements were complete. Therefore, we were unable to re-examine any shells flagged as having errors by COFECHA. Consequently, only those individuals initially validated to the master chronology (Table 1) were used in our final data set. After quality control measures were complete, only those series that were both positively and significantly ($\alpha < 0.05$) correlated to the master chronology were validated and included in the data set. The interseries correlation coefficient resulting from the validated individuals was used as a measurement of the strength of the common signal for the annual growth pattern for each population.

Using the raw, corrected growth measurements from COFECHA, data were transformed into standardized growth indices for each population using ARSTAN V1.26 (Dendrochronology Program Library; [55]). ARSTAN uses a detrending function to remove age-related growth and retain as much of the high-frequency variation in annual growth as possible. For this study, we used a negative exponential curve to detrend the growth measurements, resulting in a unitless, standardized growth index with an average value of 1 [22,55]. Growth values greater than 1 represented above average growth, whereas values less than 1 represented below average growth [56].

Comparison of internal and external processing

Analyses of covariance (JMP statistical software, version 8.0, SAS Institute, Cary, NC) were used to evaluate the age and growth estimates derived from both internal and external processing methods for the *A. ligamentina* collected from Interstate in 1994. Age was estimated by counting the number of annuli on the external and internal surface, respectively. To remove age-related variation in the growth patterns, we developed a procedure to standardize raw growth measurements: the last 5 full, consecutive years of raw growth were summed and divided by total shell length or cross-section length, depending on internal or external measurements. External measurements were based on the anterior to posterior length of annuli; thus, for external measurements the total shell length was used to standardize growth. In contrast, internal measurements were measured perpendicular to the growth lines from the umbone to shell margin; thus, for internal measurements the total height

(dorso-ventral dimension) was used to standardized growth.

Influence of hydroelectric dam operation

We evaluated long-term patterns in growth downstream of the dam by comparing growth in 1994 to growth in 2010, as this could lead to further inferences about the impact of dam operation on mussel growth. We used a Ford-Walford plot [57], plotting $length_{(t)}$ against $length_{(t+1)}$, to model the growth of each population. An analysis of covariance (JMP statistical software, version 8.0, SAS Institute, Cary, NC) was used to compare growth rates between collection dates at Interstate. Because no internal measurements were taken in 2010, this analysis was performed only with external measurements. Finally, we obtained water temperature (converted to growing degree-days; Snyder, DegDay, University of California, version 1.01) and discharge data from USGS gauge 05340500 and used regression analysis (JMP statistical software, version 8.0, SAS Institute, Cary, NC) to compare growth to each.

Competing interests
The authors claim no competing interests of this study.

Authors' contribution
BS participated in the collection and measurements of specimens, carried out the creation and measurement of thin section, and drafted the manuscript. DH coordinated the study, participated in the design of the study and collection and measurements of specimens, and helped draft the manuscript. MH participated in the study design, contributed to the collection and measurements of specimens, and made revisions to the manuscript. JK offered dendrological methods to improve analysis of growth patterns, participated in the interpretation and measurements of thin sections, and made revisions to the manuscript. All authors approved the final manuscript.

Acknowledgements
This work was supported in part by the Undergraduate Science Education Program Grant No. 52006323 from the Howard Hughes Medical Institute to Washington & Jefferson College, a Margrett A. Cargill Foundation restricted grant to the Environmental Studies Program at Washington & Jefferson College, National Park Service cooperative agreements PMIS 111881 and PMIS 131196, and Macalester College faculty travel fund. We thank the following people for help collecting mussels, and preparing and interpreting thin sections: Derek Ochi, Elise Griffin, Karen Jackson, Carl Skarbek, Kelly MacGregor, Jeff Thole, Ronald Bayline, Jamie March, and Robert East. We wish to thank both biology departments at Macalester College and Washington & Jefferson College who helped facilitate this collaborative research. Finally, thanks to two anonymous reviewers for helpful comments on the manuscript.

Author details
[1]Department of Biology, Washington and Jefferson College, Washington, PA, USA. [2]Department of Environmental Studies, Macalester College, St. Paul, MN, USA. [3]Department of Biology, Macalester College, St. Paul, MN, USA. [4]Current address: Oklahoma Biological Survey and Department of Biology, University of Oklahoma, 111 East Chesapeake St, Norman, OK, 73019, USA.

References

1. Gutierrez JL, Jones CG, Strayer DL, Iribarne OO: Mollusks as ecosystem engineers: the role of shell production in aquatic habitats. *Oikos* 2003, 101:79–90.
2. Spooner DE, Vaughn CC: Context-dependent effects of freshwater mussels on stream benthic communities. *Freshwater Biology* 2006, 51:1016–1024.
3. Vaughn CC, Nichols SJ, Spooner DE: Community and foodweb ecology of freshwater mussels. *Journal of the North American Benthological Society* 2008, 27:409–423.
4. Neves RJ: Conservation and commerce: management of freshwater mussel (Bivalvia: Unionoidea) resources in the United States. *Malacologia* 1999, 41:461–474.
5. Williams JD, Warren ML, Cummings KS, Harris JL, Neves RJ: Conservation status of fresh-water mussels of the United States and Canada. *Fisheries* 1993, 18:6–22.
6. Strayer DL: *Freshwater Mussel Ecology: A Multifactor approach to Distribution and Abundance*. Los Angeles: University of California Press; 2008.
7. Watters GT: *Freshwater mussels and water quality: a review of the effects of hydrologic and instream habitat alterations. In Proceedings of the First Freshwater Mollusk Conservation Society Symposium.* 1999:261–274.
8. Baxter RM: *Environmental effects of dams and impoundments. In Annual Review of Ecology and Systematics.* 1977:255–283.
9. Galbraith HS, Vaughn CC: Effects of reservoir management on abundance, condition, parasitism and reproductive traits of downstream mussels. *River Research and Applications* 2011, 27:193–201.
10. Tiemann JS, Dodd HR, Owens N, Wahl DH: Effects of lowhead dams on unionids in the Fox River, Illinois. *Northeastern Naturalist* 2007, 14:125–138.
11. Vaughn CC, Taylor CM: Impoundments and the decline of freshwater mussels: A case study of an extinction gradient. *Conservation Biology* 1999, 13:912–920.
12. Williams JD, Fuller SLH, Grace R: Effects of impoundments on freshwater mussels (Mollusca: Bivalvia: Unionidae) in the main channel of the Black Warrior and Tombigbee Rivers in western Alabama. *Bulletin Alabama Museum of Natural History* 1992:1–10.
13. Dennis B, Munholland PL, Scott JM: Estimation of growth and extinction parameters for endangered species. *Ecological Monographs* 1991, 61:115–143.
14. Lutz RA, Rhoads DC: Anaerobiosis and a theory of growth line formation. *Science* 1977, 198:1222–1227.
15. Bauer G: Variation in the life-span and size of the fresh-water pearl mussel. *Journal of Animal Ecology* 1992, 61:425–436.
16. Hastie LC, Young MR, Boon PJ: Growth characteristics of freshwater pearl mussels, *Margaritifera margaritifera* (L.). *Freshwater Biology* 2000, 43:243–256.
17. Negus CL: A quantitative study of growth and production of unionid mussels in River Thames at Reading. *Journal of Animal Ecology.* 1966:513–532.
18. Christian AD, Smith BN, Berg DJ, Smoot JC, Findlay RH: Trophic position and potential food sources of 2 species of unionid bivalves (Mollusca: Unionidae) in 2 small Ohio streams. *Journal of the North American Benthological Society* 2004, 23:101–113.
19. Nichols SJ, Garling D: Food-web dynamics and trophic-level interactions in a multispecies community of freshwater unionids. *Canadian Journal of Zoology-Revue Canadienne De Zoologie* 2000, 78:871–882.
20. Raikow DF, Hamilton SK: Bivalve diets in a midwestern US stream: A stable isotope enrichment study. *Limnology and Oceanography* 2001, 46:514–522.
21. Thorp JH, Delong MD, Greenwood KS, Casper AF: Isotopic analysis of three food web theories in constricted and floodplain regions of a large river. *Oecologia* 1998, 117:551–563.
22. Rypel AL, Haag WR, Findlay RH: Validation of annual growth rings in freshwater mussel shells using cross dating. *Canadian Journal of Fisheries and Aquatic Sciences* 2008, 65:2224–2232.
23. Kesler DH, Downing JA: Internal shell annuli yield inaccurate growth estimates in the freshwater mussels *Elliptio complanata* and *Lampsilis radiata*. *Freshwater Biology* 1997, 37:325–332.
24. Hanson JM, Mackay WC, Prepas EE: The effects of water depth and density on the growth of a unionid clam. *Freshwater Biology* 1988, 19:345–355.
25. Morris TJ, Corkum LD: Unionid growth patterns in rivers of differing riparian vegetation. *Freshwater Biology* 1999, 42:59–68.

26. Negishi JN, Kayaba Y: Size-specific growth patterns and estimated longevity of the unionid mussel (*Pronodularia japanensis*). *Ecological Research* 2010, 25:403–411.
27. Bolden SR, Brown KM: Role of stream, habitat, and density in predicting translocation success in the threatened Louisiana pearlshell, *Margaritifera hembeli* (Conrad). *Journal of the North American Benthological Society* 2002, 21:89–96.
28. Chamberlain T: Annual growth of fresh-water mussels. *Bulletin of the United States Fish Commission* 1930, 48:713–739.
29. Strayer DL, Downing JA, Haag WR, King TL, Layzer JB, Newton TJ, Nichols SJ: Changing perspectives on pearly mussels, North America's most imperiled animals. *Bioscience* 2004, 54:429–439.
30. Beamish RJ, McFarlane GA: The forgotten requirement for age validation in fisheries biology. *Transactions of the American Fisheries Society* 1983, 112:735–743.
31. Ghent AW, Singer R, Johnsonsinger L: Depth distributions determined with SCUBA, and associated studies of freshwater unionid clams *Elliptio-complanata* and *Anodonta-grandis* in Lake Bernard, Ontario. *Canadian Journal of Zoology-Revue Canadienne De Zoologie* 1978, 56:1654–1663.
32. Haag WR, Commens-Carson AM: Testing the assumption of annual shell ring deposition in freshwater mussels. *Canadian Journal of Fisheries and Aquatic Sciences* 2008, 65:493–508.
33. Neves RJ, Moyer SN: Evaluation of techniques for age-determination of fresh-water mussels (Unionidae). *American Malacological Bulletin* 1988, 6:179–188.
34. Black BA, Dunham JB, Blundon BW, Raggon MF, Zima D: Spatial variability in growth-increment chronologies of long-lived freshwater mussels: Implications for climate impacts and reconstructions. *Ecoscience* 2010, 17:240–250.
35. Haag WR, Rypel AL: Growth and longevity in freshwater mussels: evolutionary and conservation implications. *Biological Reviews* 2011, 86:225–247.
36. Schone BR, Dunca E, Mutvei H, Norlund U: A 217-year record of summer air temperature reconstructed from freshwater pearl mussels (*M. margarifitera*, Sweden). *Quaternary Science Reviews* 2004, 23:1803–1816.
37. Haag WR: Extreme longevity in freshwater mussels revisited: sources of bias in age estimates derived from mark-recapture experiments. *Freshwater Biology* 2009, 54:1474–1486.
38. Douglas A: Crossdating in dendrochronology. *Journal of Forestry* 1939, 39:825–831.
39. Grissino-Mayer H: Evaluating crossdating accuracy: a manual and tutorial for the computer program COFECHA. *Tree-Ring Research* 2001, 57:205–221.
40. Downing WL, Shostell J, Downing JA: Nonannual external annuli in the fresh-water mussels *Anodonta-grandis* and *Lampsilis-radiata-siliquoidea*. *Freshwater Biology* 1992, 28:309–317.
41. Black BA, Boehlert GW, Yoklavich MM: Using tree-ring crossdating techniques to validate annual growth increments in long-lived fishes. *Canadian Journal of Fisheries and Aquatic Sciences* 2005, 62:2277–2284.
42. Mutvei H, Westermark T: How environmental information can be obtained from naiad shells. *Ecology and Evolution of the Freshwater Mussels Unionoida* 2001, 145:367–379.
43. Valdovinos C, Pedreros P: Geographic variations in shell growth rates of the mussel *Diplodon chilensis* from temperate lakes of Chile: Implications for biodiversity conservation. *Limnologica* 2007, 37:63–75.
44. Moles KR, Layzer JB: Reproductive ecology of *Actinonaias ligamentina* (Bivalvia: Unionidae) in a regulated river. *Journal of the North American Benthological Society* 2008, 27:212–222.
45. Hornbach DH, Hove MC, Sansom BJ: Where have all the deertoes gone? The decline in a single species within a multi-species mussel assemblage. *Bulletin of the North American Benthological Society* 2011, 58. In Press.
46. Lytle DA, Poff NL: Adaptation to natural flow regimes. *Trends in Ecology and Evolution* 2004, 19:94–100.
47. Beechie T, Buhle E, Ruckelshaus M, Fullerton A, Holsinger L: Hydrologic regime and the conservation of salmon life history diversity. *Biological Conservation* 2006, 130:560–572.
48. Hornbach DJ, Kurth VJ, Hove MC: Variation in freshwater mussel shell sculpture and shape along a river gradient. *The American Midland Naturalist* 2010, 164:22–36.
49. Layzer JB, Gordon ME, Anderson RM: Mussels - the forgotten fauna of regulated rivers - a case-study of the Caney Fork River. *Regulated Rivers-Research & Management* 1993, 8:63–71.

50. Galbraith HS, Vaughn CC: **Temperature and food interact to influence gamete development in freshwater mussels.** *Hydrobiologia* 2009, **636:**35–47.

51. Fago D, Hatch J: **Aquatic resources of the St. Croix River Basin.** *U S Fish and Wildlife Service Biological Report* 1993, **19:**23–56.

52. Cummings KS, Mayer CA: *Field guide to freshwater mussels of the Midwest.* Illinois Natural History Survey; 1992.

53. Hornbach DJ: **Macrohabitat factors influencing the distribution and abundance of naiads in the St. Croix River, MN and WI, USA.** In *Ecological Studies 145: Ecology and Evolutionary Biology of the Freshwater Mussels Unionoidea.* Edited by Bauer G, Wachtler W. Berlin: Springer; 2001.

54. Clark GR II: **Study of molluscan shell structure & growth lines using thin sections.** *Topics in Geobiology* 1980, **1:**603–606.

55. Cook E, Holmes R: *Program ARSTAN and users manual.* Palisades, NY: Lamont Doherty Geological Observatory; 1984.

56. Fritts HC, Xiangding W: **A comparison between response-function analysis and other regression techniques.** *Tree-Ring Bulletin* 1986, **46:**31–46.

57. Anthony JL, Kessler DH, Downing WL, Downing JA: **Length-specific growth rates in freshwater mussels (Bivalvia: Unionidae): extreme longevity or generalized growth cessation?** *Freshwater Biology* 2001, **46:**1349–1359.

11

Marine crude-oil biodegradation: a central role for interspecies interactions

Terry J McGenity[*], Benjamin D Folwell, Boyd A McKew and Gbemisola O Sanni

Abstract

The marine environment is highly susceptible to pollution by petroleum, and so it is important to understand how microorganisms degrade hydrocarbons, and thereby mitigate ecosystem damage. Our understanding about the ecology, physiology, biochemistry and genetics of oil-degrading bacteria and fungi has increased greatly in recent decades; however, individual populations of microbes do not function alone in nature. The diverse array of hydrocarbons present in crude oil requires resource partitioning by microbial populations, and microbial modification of oil components and the surrounding environment will lead to temporal succession. But even when just one type of hydrocarbon is present, a network of direct and indirect interactions within and between species is observed. In this review we consider competition for resources, but focus on some of the key cooperative interactions: consumption of metabolites, biosurfactant production, provision of oxygen and fixed nitrogen. The emphasis is largely on aerobic processes, and especially interactions between bacteria, fungi and microalgae. The self-construction of a functioning community is central to microbial success, and learning how such "microbial modules" interact will be pivotal to enhancing biotechnological processes, including the bioremediation of hydrocarbons.

Keywords: Hydrocarbon, Crude oil, Salt marsh, Marine microbiology, Biodegradation, Bioremediation, Microbial interactions, Biogeochemistry, *Alcanivorax*

The problem of marine oil pollution

Our seas, oceans and coastal zones are under great stress; and pollution, particularly by crude oil, remains a major threat to the sustainability of planet Earth [1]. An estimated 1.3 million tonnes of petroleum enters the marine environment each year [2]. Acute pollution incidents cause great public concern, notably ~600,000 tonnes of crude oil released after the Deepwater Horizon explosion in the Gulf of Mexico [3] and ~63,000 tonnes from the Prestige oil-tanker [4] off the coast of north-west Spain. The fate of crude oil spilled at sea (Figure 1) depends on both the prevailing weather and the composition of the oil; but its environmental impact is exacerbated on reaching the shoreline, especially in low-energy habitats, such as lagoons and salt marshes. Acute pollution events can result in mass mortality; for example, more than 66% of total species richness (including polychaetes, molluscs, crustaceans and insects) was lost in the worst affected beaches following the

Prestige spill [5]. Hydrocarbons also contaminate the feathers and fur of marine birds and mammals, resulting in the loss of hydrophobic properties, leading to death from hypothermia [6], or lethal doses following ingestion of oil during preening.

Moreover, the impact of hydrocarbons, especially polycyclic aromatic hydrocarbons (PAHs), on wildlife and fisheries may be long-lasting; for example the Fisheries Exclusion Zone imposed after the Braer spill (Shetland Islands, United Kingdom, 1993) due to contaminated fish and shellfish, remained in place for over 6 years. Chronic pollution can cause physiological or behavioural damage at sub-lethal concentrations; and genetic damage and decreases in both growth and fecundity have been observed in fish [7,8]. Deep-sea sediments and associated biota are also chronically affected by drilling, which deposits vast amounts of oil-contaminated drill cuttings on the seafloor [9]. Even when oil-contaminated coastal sediments appear to be clean (e.g. Prince William Sound that was contaminated by the Exxon Valdez spill in 1989), toxic oil components, such as high molecular

* Correspondence: TJMcGen@Essex.ac.uk
School of Biological Sciences, University of Essex, Wivenhoe Park, Colchester CO4 3SQ, UK

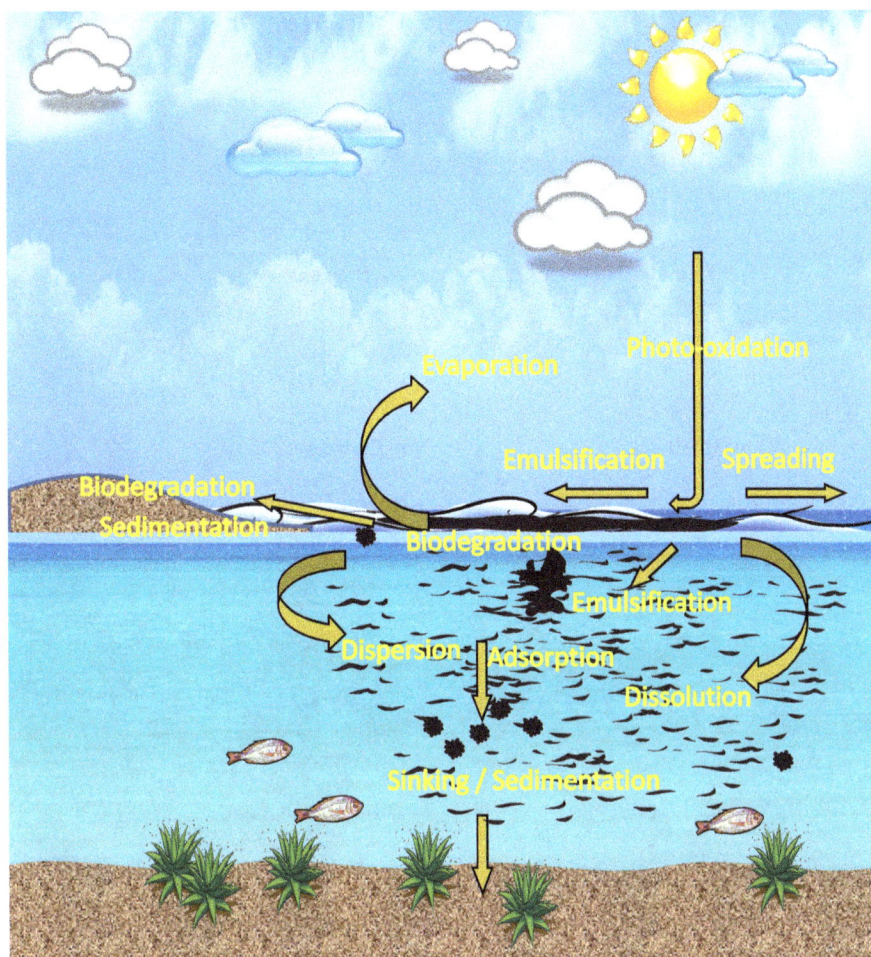

Figure 1 Fate of a marine oil spill (for a more detailed explanation, see http://www.itopf.com/marine-spills/fate/weathering-process/).
Spreading is affected by the action of winds, waves, water currents, oil type and temperature, and enhances evaporation of the volatile fractions such as low molecular weight alkanes and monoaromatic hydrocarbons. Spilt oil is broken into droplets and dispersed through the water column, enhancing the biodegradation of hydrocarbons and dissolution of water-soluble fractions of oil. Turbulent seas cause water droplets to be suspended in the oil, resulting in water-in-oil emulsions, alternatively known as chocolate mousse, which is difficult to degrade because of its high viscosity and reduced surface area. Photo-oxidation is the process by which hydrocarbons, especially PAHs, react with oxygen in the presence of sunlight, resulting in structural changes that can on the one hand lead to increased water solubility or, conversely, increased recalcitrance to biodegradation. Sedimentation will general only occur when oil adsorbs to particles owing to nearly all crude oils having a lower density than seawater.

weight (HMW) PAHs, may remain buried and sorbed to sediment particles, and can be released to the environment by bioturbation or human activities such as dredging [10].

Crude oil is a natural, heterogeneous mixture of hydro-carbons, with potentially 20,000 chemical components [11], consisting mainly of alkanes with different chain lengths and branch points, cycloalkanes, mono-aromatic and polycyclic aromatic hydrocarbons (Figure 2; [12]). Some compounds contain nitrogen, sulfur and oxygen [12]; while trace amounts of phosphorus, and heavy metals such as nickel and vanadium are also found [13]. Its composition varies widely, and each oil component has different physico-chemical properties, including viscosity,

solubility and capacity to absorb (Table 1), as well as vary-ing in its bioavailability and toxicity. Crude oil, released naturally from the geosphere to the biosphere (e.g. from cold seeps [14]) may supply up to half of the oil in the sea [2]. Although hydrocarbons are relatively stable molecules, their "fuel value" and presence in the environment for mil-lions of years have led to the evolution of many microbes able to activate and use them as a major or sole source of carbon and energy, including at least 175 genera of Bacteria [15]. Several haloarchaeal genera [16] and many Eukarya can grow on or transform hydrocarbons [17]. Biodegradation of crude oil to carbon dioxide and water is the major process by which hydrocarbon-contaminated environments are remediated.

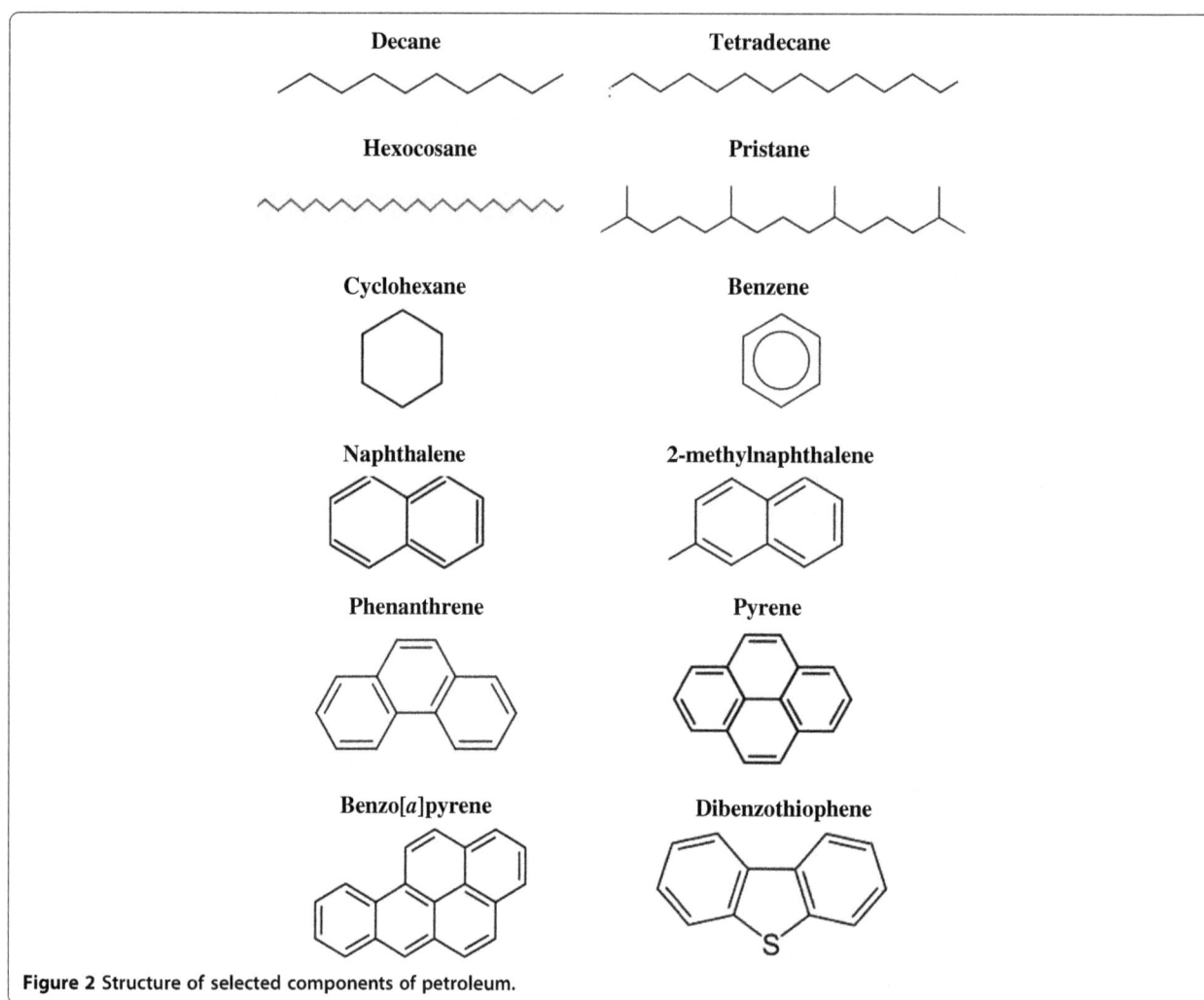

Figure 2 Structure of selected components of petroleum.

Table 1 Selected hydrocarbons and their solubility in deionised water at 25°C and hydrophobicity indicated as Log K_{ow}

Compound	Solubility (mg L^{-1})	Log K_{ow}
Decane	0.091	6.1
Tetradecane	0.009	7.2
Hexocosane	NA	14.7
Pristane	5×10^{-5}	11.4
Cyclohexane	43.0	3.2
Dibenzothiophene	2.41	4.3
Benzene	1790	2.1
Naphthalene	31.7	3.3
2-methylnaphthalene	24.6	3.9
Phenanthrene	1.29	4.5
Pyrene	0.14	5.3
Benzo[a]pyrene	0.004	6.0

The principal marine hydrocarbon degraders

The starting point in elucidating potential complex interactions involved in hydrocarbon biodegradation is to identify the microbes primarily responsible for biodegradation, and their catabolic pathways. It has long been known that the enzymatic activation of hydrocarbons by oxygen is a pivotal step in their biodegradation, and several mechanisms have been elucidated for aromatic [12,18,19] and aliphatic [12,20] compounds. However, our understanding of the catabolic processes for HMW PAHs [21] and anaerobic activation mechanisms and pathways, e.g. fumarate addition, carboxylation and O$_2$-independent hydroxylation, have emerged only recently [22-25].

The microbial response to an oil spill at sea is dependent on numerous factors, including the oil composition and degree of weathering, as well as environmental conditions, particularly temperature and nutrient concentrations. Nevertheless, there are some typical patterns; most notable is the large increase in abundance of *Alcanivorax* spp., which degrade straight-chain and branched alkanes [26-

32], followed by *Cycloclasticus* spp., which degrade PAHs [26-30,33-36].

Since the cultivation of *Alcanivorax borkumensis* [37], functional genomic, biochemical and physiological analyses have revealed the underlying basis of its success [28,38-40]. While it lacks catabolic versatility, utilising alkanes almost exclusively as carbon and energy sources, it has multiple alkane-catabolism pathways, with key enzymes including alkane hydoxylases (a non-haem diiron monooxygenase; AlkB1 and AlkB2) and three cytochrome P450-dependent alkane monooxygenases [38]. Their relative expression is influenced by the type of alkane supplied as carbon and energy source and phase of growth [38]. *Alcanivorax borkumensis* also possesses a multitude of other adaptations to access oil (e.g. synthesis of emulsifiers and biofilm formation [38]) and to survive in open marine environments (e.g. scavenging nutrients and resistance to ultraviolet light [38,40]). *Acinetobacter* spp., which are commonly isolated from oil-contaminated marine environments [41], also have a diverse array of alkane hydroxylase systems enabling them to metabolize both short- and long-chain alkanes [20,42]. For example, *Acinetobacter* strain DSM 17874 contains a flavin-binding monooxygenase, AlmA, which allows it to utilize C_{32} and C_{36} n-alkanes [43]. The *almA* gene has also been found in *Alcanivorax dieselolei* B-5 and is induced by long-chain n-alkanes of C_{22} - C_{36} [44]. A diverse array of *alk*B gene sequences, encoding alkane hydroxylase, has been detected in the environment [45,46] and in a wide range of bacteria [38,42,46], however Païssé *et al.* [47] argue that *alk*B expression may not always be a good indicator for microbial oil degradation, implying that we have not fully explored the gene diversity and/or that other hydrocarbon catabolic processes were prevalent in the environment under investigation.

In cold marine environments, the obligate alkane-degrading psychrophile, *Oleispira*, rather than *Alcanivorax* spp., are commonly associated with oil spills [29,48]; and *Alcanivorax* spp. are sometimes outcompeted by *Thalassolituus* spp. in temperate environments [34]. Such obligate hydrocarbon-degrading bacteria can constitute 90% of the microbial community in the vicinity of the oil spill and have a wide global distribution [28]. New genera of obligate alkane degraders are still being discovered, e.g. *Oleibacter* sp. [31,49], and there are likely to be many more, such as the uncharacterised Oceanospirillales strain ME113 [50], which has been detected in abundance in other oil-rich marine environments [51,52].

The role of the generalists that degrade alkanes and/or PAHs as well as non-hydrocarbons is often overlooked, yet they can constitute a significant proportion of a hydrocarbon-degrading community. For example, Buchanan and Gonzalez [53] outline eight studies in which members of the *Roseobacter* lineage, which harbours a diversity of ring-hydroxylating dioxygenases and alkane hydroxylases, increase in abundance in hydrocarbon-enriched marine waters. Other generalists, including *Acinetobacter, Marinobacter, Pseudomonas* and *Rhodococcus* spp. [54-57], contribute to hydrocarbon degradation. Sediments add to the complexity of identifying the main hydrocarbonoclastic microbes, but nearly all of the above genera are detected in the aerobic zone of marine sediments and presumed to be active in hydrocarbon degradation. It is important to recognise that within most of the genera labelled here as generalists (e.g. *Marinobacter*) there are many species, ranging from those that do not degrade hydrocarbons to specialists like *Marinobacter hydrocarbonoclasticus*, which almost exclusively utilises n-alkanes [56].

Although *Cycloclasticus* is frequently the main marine PAH-degrading microbe detected, many others from several tens of genera are known [15], and the underlying mechanisms of their interactions with, and degradation of PAHs are only beginning to be elucidated. For example, in San Diego Bay sediments, isolates able to grow on phenanthrene or chrysene were from the genera *Vibrio, Marinobacter, Cycloclasticus, Pseudoalteromonas, Marinomonas* and *Halomonas* [58]. Another marine specialist PAH degrader, named *Porticoccus hydrocarbonoclasticus*, was recently isolated [59], and strains of *Microbacterium* and *Porphyrobacter*, previously not known to be involved in PAH degradation, were isolated on benzo[*a*]pyrene after enriching for two years [60]. Based on DGGE analysis, Hilyard *et al.* [61] suggested that Planctomyces and Bacteroidetes were involved in PAH degradation, and many more species from diverse genera that are implicated in PAH degradation remain to be cultivated, particularly those growing on HMW PAHs.

Incubation of marine sediment in the presence of phenanthrene and bromodeoxyuridine (BDU), followed by analysis of BDU-labelled DNA, revealed a remarkable diversity of putative PAH degraders belonging to the genera *Exiguobacterium, Shewanella, Methylomonas, Pseudomonas, Bacteroides*, as well as Deltaproteobacteria and Gammaproteobacteria that were not closely related to cultivated organisms [62]. Some were also cultivated, including a novel *Exiguobacterium* strain, but the rest remain to be grown [62]. Similarly, stable-isotope probing (SIP) of DNA was used to identify the involvement of a novel clade of Rhodobacteraceae in biodegradation of low molecular weight (LMW) PAHs in marine algal blooms [63]. Obtaining pure cultures of the main microbes responsible for hydrocarbon biodegradation is no longer a prerequisite for their study, but it makes their investigation very much easier, allowing genomic, biochemical and physiological analyses that in turn can help to explain their *in-situ* function and interactions. It is also frequently their reliance on other microbes that prevents cultivation in the first instance, and growth in the proximity of microbes (or their diffusible products) from the same habitat [64] can be employed to improve recovery. Numerous other procedures can enhance cultivation [65], especially by increasing the bioavailability of hydrocarbons. Calvo *et al.* [66], for example, extracted extracellular polymeric substances (EPS) from *Halomonas eurihalina*, not a PAH-degrader, which enhanced the isolation of other microbes growing on PAHs.

General considerations of microbial interactions

A volume of 1 mm^3 of surface seawater, approximately equivalent to the size of a poppy-seed, contains ~600 bacteria, 150 cyanobacteria, 9 small algae, <1 protozoan [67] and ~10,000 viruses [68]. Numerous ecophsyiological investigations [69] together with modelling the co-occurrence of bacterial phylotypes [70] reveal a network of direct and indirect interactions within and between species in seawater that are vital for maintaining the microbial loop that drives marine biogeochemical cycles [71]. Some interactions exist between spatially separated species that use soluble or volatile metabolites to transmit information; while other interactions involve species in very close proximity, either as a biofilm on the same particle or physically associated to one another. Grossart [69] noted that a chain of the marine diatom, *Thalassiosira rotula*, can host up to 10^8 bacteria [72], while a single copepod can harbour up to 10^9 bacteria [73]. Surprisingly, in many studies, the attached microbiota, which is numerically equivalent to the non-attached microbiota, is removed by pre-filtration [69], and so not considered.

Microbial communities from coastal sediments vary more from one location to another than those from open waters, and have much greater community evenness [74]. Moreover, in sediments, cells are much more concentrated, resulting in a greater likelihood of interactions, which becomes even more prevalent in biofilms where cells are more densely packed. Highly productive photosynthetic microbial mats develop at the water-sediment interface. These multispecies biofilms consist of horizontally stratified layers with extremely steep gradients of light, redox potential, oxygen, sulfur species etc. The exceptionally high microbial diversity within a few microns covers a large range of metabolic groups (oxygenic and anoxygenic phototrophs, sulfate reducers, methanogens etc.) [75]. We are at an early stage in our understanding of communication mechanisms in each of these environments (open water, sediment and biofilms), where small molecules, either diffusing from cell to cell [76], or transported by vesicles [77] or via nanotubes bridging cells [78], elicit intra- and inter-species effects that could be antagonistic or beneficial.

Microbes exhibit all of the types of social behaviour (mutual benefit, selfishness, altruism and spite [79]) seen in multicellular organisms. However, it is often difficult to categorise such behaviour in complex multi-species natural environments, and so in this review we talk largely in terms of cooperation and competition, and how they are affected by hydrocarbons, and in turn influence their fate. Our knowledge gained from studying pure cultures of hydrocarbon degraders is important, but hydrocarbonoclastic bacteria rarely, if ever, function in isolation in nature. Therefore, a better understanding of crude-oil biodegradation, and thus the capability to more rationally remediate contaminated environments, requires us to consider the mechanisms of the interactions between different hydrocarbon-degrading microbes and with non-degrading organisms [27]. This review considers such interactions, with most emphasis on aerobic processes and interactions between phototrophic microalgae and hydrocarbonoclastic bacteria.

Interactions between microbes during aerobic degradation of hydrocarbons

When crude oil is added to seawater, the microbial community changes and consists of multiple co-existing species [80], which can be explained most simply by resource sharing. As indicated above, crude oil consists of a variety of chemically distinct hydrocarbons, which require specific mechanisms for activation and degradation. In seawater microcosms, each supplied with a different hydrocarbon, McKew *et al.* [34] observed that: 1) *Alcanivorax* dominated when the branched alkane, pristane, was supplied, but was not detected in other microcosms, 2) *Cycloclasticus* was dominant with most PAHs, but was undetected when fluorene was supplied, and 3) *Thalassolituus* was the dominant species when *n*-alkanes with 12 to 32 carbons were added, but was not detected when decane was the sole alkane added to seawater. Thus, it appears that the ability to be competitive in the marine / estuarine environment requires that hydrocarbonoclastic bacteria are relatively specialised. Probably the extra genetic and cellular load needed to allow bacteria to grow on a wider range of hydrocarbons would demand greater nutrient resources, making them less competitive overall, especially in oligotrophic oceans. This, in turn, requires the presence of a consortium of microbes for complete degradation of crude oil.

Competition for resources is also an important element of petroleum biodegradation: all known *Alcanivorax* spp. can degrade *n*-alkanes, yet in the above study [34] *Thalassolituus* out-competed *Alcanivorax*. Furthermore, in a follow-up study *Alcanivorax* was undetected in the microcosms to which *Thalassolituus oleivorans* had been added previously, whereas it grew in all other microcosms, though its abundance was negatively correlated with that of *Thalassolituus* [30]. The nature of this competition deserves more detailed study. It could simply be competition for common resources, such as nutrients, but the idea that *Thalassolituus* actively releases bioactive compounds to inhibit competitors must be considered. Nevertheless, as noted above, in most oil-amended experiments and environmental surveys *Alcanivorax* is the dominant microbe, so it is pertinent to consider whether it produces antibacterial molecules. *Alcanivorax jadensis* produces an antibiotic which has been termed "alcanivorone" [81], but the impact of this antibiotic on other microorganisms during hydrocarbon degradation is still unknown. In a two-species experiment, *Alcanivorax borkumensis* outcompeted *Acinetobacter venetianus*, but the filtered spent medium from *Alcanivorax borkumensis* did not influence the growth of *Acinetobacter venetianus*, rather Hara *et al.* [82] proposed the former's ability to use branched alkanes as a key factor. However, such branched alkanes are a relatively minor component of crude oil, and so the extra carbon and energy available to *Alcanivorax borkumensis* may be just one of several possible explanations.

Even when a single hydrocarbon is added to seawater microcosms, multiple species are always detected [34,36,80,83], and frequently mixed cultures outperform single species isolated from a consortium [83]. For example, the dominant benzo[*a*]pyrene-degrading bacteria from a marine enrichment were isolated, and faster degradation was seen when the three strains (*Ochrabactrum*, *Stenotrophomonas* and *Pseudomonas* spp.) were combined than when tested individually [84]. Both *Cycloclasticus* and *Pseudomonas* were abundant in estuarine waters enriched with naphthalene, but *Pseudomonas* appeared in the latter stages of the enrichment [36]. Perhaps the most compelling explanation for multiple species growing on one carbon and energy source, is that a measurable amount of the PAH is not completely oxidized to CO_2 and H_2O by one organism, resulting in oxidation products being liberated into the environment. Numerous microbes may take advantage of this so-called epimetabolome [85,86] as sources of carbon and energy [87,88].

It is becoming apparent that metabolite sharing is widespread in nature and in the laboratory as shown using auxotrophic mutants of *Escherichia coli* that complemented each other's growth by cross-feeding essential metabolites [89]. The cooperative behaviour of microbes to self-construct a functioning community is central to their success, and learning how such "microbial modules" interact will be pivotal to enhancing biotechnological processes, including the bioremediation of hydrocarbons. However, few studies have tracked the flow of hydrocarbon-derived metabolites between microbes in a consortium, and many interesting metabolites are transient and therefore difficult to detect. Pelz *et al.* [87] tracked the biodegradation of 4-chlorosalicylate through a three-member consortium of *Pseudomonas* MT1, *Pseudomonas* MT4 and *Achromobacter* MT3 using ^{13}C-labelled substrates. Analysis revealed a network of carbon sharing: strain MT1, the only member

able to degrade 4-chlorosalicylate, provided carbon skeletons to the other strains (MT3 and MT4), while they degraded toxic metabolites that inhibited strain MT1 if allowed to accumulate [87]. One of the toxic intermediates (4-chlorocatechol) was partially taken up by strain MT3 and further degraded [87]; and a proteomic and metabolite analysis of a co-culture of strains MT1 and MT3 revealed the importance of strain MT3, not only in consuming the toxic intermediate but also in reducing the degradation rate of the parent compound by strain MT1; both of which minimized the stress experienced by strain MT1 as judged by negligible detection of stress-response proteins in the mixed culture compared with the pure culture [90].

Reducing the stress imposed by metabolites may also be a typical feature in bacterial members of consortia degrading PAHs. However, our current knowledge of the catabolic routes for PAH degradation requires considerable development as diverse novel metabolites are produced by PAH-degrading microbes [43,91]; for example *Cycloclasticus* strain P1, derived from a deep-sea pyrene-degrading consortium, produced three metabolites, two of which could be identified as cyclopenta[*d,e,f*]phenanthreone and 4-phenanthrenol [83]. These metabolites are unusual as they involve the creation of a pentagonal ring suggesting a novel catabolic pathway is adopted by strain P1 [83].

Chen and Aitken [92] showed that salicylate, an intermediate produced by a *Pseudomonas* sp. pre-grown on phenanthrene as a sole source of carbon and energy, induced production of a PAH dioxygenase leading to degradation of HMW PAHs that the isolate could not use for growth [92]. The importance of metabolites as inducers of co-metabolic degradation may be significant also in natural communities.

A wide variety of fungi are known to be important in initiating biodegradation of HMW PAHs in terrestrial environments by co-metabolism using a battery of enzymes (e.g. lignin peroxidases, manganese peroxidases, laccases and epoxide hydrolases) that probably evolved to breakdown other compounds such as lignin, but which fortuitously degrade PAHs [91,93-95]. Extracellular enzymes and radicals produced by ligninolytic fungi are not constrained by slow desorption and mass transfer which limit the activity of those microbes that need PAHs to enter the cell. Moreover, these metabolites are generally more polar, and so more bioavailable, than the parent compounds [96]. An increase in bioavailability of polar metabolites was demonstrated by experiments undertaken with the white rot fungus *Bjerkandera* strain BOS55 [97]. As a pure culture it was able to degrade 74% of ^{14}C-benzo[*a*]pyrene but only produced a limited amount of ^{14}CO$_2$. The addition of soil, sludge or LMW PAH enrichment cultures led to a rapid increase in ^{14}CO$_2$ production as the polar metabolites produced by the fungus were mineralised, but only up to 34%, indicating that some [14]C- benzo[a]pyrene fungal metabolites were readily biodegraded while others persisted [97]. This has also been demonstrated with fungal-bacterial co-cultures containing the non-ligninolytic fungus *Penicillium janthinelum* VUO 10,201, which showed significant degradation of a range of HMW PAHs including pyrene and benzo[a]pyrene compared with either the fungal or bacterial species incubated alone [98]. Twenty-five percent of benzo[a]pyrene was mineralised to CO_2 over 49 days by the co-cultures, accompanied by the detection of transient intermediates [98].

Figure 3 provides a schematic illustration of some of the interactions involved in hydrocarbon biodegradation. When present in mixtures, PAHs have the capacity to negatively influence the rate and extent of biodegradation of other components in the mixture [99]. Some metabolites may not be degraded further in a particular environment (dead-end metabolites), and while they are usually less toxic than the parent compound, some are more toxic, and so it is important to monitor production of metabolites and the overall toxicity during bioremediation processes. For example, metabolites, such as pyrene-4,5-dione derived from pyrene transformation have the potential to accumulate in PAH-contaminated systems and significantly inhibit the biodegradation of other PAHs [100].

Although fungi are considered to be largely terrestrial, they have been found in marine mats [101] and it is known that many can function in saline conditions [102], but in general salt-adapted fungi have received little attention despite a potentially major role in coastal PAH degradation. The ubiquitous co-existence of bacteria and fungi in soil and sediments [103] and their known catabolic co-operation suggests that physical interactions between them may be of importance for PAH degradation. There is also evidence that filamentous fungal networks may facilitate the movement of hydrocarbon-degrading bacteria through soils and sediments – the so-called "fungal highway" – by providing continuous liquid films in which gradients of chemo-attractants can form and chemotactic swimming can take place, thus greatly increasing the accessibility to pollutants [104].

Biosurfactants and the interactions between hydrocarbon-degrading microbes and their environment

PAHs are usually found mixed with other organic pollutants (commonly petroleum and derived products) in contaminated sites, which may alter their fate and transport. This is of particular relevance when considering aged or weathered oils, in which PAHs will be less bioavailable because they are more effectively partitioned within the residual oil phase [105]. PAHs, particularly HMW PAHs, adsorb strongly to minerals and their associated organic matter [106], further diminishing their bioavailability. Owing to the low solubility and high levels of adsorption of PAHs, many microbes have evolved mechanisms to access them more readily. For

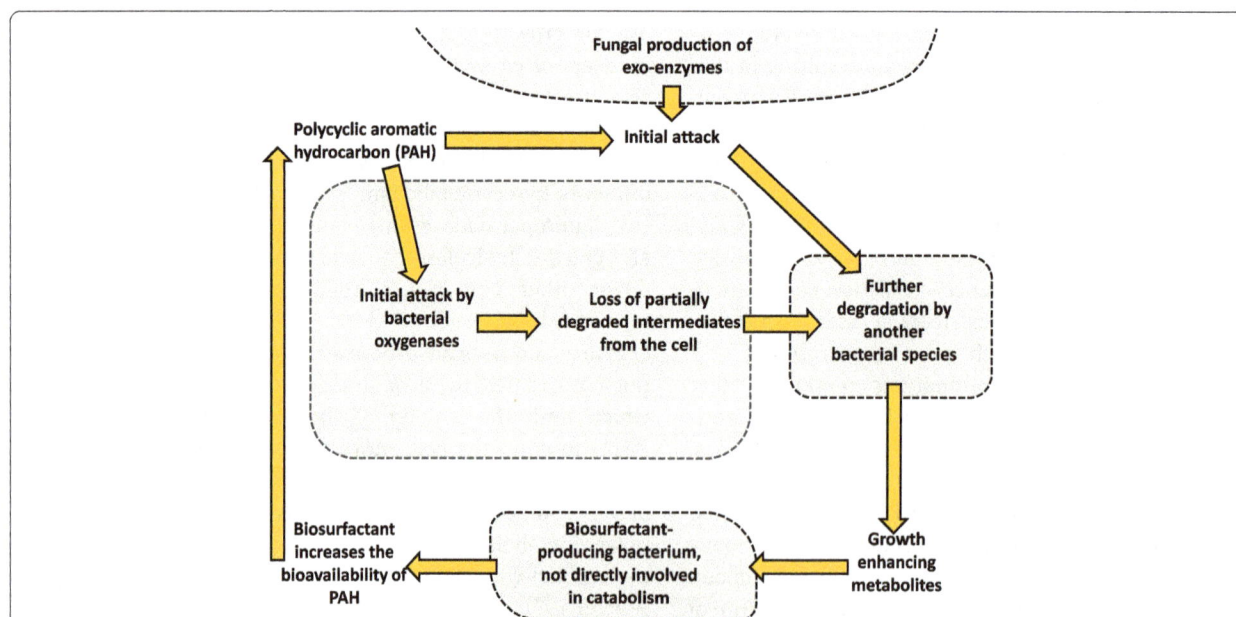

Figure 3 Schematic illustration of some of the interactions seen in a microbial consortium degrading polycyclic aromatic hydrocarbons (PAHs). Different microbial cells are represented by shaded shapes surrounded by a dashed line. Elements of these interactions have been seen in several studies (see text for details). Further complexity can be built into this simple schematic if one considers multiple PAHs invoking several pathways in one or more microbes, as well as co-metabolic degradation.

example some PAH-degrading microbes have high-affinity uptake systems that efficiently reduce the PAH concentration close to the cell surface, thereby enhancing diffusive flux [107,108]. Living on the mineral surfaces to which PAHs are adsorbed is another strategy that reduces diffusion time [109] by physically reducing the distance between cells and substrate. During such interactions, the nature of the cell surface is extremely important; for example the mycolic acids of mycobacteria and related Actinobacteria enhance cell-surface hydrophobicity which serves to encourage biofilm formation and uptake of lipophilic compounds into the cell [110]. The production of extracellular polymeric substances (EPS) has also been shown to be an important mechanism in allowing attachment of *Pseudomonas putida* to solid PAHs [111]. Vaysse *et al.* [112] showed that *Marinobacter hydrocarbonoclasticus* exhibited a major change in the proteome of cells freshly detached from hexadecane compared with those attached to hexadecane. Their mobilization may be fuelled by intracellular wax esters accumulated while growing as a biofilm on hexadecane, and the dispersed cells demonstrated a high capacity to reattach to the *n*-alkane [112]. Thus, the ability to readily attach to hydrocarbons and then move to a new patch appears to be essential for many hydrocarbon-degrading bacteria. During this process the hydrocarbon surface will be modified by excreted microbial products, and would thus be expected to lead to colonization by a succession of different microbes, however we are not aware of any studies exploring this in detail. Wouters *et al.* [113] used differential fluorescence staining to analyse a model, three-species community on the surface of PAH crystals, which looks like a promising tool to investigate their interactions and succession on the hydrocarbon surface.

Another mechanism for increasing the bioavaliabilty of these compounds is the production of biosurfactants (biological surface-active agents that have both hydrophilic and hydrophobic moieties). Some biosurfactants are known to inhibit certain microbes, while at the same time benefiting others by increasing the bioavailability of hydrophobic compounds that can serve as a carbon and energy source, thus acting as "common goods" [79]. Numerous studies have shown that production of biosurfactants, by either degrading or non-degrading microbes, is essential in enhancing the bioavailability of poorly soluble and adsorbed hydrocarbons [114,115]. Low-molecular-weight biosurfactant molecules are mostly glycolipids, including rhamnolipid, trehalose lipids and sophorolipids, or lipopeptides such as surfactin, gramicidin S, and polymyxin [114,115]. High molecular weight EPS can also act as a biosurfactant, and represents a heterogeneous range of polymers composed of polysaccharides, proteins, lipopolysaccharides, lipoproteins or complex mixtures of these biopolymers [114,115]. Biosurfactants preferentially partition at the interface between polar and apolar molecules (e.g. hydrocarbons and water), producing micro-emulsions which in many cases enhance bioavailability and desorption of the hydrocarbon [115].

McKew *et al.* [30] demonstrated that the addition of *Alcanivorax borkumensis* to seawater microcosms containing crude oil, increased PAH-degradation rates despite the fact that *A. borkumensis* does not mineralise PAHs. *A. borkumensis* is known to produce biosurfactants, which enhance uptake of alkanes, its main source of carbon and energy [38]. It is probable that such biosurfactants produced by *A. borkumensis* fortuitously increase the availability of PAHs thereby enhancing their biodegradation by other microbes in the seawater [30]. The release of such "common goods" may benefit *A. borkumensis* by reducing the concentration of stress-inducing PAHs, however those PAH-degraders will be competing for nitrogen and phosphorous that are commonly limiting nutrients in petroleum-contaminated environments. Furthermore, the biosurfactants may benefit other alkane degraders competing directly with *A. borkumensis* for alkanes. *Rhodanobacter* strain BPC1 from an eight-strain consortium degrading benzo[*a*]pyrene in a mixture of diesel fuel components, was found to be the pivotal organism in making benzo[*a*]pyrene ~500 times more soluble, thus enhancing its degradation [116]. Strain BPC1 was unable to grow on the mixture, but grew in the presence of the other microbes, indicating that it was probably utilizing metabolites produced by other consortium members [116]. Similarly, the addition to seawater of EPS from *Rhodococcus rhodochrous*. S-2, that serves to protect this strain from aromatic-hydrocarbon-induced stress, enhanced crude oil degradation and stimulated the growth of *Alcanivorax* and especially *Cycloclasticus* spp. [117]. Although *Cycloclasticus* spp. grow in pure culture, they are frequently difficult to maintain, which together with the above observations [30,117], suggests that in nature they may typically take advantage of biosurfactants produced by other microbes.

Biosurfactants may also serve an antagonistic role – they are after all important virulence factors in many pathogens – and their effects will be dose- and species-dependent. Rhamnolipid generally enhances hydrocarbon bioavailability and degradation [30,118], but Shin *et al.* [119] reported that it inhibited degradation of phenanthrene by a two-species consortium of *Sphingomonas* and *Paenibacillus* sp., even though in pure culture the rhamnolipid inhibited only *Sphingomonas* sp. It was therefore suggested that the increased stress caused by the solubilized phenanthrene, or the rhamnolipid in the presence of solubilized phenanthrene, was responsible for inhibition of *Paenibacillus* sp. It is also important to consider the potential synergistic role of multiple biosurfactants. Rambeloarisoa et al. [120] studied an eight-strain microbial consortium from the French coast, and found that biosurfactants produced by a pure strain did not emulsify crude oil, whereas those produced by the whole bacterial community did emulsify oil and led to rapid hydrocarbon degradation [120]. The extent to which such multi-species synthesis of biosurfactants may be coordinated remains to be discovered. Microbial, petroleum and clay interactions are important but very poorly understood. Chaerun et al. [121], for example, showed that montmorillonite and kaolinite enhanced growth on heavy oil, acting as supports for microbes producing EPS, as well as buffering the pH. Degradation of adsorbed PAHs involves specific adaptations that are still not well understood, and some microbes specialise in accessing and degrading adsorbed PAHs [107]. Vacca et al. [122] showed that none of the 25 soil strains isolated with non-sorbed phenanthrene could mineralise humic-acid sorbed phenanthrene (HASP), whereas all three strains that were enriched on HASP were proficient at mineralising it, clearly indicating that different capacities are needed for the biodegradation of adsorbed PAHs.

Microbial interactions during anaerobic degradation of hydrocarbons

Biodegradation of hydrocarbons in anoxic marine sediments is slower than in oxic zones, and it is generally assumed that the primary mechanism of hydrocarbon degradation even in marine sediments is aerobic respiration [123]. Despite the absence of oxygen to activate hydrocarbons, other mechanisms [124] can lead to the initiation of their degradation by a wide range of anaerobic species utilising diverse terminal electron acceptors [124]. In the environment, anaerobic hydrocarbon biodegradation is most likely to involve syntrophic consortia. Conversion of n-hexadecane to methane in an anaerobic enrichment culture was shown to involve a consortium of microorganisms, which on the basis of phylogenetic affiliation had the following putative phenotypes: syntrophs belonging to the Syntrophaceae (called Syntrophus but probably Smithella [125]) that convert n-hexadecane to acetate, hydrogen and CO_2; methanogens that convert acetate to methane and CO_2; other methanogens that convert hydrogen and CO_2 to methane; and a Desulfovibrio sp. that may couple hydrogen and CO_2 consumption with sulfate reduction [126]. However, a fermentative, syntrophic role for Desulfovibrio sp. must be considered given its metabolic flexibility [127]. A methanogenic consortium with a remarkably similar structure was also found to degrade toluene [128]. Such microbial teamwork is common in the anaerobic mineralisation of structurally complex compounds. The syntrophic association is important because the methanogens lower the concentrations of hydrogen and acetate, which makes the breakdown of the alkane energetically favourable. It will be

important to elucidate the precise nature of such interactions involved in the thermodynamically challenging anaerobic degradation of hydrocarbons, particularly identifying the microbes responsible for the initial activation and their mode of action [129]. The extremely high level of enrichment in methanogenic hydrocarbon-degrading consortia provides strong evidence [125,126] that Smithella spp. play this role. Better means of identifying and tracking intermediate metabolites will also be essential to better understanding the mechanism of these closely coupled syntrophic consortia.

It is important to consider that in many environments a gradient of oxygen concentrations can be found, with consequent microbial adaptations to a microaerobic lifestyle. Benzene degradation, for example, has been shown to occur at 0.05 mg l^{-1} of oxygen [130]. Moreover, aerobes and anaerobes can co-exist in chemostats [131,132]. For example, the strict aerobe, Comamonas testosteroni, and strict anaerobe, Methanosarcina barkeri, grew together, with the aerobe consuming the oxygen and maintaining it at a sub-inhibitory concentration for the methanogen [132]. Similar mixed cultures were detected in a benzene-contaminated aquifer [133], but the nature of the interaction in situ remains to be elucidated. Diurnal fluctuation in photosynthetically derived oxygen is an important consideration in coastal biofilms, and sequential aerobic-anaerobic hydrocarbon degradation may be an important mechanism. For instance, Chayabutra and Ju [134] investigated the sequential degradation of n-hexadecane by Pseudomonas aeruginosa using aerobic resting cells in the initial aerobic mineralization and inducing nitrate-reducing conditions for subsequent anaerobic degradation of oxidized metabolites. Providing oxic-anoxic transitions for the treatment of oily sludge proved as effective as oxic conditions alone in the degradation of PAHs by a microbial community dominated by Pseudomonas spp. [135]. Rocchetti et al. [136] also compared microbial degradation of hydrocarbons under both oxic and anoxic conditions in addition to sequential oxic-anoxic treatment in microcosms containing contaminated sediments. They reported that hydrocarbon degradation was significantly enhanced via sequential anaerobic-aerobic degradation involving sulfate-reducing bacteria in the anaerobic step, compared to degradation under either aerobic or anaerobic conditions. A more thorough review of this topic that describes other outcomes as well as the effect of the starting conditions (oxic or anoxic) is provided by Cravo-Laureau et al. [137].

Phototroph-heterotroph interactions

Marine phototrophs (primarily eukaryotic microalgae and cyanobacteria) contribute half the Earth's primary production and half of the oxygen liberated to the atmosphere [138]. However, they do not exist in isolation, and their phycosphere (loosely defined as the zone around algal cells in which bacteria feed on algal products) constitutes an important habitat that is colonised by an abundant and diverse community of heterotrophic bacteria [72,139]. Bacteria are also found living inside microalgal cells - many with unknown function [140]. The composition of free-living marine microbial communities is frequently very different from those attached to microalgae [141], with certain groups often preferring the attached lifestyle [142] and showing higher levels of activity [143]. Moreover, different species of microalgae host distinct bacterial communities that change with time and environmental conditions [72,144]. However, there is likely to be a large spectrum of bacterial heterotroph-phototroph specificity [145], and certainly many attached bacteria can also live in the absence of a microalgal or cyanobacterial host [146]. While antagonistic interactions occur between marine phototrophs and their attached microbiota [147,148], mutualistic interactions are common, with the host supplying carbon and energy sources [149], as well as potential protection from desiccation and grazing via their EPS; while the bacteria have been shown to provide iron [150], haem [151], vitamin B12 [152], to consume oxygen [153] and provide protection from reactive oxygen species [154]. Symbiotic cyanobacteria supply fixed nitrogen to diatoms [155] and other algae and protists [156], and heterotrophic N_2-fixing bacteria may also be important in interactions with microalgae, as evidenced by the abundance of alphaproteobacterial diazotrophs in seawater size fractions of >10 μm [157]. Attached bacteria can affect microalgal morphogenesis [158], the composition of their EPS [159] and enhance aggregate formation [160]. Indeed, many microalgae function less efficiently or do not even grow as axenic cultures [161]. Bruckner et al. [162] showed that a complex network of chemical cues, including amino acids and EPS, may be involved in regulation of diatom-bacteria biofilms. The variety of metabolites released from both microalgal and bacterial cells is immense [163], and dissecting out those that are important or essential for nurturing specific or general interactions is a major task for the marine biochemist.

Such heterotroph-phototroph interactions are of direct relevance to hydrocarbon degradation, not least because oil has most environmental impact where it floats on the sea surface and especially intertidal areas where microalgal biofilms are usually dominant. Although the water-soluble fraction from oil was shown to reduce the abundance of marine phytoplankton (primarily *Prochlorococcus*), the effect on coastal planktonic diatoms was stimulatory for small (<20 μm) species and either inhibitory or stimulatory depending on the concentration for larger diatoms [164]. Many marine phototrophs can withstand high concentrations of crude oil, and some cyanobacteria appear to accumulate hydrocarbons

without degrading them in inter-thylakoid spaces [165]. Coastal biofilms are particularly resistant to oil pollution, which can even result in enhanced photosynthetic activity [166]. The cyanobacterial genus, *Oscillatoria*, is particularly common in oil-polluted mats [167-169]. Diatoms too are often abundant in diverse oil-polluted sediments, including a chronically oil-polluted lagoon in which diatom chloroplast 16S rRNA gene sequences constitute up to 21% of the sequences from the surface sediment [170].

Although there are many reports of hydrocarbon degradation directly by microalgal species, primarily chlorophytes and diatoms (as summarised by Prince [17]), but also cyanobacteria, it is questionable whether microalgae would be competitive with specialist aerobic heterotrophs, and they are probably involved only in partial oxidation [171-174]. For example, Todd et al. [174] showed that the chlorophyte, *Chlorella vulgaris* slowly metabolized naphthalene to 1- naphthol. However, other evidence implicates photo-(mixo)trophs in complete hydrocarbon oxidation. For example, fatty acid analysis of cyanobacteria grown with and without hydrocarbons, suggests that they are incorporated into biomass [175]. Also, Lei et al. [176] reported that six strains from diverse microalgal genera, including *Chlamydomonas, Chlorella, Scenedesmus, Selenastrum and Synechocystis*, could degrade 34 to 100% of the supplied pyrene in 7 days.

It is difficult to obtain axenic cultures of microalgae, and so in some reports of more complete and rapid hydrocarbon degradation by phototrophs the degradation could have been performed wholly or partly by associated microbes [177]. For example, the medium used to check for the absence of heterotrophic bacteria in cyanobacterial cultures that degraded 50% of hexadecane and up to 90% of PAHs in 10 days [178] contained peptone-glucose that would not have allowed *Alcanivorax* spp. to grow, and so they would evade detection. De Oteyza et al. [179] have shown that while cyanobacterial filaments surround oil droplets, biodegradation was most likely due to associated heterotrophic bacteria. Cohen [168] found rapid degradation in cyanobacterial mats, whereas pure cyanobacterial cultures could not degrade hydrocarbons. Therefore, while cyanobacteria-dominated mats can degrade hydrocarbons, it is the heterotrophic bacteria that are mainly responsible for the degradation [166,168,177,180-182]. However, it is important to determine the extent to which microalgal biodegradation of hydrocarbons and their metabolites [173,183] is relevant in the marine environment.

Phototroph-heterotroph interactions are very important to hydrocarbon biodegradation. Many algae produce hydrocarbons [184,185], and nearly all produce the volatile hydrocarbon, isoprene [186,187], which could serve to sustain hydrocarbon-degrading communities in the absence of an oil spill [188], and may explain why hydrocarbon-degrading bacteria, such as *Alcanivorax* spp., are often associated with micro-[189] and macro-[190] algae. PAHs adsorb to the cell surface of marine microalgae at relatively high concentrations [191], and have been shown to be transported by phytoplankton cells from the surface layers of the Southern Baltic to the sea floor [192]. Thus, exogenous hydrocarbons may also support hydrocarbonoclastic bacteria attached to algae. Other bacterial genera that have many species with the capacity for hydrocarbon degradation, such as *Marinobacter* and *Roseobacter*, are also commonly associated with algae [144,160,168,189,193]; however both are nutritionally versatile and so could use diverse sources of carbon and energy supplied by their hosts. Gutierrez *et al.* [59] isolated a new species of specialist PAH degrader, named *Porticoccus hydrocarbonoclasticus*, from the marine dinoflagellate *Lingulodinium polyedrum*, and also used quantitative PCR to show that it was associated with other phytoplankton.

Figure 4 shows some of the means by which algae and associated bacteria collectively interact, as discussed previously. These associations may be enhanced by the presence of hydrocarbons; for example oxygen liberated by photosynthesis is likely to be very important in activating hydrocarbons and serving as an electron acceptor in aerobic respiration [75,168,179,194]. In turn, the locally increased concentration of CO_2 produced by the heterotrophs, will generally allow enhanced photosynthesis. Abed [194] studied the interactions between cultivated cyanobacteria and aerobic heterotrophic bacteria in the degradation of hydrocarbons, showing an increase in growth of the bacteria and enhanced hydrocarbon degradation in the presence of cyanobacterial organic exudates. Similarly, extracts from a chlorophyte enhanced benzo[*a*]pyrene degradation by a *Mycobacterium* sp. and *Sphingomonas* sp. [195]. The consortium constructed by Tang *et al.* [196] consisted of an

alga, *Scenedesmus obliquus* GH2, that could not degrade petroleum hydrocarbons but promoted the degradation of both aliphatic and aromatic hydrocarbons (especially HMW PAHs) by the added bacterial members of the consortium. In this interesting study it was also observed that when a unialgal, but non-axenic, culture of *Scenedesmus obliquus* GH2 was added to the consortium, degradation was inhibited, implying that unidentified non-hydrocarbon-degrading bacteria associated with the alga outcompeted the added oil-degrading bacteria.

The organic compounds produced by algae may influence hydrocarbon degradation in different ways. Algal EPS could serve to emulsify hydrocarbons as suggested by Cohen [168]. Additionally, EPS together with excreted amino acids and sugars provide a source of carbon and energy for associated bacteria (as well as the microbial community beyond the phycosphere) [197]. It is not known what effect algal dissolved organic matter (DOM) might have on hydrocarbon biodegradation, but in other environments the addition of organic compounds led to both increased and decreased hydrocarbon consumption [198-200]. Such simple organic compounds significantly enhance microbial populations, a proportion of which may also have the capacity to degrade hydrocarbons. (This is the explanation often given for the success of phytoremediation of polluted land, where plant-root exudates stimulate microbial growth). Alternatively, the stimulated populations may out-compete hydrocarbon-degrading bacteria, especially obligate hydrocarbonoclastic species. In summary, we do not yet have a mechanistic explanation for the above [194-196] observations of stimulation of hydrocarbon degradation by algal exudates. The possibility should also be considered that algae produce secondary metabolites to specifically nurture hydrocarbonoclastic

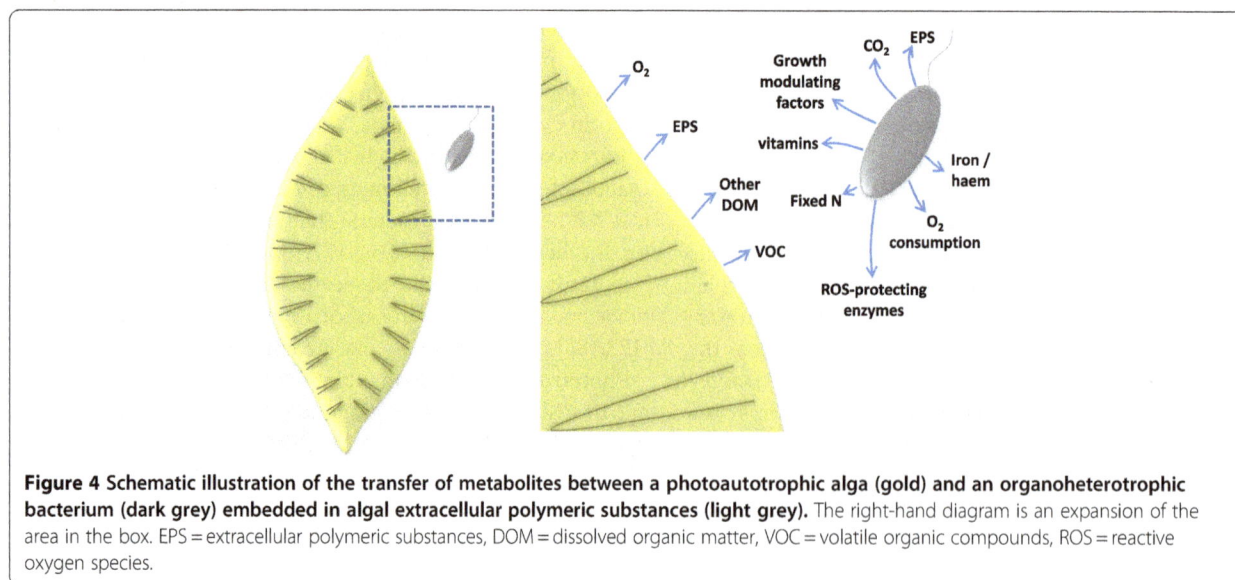

Figure 4 Schematic illustration of the transfer of metabolites between a photoautotrophic alga (gold) and an organoheterotrophic bacterium (dark grey) embedded in algal extracellular polymeric substances (light grey). The right-hand diagram is an expansion of the area in the box. EPS = extracellular polymeric substances, DOM = dissolved organic matter, VOC = volatile organic compounds, ROS = reactive oxygen species.

bacteria, as removal of stressful hydrocarbons would bene-fit the host.

Hydrocarbon-degrading bacteria could supply the algae with the benefits outlined in Figure 4. *Alcanivorax* and *Marinobacter* spp., for example, are well adapted to sequestering iron [40,150]. Most importantly, hydrocarbonoclastic bacteria will decrease the concentration and toxicity of hydrocarbons in the immediate vicinity of algal cells. There are several studies that demonstrate the benefit of such a co-culture; for example Abed [194] showed that the cyanobacterium *Synechocystis* sp. grew best in the presence of aerobic hydrocarbon-degrading bacteria and hexadecane. *Alcanivorax* spp., which have been shown to inhabit the phycosphere of algae such as the dinoflagellate *Gymnodinium catenatum* [189], can reduce the lag phase and enhance the maximum chlorophyll fluorescence of the cyanobacterium *Prochlorococcus* by means of diffusible molecules [201].

Nitrogen often becomes limiting in petroleum-contaminated environments [202], yet there are few studies on the impact of hydrocarbons on fixation of atmospheric nitrogen and in turn how this may influence biodegradation. Oil had little effect on nitrogen fixation in Arctic marine sediments [203] and marine-sediment microcosms [204], and had variable impact in salt-marsh sediments [205]. However, nitrogen limitation in other oil-polluted habitats can be overcome by dinitrogen fixation [206,207]. Musat *et al.* [204] demonstrated that cyanobacteria were the most active dinitrogen fixers in nitrogen-limited pristine and oil-polluted marine sediments reconstructed in aquaria, by combining acetylene-reduction assays with light–dark incubations and sequence analysis of expressed *nif*H genes. The capacity to fix atmospheric nitrogen and solubilise phosphate should be advantageous for microbes that rely largely on a diet of hydrocarbons. Also, the ability to scavenge iron, a major component of hydrocarbon-activating oxygenases, would be important in oligotrophic environments. There was little data suggesting that these capabilities may be widespread in hydrocarbon degraders until two recent studies showed that many hydrocarbon-degrading bacterial isolates potentially [208] or actually [190] fix nitrogen, and 84% of isolates produced sidero-phores to access iron and 51% solubilised phosphate [208].

Grazers and viruses

In order to better understand natural attenuation and determine the potential for bioaugmentation of oil-contaminated marine environments, it is essential to understand the effect of oil on grazers [27]. Grazing organisms play a role in the transfer of hydrocarbons or their metabolites to higher trophic levels, and also affect degradation rates, both positively and negatively [209]. It is pertinent to ask whether hydrocarbonoclastic bacteria

forming biofilms on oil droplets are grazed by protozoa (e.g. ciliates and flagellates) or meiofauna (e.g. nema-todes, copepods and ostracods) to the same extent as other bacteria. The grazer would have to avoid co-ingestion of oil or subsequently tolerate or expel it. Stoeck and Edgcomb [209], summarising the rather scant literature on this topic, state that defence mechanisms include release of protective mucous and complexation of hydrocarbons with lipids. Many grazers are resistant to crude-oil components, for example Gertler *et al.* [210] found an abundant, fluctuating protozoal community alongside an abundant, inversely fluctuating and active hydrocarbon-degrading bacterial community in a marine mesocosm. The main protozoal species changed over time, with selection in the oiled mesocosm of *Scuticocili-tia* spp. initially and *Euplotes* spp. later, both of which had been found by other researchers in polluted environments [210]. Also, Dalby *et al.* [211] concluded that cosmopolitan generalist protozoa could effectively graze bacteria in crude-oil amended microcosms. In the presence of oil, the flagellate, *Paraphysomonas foraminifera*, became dominant (48-82% of 18S rRNA phylotypes), keeping the bacterial population below 10^7 cells ml^{-1}.

Grazing frequently leads to enhanced rates of organic matter mineralisation by releasing nutrients and/or maintaining heterotrophic populations in exponential growth phase [209]. However, there are few studies investigating the effects on hydrocarbon mineralisation, and the outcomes are sometimes conflicting, perhaps as a consequence of environmental differences or technical approaches. Using eukaryote inhibitors, Tso and Taghon [212] showed that grazing had a beneficial effect on naphthalene degradation in estuarine sediments, possibly because the protozoa selectively grazed those bacteria that were not attached to naphthalene, thus allowing attached naphthalene-degrading bacteria to flourish by reducing competition for nutrients and other resources. Mattison and Harayama [213] reported a four-fold increase in toluene mineralization by a *Pseudomonas* sp. in the presence of the bacterivorous flagellate *Heteromita globosa* than in its absence, though *Pseudomonas* numbers reduced to 60% of the original biomass in the presence of the flagellate. In this case it was suggested that, in addition to selectively grazing the less-active bacteria, *H. globosa* enhanced naphthalene degradation by excreting growth-stimulating metabolites or ammonium and phosphate. Rogerson and Berger [214] proposed that stimulation of crude-oil degradation by *Colpidium col-poda* may additionally have been due to increasing oxygen flow caused by the swimming action of the ciliate and/or production of oil-emulsifying mucus that may have enhanced hydrocarbon bioavailability. Stoeck and Edgcomb [209] provide examples of other indirect bene-fits of protozoa to oil biodegradation. In contrast,

Näslund *et al.* [215] found that meiofaunal grazers reduced naphthalene degradation in marine sediments. By reducing the number of larger grazers, oil pollution can result in microalgal blooms [216,217]. Although the benefits of phototrophs have been outlined earlier, such a bloom may be disadvantageous because of algal competition for nutrients with hydrocarbon-degrading bacteria. More systematic studies investigating the role of different types of grazers under defined scenarios with varying levels of complexity are required to provide a clearer understanding of the nature of the interactions involved and the impact of grazers on hydrocarbon degradation.

Bacteriovorax spp. are obligate predatory bacteria that prey on other bacteria, but information regarding their potential role in oil-degrading communities is limited and conflicting. During hydrocarbon-degradation mesocosm experiments, *Bacteriovorax* were detected in microbial communities between days 21 and 35 [218] and days 21 and 28 [210]. However, in a similar experiment *Bacteriovorax* represented 11% of the bacterial community at day 0, but by day 15 none were detected [219].

Bacteriophages might also affect microbial oil degradation either positively or negatively. Pollutants can induce prophage [27,220], and the resultant bacteriophage-induced lysis of bacterial cells, unlike grazing, releases all cellular components back into the marine environment for reuse by other microbes. Such a phage-driven microbial-loop was implicated in enhancing total organic carbon removal in reactors treating oil-contaminated waters [221]. Rosenberg *et al.* [221] found extremely high densities of bacteria and phages in these reactors, and they isolated phages, including one that infected a strain of *Marinobacter* cultured from the same location. Using the GeoChip-based high-throughput microarray, Lu *et al.* [222] observed significantly higher numbers of bacteriophage replication genes in the Deepwater Horizon deep-sea oil plume samples than in non-plume control samples collected at the same depth. Because previous studies had reported a significant increase in biomass in the plume samples [223], it was surmised that the bacteriophages provided a constant supply of nutrients needed for bacterial hydrocarbon degradation through phage-mediated biomass turnover. Furthermore, phages, together with various mobile genetic elements, are important in dissemination of valuable genetic material, including hydrocarbon-degradation genes and in the generation of new catabolic pathways via lateral gene transfer [224,225].

A brief overview of microbial interactions with macrofauna and plants

There exists substantial evidence that bioturbation by larger fauna has a significant impact on the degradation of petroleum hydrocarbons in oil-contaminated sediments. By selective-removal experiments, Cuny *et al.* [226] found

that the marine polychaete, *Nereis diversicolor*, increased the abundance of bacteria known to play important roles in aerobic hydrocarbon degradation. It was suggested that digestive solubilizers produced by the polychaete via feeding might have enhanced the bioavailability of the hydrocarbons and/or burrowing activities enhanced oxygen transfer to hydrocarbon-degrading bacteria. Gilbert *et al.* [227] had demonstrated previously that the digestive process of the polychaete *Nereis virens* altered the composition and reduced the concentration of ingested aliphatic hydrocarbons. It was therefore surmised that surfactant production in the gut of the worm led to these changes in the hydrocarbons. In addition to aerating deeper sediments, burrowing animals may transport pollutants or degrading bacteria deeper into sediments or return buried pollutants back to the surface [228,229].

Plant roots oxygenate their rhizosphere and provide sugars and other compounds that stimulate microbial activity; and ultimately their major polymers, such as lignin, upon entering the soil will be attacked by a suite of (fungal) extracellular enzymes, which will initiate fungal degradation of PAHs. Phytoremediation, which exploits these features, has been employed in terrestrial soils, but only trials have been carried out in coastal zones [230]. For example, Lin and Mendelssohn [231] investigated both tolerance limit to crude oil and phytoremediation potentials of the salt-marsh grass *Spartina patens*. It could survive at concentrations up to 320 mg oil g^{-1} dry sediment, and at oil doses of between 40 and 160 mg g^{-1} oil degradation was significantly higher than in unplanted sediments. The rhizosphere of mangrove species were shown to harbour a variety of bacteria that both degraded oil and potentially stimulated plant growth [208]. As with algal-bacterial interactions, a more complete understanding of the molecular interactions between plants and associated bacteria and fungi will only improve the possibility of this technology being rationally applied to remove oil in the coastal zone [232].

Concluding remarks and prospects for using interacting microbes for oil-spill cleanup

There has been a lot of debate about the validity of bioaugmentation, specifically supplementing the environment with microbes to enhance biodegradation or detoxification of pollutants. Examples of success and failure abound. The key reasons for failure include: use of a single organism, focus on biodegrading strains only, microbes not adapted to the environment, inadequate dispersion/access to the pollutant, lack of protection (e.g. from grazers), other factors limiting biodegradation (e.g. nutrients). Now, there is overwhelming evidence that using a consortium of microbes rather than a single strain greatly enhances the chances of successful bioaugmentation.

A well designed microbial consortium will have complementary catabolic pathways, as well as the potential to disperse and make the hydrocarbons readily bioavailable. Gallego *et al.* [233], for example, demonstrated the vastly superior efficacy of a designed four-species consortium over individual species in the bioremediation of oil-tank sludge. A six-species manufactured consortium, including a fungus, *Fusarium* sp., mineralised 78% of the PAHs from soil in 70 days, compared with negligible mineralization in an uninoculated control, and much lower degradation with single-species inocula [234]. Successful bioaugmentation is also a function of the competition between the introduced microorganisms and the autochthonous microbial community, and the study of this biotic pressure requires more attention.

Despite the improved biodegradation of hydrocarbons in bacterial co-cultures with microalgae, there have been few attempts to exploit this in the remediation of petroleum contamination. Munoz and Guieysse [235] describe *ex-situ* bioremediation using photobioreactors, but for marine pollution an *in-situ* approach is preferred owing to the large volume of polluted material. The critical phase of crude-oil contamination of the shoreline is the first few days. If the oil is not rapidly degraded then it will start to sink into the sediment where it can remain for decades. While it is true that hydrocarbonoclastic microbes will emerge from the native community, this process may take days. Thus, there is a role for bioaugmentation to bolster the *in-situ* hydrocarbon-degrading community in this crucial period. The potential to apply relevant hydrocarbonoclastic bacteria with or without associated microalgae should be investigated further.

Clearly there are many fundamental gaps in our understanding of microbial interactions; however, by a combination of reductionist experiments through to modelling the co-occurrence of microbial communities on a large scale, the field is advancing. The nature of interactions can be captured by single-cell and *in-situ*-metabolism imaging techniques such as Raman-FISH [236] and Nano-SIMS [237], as well as co-localisation studies using Magneto-FISH [238]. The requisite tools are constantly being developed, such that we can characterise and analyse in more depth the function of diverse components of DOM or the epimetabolome, as well as the volatile organic compounds, including the all-important signalling molecules. It is essential to make greater sense of metabolomics and protein and gene expression analyses in microbial consortia via the tools of systems biology [86,239]. A better understanding of microbial community metabolic networks will arise from recreating natural consortia in which modifications can be made a gene at a time. The result will be a clearer picture of microbial interactions and thus the functioning of global biogeochemical cycles, with potential practical offshoots, not least a more rational approach to the remediation of marine pollution.

Competing interests
The authors declare that they have no competing interests.

Authors' contributions
TJM conceived the review and wrote the first draft. All authors contributed to the writing and read and approved the final manuscript.

Acknowledgements
TJM is grateful to members of the European Community Sixth Framework project FACEiT (project n°018391) for many interesting discussions. GOS and BDF would like to acknowledge the Petroleum Technology Development Fund (PTDF), Nigeria, and NERC, UK, respectively for funding their PhD Programmes. Particular thanks go to Professors Ken Timmis and Graham Underwood for inspirational and entertaining discussions about microbial interactions.
Invited review for Aquatic Biosystems.

References

1. Halpern BS, Walbridge S, Selkoe KA, Kappel CV, Micheli F, D'agrosa C, Bruno JF, Casey KS, Ebert C, Fox HE, Fujita R, Heinemann D, Lenihan HS, Madin EM, Perry MT, Selig ER, Spalding M, Steneck R, Watson R: A global map of human impact on marine ecosystems. *Science* 2008, 319:948–952.
2. National Research Council (NRC): *Oil in the Sea III: Inputs, Fates, and Effects.* Washington, D.C: National Academy Press; 2003.
3. Crone TJ, Tolstoy M: Magnitude of the 2010 Gulf of Mexico oil leak. *Science* 2010, 330:634.
4. ITOPF: *Oil Tanker Spill Statistics: 2005.* London: The International Tanker Owners Pollution Federation Ltd; 2006.
5. de la Huz R, Lastra M, Junoy J, Castellanos C, Vieitez JM: Biological impacts of oil pollution and cleaning in the intertidal zone of exposed sandy beaches: Preliminary study of the "Prestige" oil spill. *Estuarine Coastal Shelf Sci* 2005, 65:19–29.
6. Mason C: *Biology of Freshwater Pollution.* Essex: Prentice Hall; 2002.
7. Carls MG, Rice SD, Hose JE: Sensitivity of fish embryos to weathered crude oil: I. Low level exposure during incubation causes malformations, genetic damage, and mortality in larval Pacific herring (*Clupea pallasi*). *Environ Toxicol Chem* 1999, 18:481–493.
8. Heintz RA: Chronic exposure to polynuclear aromatic hydrocarbons in natal habitats leads to decreased equilibrium size, growth, and stability of pink salmon populations. *Integr Environ Assess Manag* 2007, 3:351–363.
9. Davies JM, Addy JM, Blackman RA, Blanchard JR, Ferbrache JE, Moore DC, Somerville HJ, Whitehead A, Wilkinson T: Environmental effects of the use of oil-based drilling muds in the North Sea. *Mar Pollut Bull* 1984, 15:363–370.
10. Venosa AD, Campo P, Suidan MT: Biodegradability of lingering crude oil 19 years after the Exxon Valdez oil spill. *Environ Sci Technol* 2010, 44:7613–7621.
11. Marshall AG, Rodgers RP: Petroleomics: the next grand challenge for chemical analysis. *Accounts of Chemical Research* 2004, 37:53–59.
12. Harayama S, Kishira H, Kasai Y, Shutsubo K: Petroleum biodegradation in marine environments. *J Mol Microbiol Biotechnol* 1999, 1:63–70.
13. van Hamme JD, Singh A, Ward OP: Recent advances in petroleum microbiology. *Microbiol Mol Biol Rev* 2003, 67:503–549.
14. Suess E: Marine cold seeps. In *Handbook of Hydrocarbon and Lipid Microbiology.* Edited by Timmis KN, McGenity TJ, van der Meer JR, de Lorenzo V. Berlin Heidelberg: Springer; 2010:187–203.
15. Prince RC, Gramain A, McGenity TJ: Prokaryotic hydrocarbon degraders. In *Handbook of Hydrocarbon and Lipid Microbiology.* Edited by Timmis KN, McGenity TJ, van der Meer JR, de Lorenzo V. Berlin Heidelberg: Springer; 2010:1671–1692.
16. Al-Mailem DM, Sorkhoh NA, Al-Awadhi H, Eliyas M, Radwan SS: Biodegradation of crude oil and pure hydrocarbons by extreme halophilic archaea from hypersaline coasts of the Arabian Gulf. *Extremophiles* 2010, 14:321–328.

17. Prince RC: **Eukaryotic hydrocarbon degraders**. In *Handbook of Hydrocarbon and Lipid Microbiology*. Edited by Timmis KN, McGenity TJ, van der Meer JR, de Lorenzo V. Berlin Heidelberg: Springer; 2010:2065–2078.

18. Pérez-Pantoja D, González B, Pieper DH: **Aerobic degradation of aromatic hydrocarbons**. In *Handbook of Hydrocarbon and Lipid Microbiology*. Edited by Timmis KN, McGenity TJ, van der Meer JR, de Lorenzo V. Berlin Heidelberg: Springer; 2010:799–837.

19. Fuchs G, Boll M, Heider J: **Microbial degradation of aromatic compounds - from one strategy to four**. *Nat Rev Microbiol* 2011, 9:803–816.

20. Rojo F: **Degradation of alkanes by bacteria**. *Environ Microbiol* 2009, 11:2477–2490.

21. Kanaly RA, Harayama S: **Advances in the field of high-molecular-weight polycyclic aromatic hydrocarbon biodegradation by bacteria**. *Microb Biotechnol* 2010, 3:136–164.

22. Widdel F, Grundmann O: **Biochemistry of the anaerobic degradation of non-methane alkanes**. In *Handbook of Hydrocarbon and Lipid Microbiology*. Edited by Timmis KN, McGenity TJ, van der Meer JR, de Lorenzo V. Berlin Heidelberg: Springer; 2010:909–924.

23. Tierney M, Young LY: **Anaerobic degradation of aromatic hydrocarbons**. In *Handbook of Hydrocarbon and Lipid Microbiology*. Edited by Timmis KN, McGenity TJ, van der Meer JR, de Lorenzo V. Berlin Heidelberg: Springer; 2010:925–934.

24. Vogt C, Kleinsteuber S, Richnow H-H: **Anaerobic benzene degradation by bacteria**. *Microb Biotechnol* 2011, 4:710–724.

25. Zedelius J, Rabus R, Grundmann O, Werner I, Brodkorb D, Schreiber F, Ehrenreich P, Behrends A, Wilkes H, Kube M, Reinhardt R, Widdel F: **Alkane degradation under anoxic conditions by a nitrate-reducing bacterium with possible involvement of the electron acceptor in substrate activation**. *Environmental Microbiology Reports* 2011, 3:125–135.

26. Harayama S, Kasai Y, Hara A: **Microbial communities in oil-contaminated seawater**. *Curr Opin Biotechnol* 2004, 15:205–214.

27. Head IM, Jones DM, Röling WFM: **Marine microorganisms make a meal of oil**. *Nat Rev Microbiol* 2006, 4:173–182.

28. Yakimov MM, Timmis KN, Golyshin PN: **Obligate oil-degrading marine bacteria**. *Curr Opin Biotechnol* 2007, 18:257–266.

29. Coulon F, McKew BA, Osborn AM, McGenity TJ, Timmis KN: **Effects of temperature and biostimulation on oil-degrading microbial communities in temperate estuarine waters**. *Environ Microbiol* 2007, 9:177–186.

30. McKew BA, Coulon F, Yakimov MM, Denaro R, Genovese M, Smith CJ, Osborn AM, Timmis KN, McGenity TJ: **Efficacy of intervention strategies for bioremediation of crude oil in marine systems and effects on indigenous hydrocarbonoclastic bacteria**. *Environ Microbiol* 2007, 9:1562–1571.

31. Teramoto M, Suzuki M, Okazaki F, Hatmanti A, Harayama S: *Oceanobacter*-related bacteria are important for the degradation of petroleum aliphatic hydrocarbons in the tropical marine environment. *Microbiology* 2009, 155:3362–3370.

32. Kostka JE, Prakash O, Overholt WA, Green SJ, Freyer G, Canion A, Delgardio J, Norton N, Hazen TC, Huettel M: **Hydrocarbon-degrading bacteria and the bacterial community response in Gulf of Mexico beach sands impacted by the Deepwater Horizon oil spill**. *Appl Environ Microbiol* 2011, 77:7962–7974.

33. Teira E, Lekunberri I, Gasol JM, Nieto-Cid M, Alvarez-Salgado XA, Figueiras FG: **Dynamics of the hydrocarbon-degrading *Cycloclasticus* bacteria during mesocosm-simulated oil spills**. *Environ Microbiol* 2007, 9:2551–2562.

34. McKew BA, Coulon F, Osborn AM, Timmis KN, McGenity TJ: **Determining the identity and roles of oil-metabolizing marine bacteria from the Thames estuary, UK**. *Environ Microbiol* 2007, 9:165–176.

35. Cui ZS, Lai QL, Dong CM, Shao ZZ: **Biodiversity of polycyclic aromatic hydrocarbon-degrading bacteria from deep sea sediments of the Middle Atlantic Ridge**. *Environ Microbiol* 2008, 10:2138–2149.

36. Niepceron M, Portet-Koltalo F, Merlin C, Motelay-Massei A, Barray S, Bodilis J: **Both *Cycloclasticus* spp. and *Pseudomonas* spp. as PAH-degrading bacteria in the Seine estuary (France)**. *FEMS Microbiol Ecol* 2010, 71:137–147.

37. Yakimov MM, Golyshin PN, Lang S, Moore ERB, Abraham WR, Lunsdorf H, Timmis KN: *Alcanivorax borkumensis* gen. nov., sp. nov., a new, hydrocarbon-degrading and surfactant-producing marine bacterium. *Int J Syst Bacteriol* 1998, 48:339–348.

38. Schneiker SVA, dos Santos PM, Bartels D, Bekel T, Brecht M, Buhrmester J, Chernikova TN, Denaro R, Ferrer M, Gertler C, Goesmann A, Golyshina OV, Kaminski F, Khachane AN, Lang S, Linke B, McHardy AC, Meyer F, Nechitaylo T, Pühler A, Regenhardt D, Rupp O, Sabirova JS, Selbitschka W, Yakimov MM,

Timmis KN, Vorhölter FJ, Weidner S, Kaiser O, Golyshin PN: **Genome sequence of the ubiquitous hydrocarbon-degrading marine bacterium *Alcanivorax borkumensis***. *Nat Biotechnol* 2006, 24:997–1004.

39. Sabirova JS, Ferrer M, Regenhardt D, Timmis KN, Golyshin PN: **Proteomic insights into metabolic adaptations in *Alcanivorax borkumensis* induced by alkane utilization**. *J Bacteriol* 2006, 188:3763–3773.

40. Sabirova JS, Chernikova TN, Timmis KN, Golyshin PN: **Niche-specificity factors of a marine oil-degrading bacterium *Alcanivorax borkumensis* SK2**. *FEMS Microbiol Lett* 2008, 285:89–96.

41. Ron E, Rosenberg E: *Acinetobacter* and Alkanindiges. In *Handbook of Hydrocarbon and Lipid Microbiology*. Edited by Timmis KN, McGenity TJ, van der Meer JR, de Lorenzo V. Berlin Heidelberg: Springer; 2010:1800–1803.

42. van Beilen JB, Funhoff EG: **Alkane hydroxylases involved in microbial alkane degradation**. *Appl Microbiol Biotechnol* 2007, 74:13–21.

43. Throne-Holst M, Wentzel A, Ellingsen TE, Kotlar HK, Zotchev SB: **Identification of novel genes involved in long-chain *n*-alkane degradation by *Acinetobacter* sp. strain DSM 17874**. *Appl Environ Microbiol* 2007, 73:3327–3332.

44. Liu CL, Wang WP, Wu YH, Zhou ZW, Lai QL, Shao ZZ: **Multiple alkane hydroxylase systems in a marine alkane degrader, *Alcanivorax dieselolei* B-5**. *Environ Microbiol* 2011, 13:1168–1178.

45. Wasmund K, Burns KA, Kurtboke DI, Bourne DG: **Novel alkane hydroxylase gene (*alk*B) diversity in sediments associated with hydrocarbon seeps in the Timor Sea, Australia**. *Appl Environ Microbiol* 2009, 75:7391–7398.

46. Wang W-P, Wang L-P, Shao Z: **Diversity and abundance of oil-degrading bacteria and alkane hydroxylase (*alk*B) genes in the subtropical seawater of Xiamen Island**. *Microb Ecol* 2010, 60:429–439.

47. Païssé S, Duran R, Coulon F, Goñi-Urriza M: **Are alkane hydroxylase genes (*alk*B) relevant to assess petroleum bioremediation processes in chronically polluted coastal sediments?** *Appl Microbiol Biotechnol* 2011, 92:835–844.

48. Golyshin PN, Ferrer M, Chernikova TN, Golyshina OV, Yakimov MM: *Oleispira*. In *In Handbook of Hydrocarbon and Lipid Microbiology*. Edited by Timmis KN, McGenity TJ, van der Meer JR, de Lorenzo V. Berlin Heidelberg: Springer; 2010:1755–1763.

49. Teramoto M, Ohuchi M, Hatmanti A, Darmayati Y, Widyastuti Y, Harayama S, Fukunaga Y: *Oleibacter marinus* gen. nov., sp. nov., a bacterium that degrades petroleum aliphatic hydrocarbons in a tropical marine environment. *Int J Syst Evol Microbiol* 2011, 61:375–380.

50. Golyshin PN, Chernikova TN, Abraham WR, Lunsdorf H, Timmis KN, Yakimov MM: **Oleiphilaceae fam. nov., to include *Oleiphilus messinensis* gen. nov., sp nov., a novel marine bacterium that obligately utilizes hydrocarbons**. *Int J Syst Evol Microbiol* 2002, 52:901–911.

51. Lanfranconi MP, Bosch R, Nogales B: **Short-term changes in the composition of active marine bacterial assemblages in response to diesel oil pollution**. *Microb Biotechnol* 2010, 3:607–621.

52. Coulon F, Chronoupolou P-M, Fahy A, Païssé S, Goñi-Urriza MS, Peperzak L, Acuña-Alvarez L, McKew BA, Brussard C, Underwood GJC, Timmis KN, Duran R, McGenity TJ: **Central role of dynamic tidal biofilms dominated by aerobic hydrocarbonoclastic bacteria and diatoms in the biodegradation of hydrocarbons in coastal mudflats**. *Appl Environ Microbiol* 2012, 78:3638–3648.

53. Buchanan A, Gonzalez JM: *Roseobacter*. In *Handbook of Hydrocarbon and Lipid Microbiology*. Edited by Timmis KN, McGenity TJ, van der Meer JR, de Lorenzo V. Berlin Heidelberg: Springer; 2010:1336–1343.

54. Gerdes B, Brinkmeyer R, Dieckmann G, Helmke E: **Influence of crude oil on changes in bacterial communities in Arctic sea-ice**. *FEMS Microbiol Ecol* 2005, 53:129–139.

55. Yu SH, Ke L, Wong YS, Ta NFY: **Degradation of polycyclic aromatic hydrocarbons by a bacterial consortium enriched from mangrove sediments**. *Environ Int* 2005, 31:149–154.

56. Duran R: *Marinobacter*. In *Handbook of Hydrocarbon and Lipid Microbiology*. Edited by Timmis KN, McGenity TJ, van der Meer JR, de Lorenzo V. Berlin Heidelberg: Springer; 2010:1725–1735.

57. Vila J, Nieto JM, Mertens J, Springael D, Grifoll M: **Microbial community structure of a heavy fuel oil-degrading marine consortium: linking microbial dynamics with polycyclic aromatic hydrocarbon utilization**. *FEMS Microbiol Ecol* 2010, 73:349–362.

58. Melcher RJ, Apitz SE, Hemmingsen BB: **Impact of irradiation and polycyclic aromatic hydrocarbon spiking on microbial populations in marine sediment for future aging and biodegradability studies**. *Appl Environ Microbiol* 2002, 68:2858–2868.

59. Gutierrez T, Nichols PD, Whitman WB, Aitken MD: *Porticoccus hydrocarbonoclasticus* sp. nov., an aromatic hydrocarbon-degrading bacterium identified in laboratory cultures of marine phytoplankton. *Appl Environ Microbiol* 2012, **78**:628–637.

60. Gauthier E, Déziel E, Villemur R, Juteau P, Lépine F, Beaudet R: Initial characterization of new bacteria degrading high-molecular weight polycyclic aromatic hydrocarbons isolated from a 2-year enrichment in a two-liquid-phase culture system. *J Appl Microbiol* 2003, **94**:301–311.

61. Hilyard EJ, Jones-Meehan JM, Spargo BJ, Hill RT: Enrichment, isolation and phylogenetic identification of polycyclic aromatic hydrocarbon-degrading bacteria from Elizabeth River sediments. *Appl Environ Microbiol* 2008, **74**:1176–1182.

62. Edlund A, Jansson JK: Use of bromodeoxyuridine immunocapture to identify psychrotolerant phenanthrene-degrading bacteria in phananthrene-enriched polluted Baltic Sea sediments. *FEMS Microbiol Ecol* 2008, **65**:513–525.

63. Gutierrez T, Singleton DR, Aitken MD, Semple KT: Stable isotope probing of an algal bloom to identify uncultivated members of the Rhodobacteraceae associated with low-molecular-weight polycyclic aromatic hydrocarbon degradation. *Appl Environ Microbiol* 2011, **77**:7856–7860.

64. Kaeberlein T, Lewis K, Epstein SS: Isolating "uncultivable" microorganisms in pure culture in a simulated natural environment. *Science* 2002, **296**:1127–1129.

65. McGenity TJ, Gramain A: Cultivation of halophilic hydrocarbon degraders. In *Handbook of Hydrocarbon and Lipid Microbiology*. Edited by Timmis KN, McGenity TJ, van der Meer JR, de Lorenzo V. Berlin Heidelberg: Springer; 2010:3847–3854.

66. Calvo C, Martínez-Checa F, Toledo FL, Porcel J, Quesada E: Characteristics of bioemulsifiers synthesized in crude oil media by *Halomonas eurihalina* and their effectiveness in the isolation of bacteria able to grow in the presence of hydrocarbons. *Appl Microbiol Biotechnol* 2002, **60**:347–351.

67. Zubkov MV, Mary I, Woodward EMS, Warwick PE, Fuchs BM, Scanlan DJ, Burkill PH: Microbial control of phosphate in the nutrient-depleted North Atlantic subtropical gyre. *Environ Microbiol* 2007, **9**:2079–2089.

68. Suttle CA: Marine viruses - major players in the global ecosystem. *Nat Rev Microbiol* 2007, **5**:801–812.

69. Grossart HP: Ecological consequences of bacterioplankton lifestyles: changes in concepts are needed. *Environmental Microbiology Reports* 2010, **2**:706–714.

70. Fuhrman JA: Microbial community structure and its functional implications. *Nature* 2009, **459**:193–199.

71. Azam F, Malfatti F: Microbial structuring of marine ecosystems. *Nat Rev Microbiol* 2007, **5**:782–791.

72. Grossart HP, Levold F, Allgaier M, Simon M, Brinkhoff T: Marine diatom species harbour distinct bacterial communities. *Environ Microbiol* 2005, **7**:860–873.

73. Carman KR, Dobbs FC: Epibiotic microorganisms on copepods and other aquatic crustaceans. *Microscope Research Technique* 1997, **37**:116–135.

74. Zinger L, Amaral-Zettler LA, Fuhrman JA, Horner-Devine MC, Huse SM, Welch DB, Martiny JB, Sogin M, Boetius A, Ramette A: Global patterns of bacterial beta-diversity in seafloor and seawater ecosystems. *PLoS One* 2011, **6**:e24570.

75. Duran R, Goñi-Urriza MS: Impact of pollution on microbial mats. In *Handbook of Hydrocarbon and Lipid Microbiology*. Edited by Timmis KN, McGenity TJ, van der Meer JR, de Lorenzo V. Berlin Heidelberg: Springer; 2010:2339–2348.

76. Shank EA, Kolter R: New developments in microbial interspecies signaling. *Curr Opin Microbiol* 2009, **12**:205–214.

77. Mashburn-Warren LM, Whiteley M: Special delivery: vesicle trafficking in prokaryotes. *Mol Microbiol* 2006, **61**:839–846.

78. Dubey GP, Ben-Yehuda S: Intercellular nanotubes mediate bacterial communication. *Cell* 2011, **144**:590–600.

79. Diggle SP: Microbial communication and virulence: lessons from evolutionary theory. *Microbiology* 2010, **156**:3503–3512.

80. Yakimov MM, Denaro R, Genovese M, Cappello S, D'Auria G, Chernikova TN, Timmis KN, Golyshin PN, Giuliano L: Natural microbial diversity in superficial sediments of Milazzo Harbor (Sicily) and community successions during microcosm enrichment with various hydrocarbons. *Environ Microbiol* 2005, **7**:1426–1441.

81. Kanoh K, Adachi K, Katsuta A, Shizuri Y: Structural determination and proposed biosynthesis of alcanivorone, a novel alpha-pyrone produced by *Alcanivorax jadensis*. *Journal of Antibiotics (Tokyo)* 2008, **61**:70–74.

82. Hara A, Syutsubo K, Harayama S: *Alcanivorax* which prevails in oil-contaminated seawater exhibits broad substrate specificity for alkane degradation. *Environ Microbiol* 2003, **5**:746–753.

83. Wang BJ, Lai QL, Cui ZS, Tan TF, Shao ZZ: A pyrene-degrading consortium from deep-sea sediment of the West Pacific and its key member *Cycloclasticus* sp P1. *Environ Microbiol* 2008, **10**:1948–1963.

84. Luo YR, Tian Y, Huang X, Yan CL, Hong HS, Lin GH, Zheng TL: Analysis of community structure of a microbial consortium capable of degrading benzo[*a*]pyrene by DGGE. *Mar Pollut Bull* 2009, **58**:1159–1163.

85. de Lorenzo V: Systems biology approaches to bioremediation. *Curr Opin Biotechnol* 2008, **19**:579–89.

86. de Lorenzo V, Fraile S, Jiménez JI: Emerging systems and synthetic biology approaches to hydrocarbon biotechnology. In *Handbook of Hydrocarbon and Lipid Microbiology*. Edited by Timmis KN, McGenity TJ, van der Meer JR, de Lorenzo V. Berlin Heidelberg: Springer; 2010:1411–1435.

87. Pelz O, Tesar M, Wittich RM, Moore ERB, Timmis KN, Abraham WR: Towards elucidation of microbial community metabolic pathways: unravelling the network of carbon sharing in a pollutant-degrading bacterial consortium by immunocapture and isotopic ratio mass spectrometry. *Environ Microbiol* 1999, **1**:167–174.

88. López Z, Vila J, Ortega-Calvo JJ, Grifoll M: Simultaneous biodegradation of creosote-polycyclic aromatic hydrocarbons by a pyrene-degrading *Mycobacterium*. *Appl Microbiol Biotechnol* 2008, **78**:165–172.

89. Wintermute EH, Silver PA: Emergent cooperation in microbial metabolism. *Mol Syst Biol* 2010, **6**:407.

90. Bobadilla Fazzini RAB, Preto MJ, Quintas ACP, Bielecka A, Timmis KN, dos Santos VAPM: Consortia modulation of the stress response: proteomic analysis of single strain versus mixed culture. *Environ Microbiol* 2010, **12**:2436–2449.

91. Cerniglia CE, Sutherland JB: Degradation of polycyclic aromatic hydrocarbons by fungi. In *Handbook of Hydrocarbon and Lipid Microbiology*. Edited by Timmis KN, McGenity TJ, van der Meer JR, de Lorenzo V. Berlin Heidelberg: Springer; 2010:2080–2110.

92. Chen SH, Aitken MD: Salicylate stimulates the degradation of high molecular weight polycyclic aromatic hydrocarbons by *Pseudomonas saccharophila P15*. *Environ Sci Technol* 1999, **33**:435–439.

93. da Silva M, Esposito E, Joanna D, Canhos V, Cerniglia C: Metabolism of aromatic hydrocarbons by the filamentous fungus *Cyclothyrium* sp. *Chemosphere* 2004, **57**:943–952.

94. Prenafeta-Boldú FX, Summerbell R, de Hoog GS: Fungi growing on aromatic hydrocarbons: biotechnology's unexpected encounter with biohazard? *FEMS Microbiol Rev* 2006, **30**:109–130.

95. Harms H, Schlosser D, Wick LY: Untapped potential: exploiting fungi in bioremediation of hazardous chemicals. *Nat Rev Microbiol* 2011, **9**:177–192.

96. Johnsen A, Wick L, Harms H: Principles of microbial PAH-degradation in soil. *Environ Pollut* 2005, **133**:71–84.

97. Kotterman M, Vis E, Field J: Successive mineralization and detoxification of benzo[*a*]pyrene by the white rot fungus *Bjerkandera* sp. strain BOS55 and indigenous microflora. *Appl Environ Microbiol* 1998, **64**:2853–2858.

98. Boonchan S, Britz ML, Stanley GA: Degradation and mineralization of high-molecular-weight polycyclic aromatic hydrocarbons by defined fungal-bacterial cocultures. *Appl Environ Microbiol* 2000, **66**:1007–1019.

99. Haritash AK, Kaushik CP: Biodegradation aspects of polycyclic aromatic hydrocarbons (PAHs): A review. *J Hazard Mater* 2009, **169**:1–15.

100. Kazunga C, Aitken M: Products from the incomplete metabolism of pyrene by polycyclic aromatic hydrocarbon-degrading bacteria. *Appl Environ Microbiol* 2000, **66**:4598–4598.

101. Allen MA, Goh F, Burns BP, Neilan BA: Bacterial, archaeal and eukaryotic diversity of smooth and pustular microbial mat communities in the hypersaline lagoon of Shark Bay. *Geobiology* 2009, **7**:82–96.

102. Valentin L, Feijoo G, Moreira MT, Lema JM: Biodegradationof polycyclic aromatic hydrocarbon in forest and salt marsh soils by white rot fungi. *IntBiodeter Biodegr* 2006, **58**:15–21.

103. Frey-Klett P, Burlinson P, Deveau A, Barret M, Tarkka M, Sarniguet A: Bacterial-fungal interactions: hyphens between agricultural, clinical, environmental, and food microbiologists. *Microbiol Mol Biol Rev* 2011, **75**:583–609.

104. Furuno S, Pazolt K, Rabe C, Neu TR, Harms H, Wick LY: Fungal mycelia allow chemotactic dispersal of polycyclic aromatic hydrocarbon-degrading bacteria in water-unsaturated systems. *Environ Microbiol* 2010, **12**:1391–1398.

105. Woolgar PJ, Jones KC: **Studies on the dissolution of polycyclic aromatic hydrocarbons from contaminated materials using a novel dialysis tubing experimental method.** *Environ Sci Technol* 1999, **33**:2118–2126.

106. Semple KT, Doick CJ, Wick LY, Harms H: **Microbial interactions with organic contaminants in soil: Definitions, processes and measurement.** *Environ Pollut* 2007, **150**:166–176.

107. Bastiaens L, Springael D, Wattiau P, Harms H, de Wachter R, Verachtert H, Diels L: **Isolation of adherent polycyclic aromatic hydrocarbon (PAH)-degrading bacteria using PAH-sorbing carriers.** *Appl Environ Microbiol* 2000, **66**:1834–1843.

108. Harms H, Smith KEC, Wick LY: **Microorganism-hydrophobic compound interactions.** In *Handbook of Hydrocarbon and Lipid Microbiology.* Edited by Timmis KN, McGenity TJ, van der Meer JR, de Lorenzo V. Berlin Heidelberg: Springer; 2010:1479–1490.

109. Guerin WF, Boyd SA: **Differential bioavailability of soil-sorbed naphthalene to two bacterial species.** *Appl Environ Microbiol* 1992, **58**:1142–1152.

110. Wick LY, Wattiau P, Harms H: **Influence of the growth substrate on the mycolic acid profiles of mycobacteria.** *Environ Microbiol* 2002, **4**:612–616.

111. Rodrigues AC, Wuertz S, Brito AG, Melo LF: **Fluorene and phenanthrene uptake by *Pseudomonas putida* ATCC 17514: Kinetics and physiological aspects.** *Biotechnol Bioeng* 2005, **90**:281–289.

112. Vaysse PJ, Sivadon P, Goulas P, Grimaud R: **Cells dispersed from *Marinobacter hydrocarbonoclasticus* SP17 biofilm exhibit a specific protein profile associated with a higher ability to reinitiate biofilm development at the hexadecane-water interface.** *Environ Microbiol* 2011, **13**:737–746.

113. Wouters K, Maes E, Spitz J-A, Roeffaers MBJ, Wattiau P, Hofkens J, Springael D: **A non-invasive fluorescent staining procedure allows Confocal Laser Scanning Microscopy based imaging of *Mycobacterium* in multispecies biofilms colonizing and degrading Polycyclic Aromatic Hydrocarbons.** *J Microbiol Methods* 2010, **83**:317–325.

114. Pacwa-Plociniczak M, Plaza GA, Piotrowska-Seget Z, Cameotra SS: **Environmental applications of biosurfactants: recent advances.** *Int J Mol Sci* 2011, **12**:633–654.

115. Perfumo A, Smyth TJP, Marchant R, Banat IM: **Production and roles of biosurfactants and bioemulsifiers in accessing hydrophobic substrates.** In *Handbook of Hydrocarbon and Lipid Microbiology.* Edited by Timmis KN, McGenity TJ, van der Meer JR, de Lorenzo V. Berlin Heidelberg: Springer; 2010:1501–1512.

116. Kanaly RA, Harayama S, Watanabe K: ***Rhodanobacter* sp. strain BPC1 in a benzo[*a*]pyrene-mineralizing microbial consortium.** *Appl Environ Microbiol* 2002, **68**:5826–5833.

117. Iwabuchi N, Sunairi M, Urai M, Itoh C, Anzai H, Nakajima M, Harayama S: **Extracellular polysaccharides of *Rhodococcus rhodochrous* S-2 stimulate the degradation of aromatic components in crude oil by indigenous marine bacteria.** *Appl Environ Microbiol* 2002, **68**:2337–2343.

118. Abalos A, Viñas M, Sabaté J, Manresa MA, Solanas AM: **Enhanced biodegradation of Casablanca crude oil by a microbial consortium in presence of a rhamnolipid produced by *Pseudomonas aeruginosa* AT10.** *Biodegradation* 2004, **15**:249–260.

119. Shin KH, Ahn Y, Kim KW: **Toxic effect of biosurfactant addition on the biodegradation of phenanthrene.** *Environ Toxicol Chem* 2005, **24**:2768–2774.

120. Rambeloarisoa E, Rontani JF, Giusti G, Duvnjak Z, Bertrand JC: **Degradation of crude oil by a mixed population of bacteria isolated from sea-surface foams.** *Mar Biol* 1984, **83**:69–81.

121. Chaerun SK, Tazaki K, Asada R, Kogure K: **Interaction between clay minerals and hydrocarbon-utilizing indigenous microorganisms in high concentrations of heavy oil: implications for bioremediation.** *Clay Minerals* 2005, **40**:105–114.

122. Vacca DJ, Bleam WF, Hickey WJ: **Isolation of soil bacteria adapted to degrade humic acid-sorbed phenanthrene.** *Appl Environ Microbiol* 2005, **71**:3797–3805.

123. Shin WS, Pardue JH, Jackson WA: **Oxygen demand and sulfate reduction in petroleum hydrocarbon contaminated salt marsh soils.** *Water Res* 2000, **34**:1345–1353.

124. Widdel F, Knittel K, Galushko A: **Anaerobic hydrocarbon-degrading microorganisms: an overview.** In *Handbook of Hydrocarbon and Lipid Microbiology.* Edited by Timmis KN, McGenity TJ, van der Meer JR, de Lorenzo V. Berlin Heidelberg: Springer; 2010:1997–2021.

125. Gray ND, Sherry A, Grant RJ, Rowan AK, Hubert CRJ, Callbeck CM, Aitken CM, Jones DM, Adams JJ, Larter SR, Head IM: **The quantitative significance of Syntrophaceae and syntrophic partnerships in methanogenic degradation of crude oil alkanes.** *Environ Microbiol* 2011, **13**:2957–2975.

126. Zengler K, Richnow H, Rosselló-Mora R, Michaelis W, Widdel F: **Methane formation from long-chain alkanes by anaerobic microorganisms.** *Nature* 1999, **401**:266–269.

127. Walker CB, He Z, Yang ZK, Ringbauer JA Jr, He Q, Zhou J, Voordouw G, Wall JD, Arkin AP, Hazen TC, Stolyar S, Stahl DA: **The electron transfer system of syntrophically grown *Desulfovibrio vulgaris*.** *J Bacteriol* 2009, **191**:5793–5801.

128. Ficker M, Krastel K, Orlicky S, Edwards E: **Molecular characterization of a toluene-degrading methanogenic consortium.** *Appl Environ Microbiol* 1999, **65**:5576–5585.

129. McInerney MJ, Sieber JR, Gunsalus RP: **Syntrophy in anaerobic global carbon cycles.** *Curr Opin Biotechnol* 2009, **20**:623–632.

130. Yerushalmi L, Lascourreges JF, Rhofir C, Guiot SR: **Detection of intermediate metabolites of benzene biodegradation under microaerophilic conditions.** *Biodegradation* 2001, **12**:379–391.

131. Wimpenny JWT, Abdollahi H: **Growth of mixed cultures of *Paracoccus denitrificans* and *Desulfovibrio desulfuricans* in homogeneous and in heterogeneous culture systems.** *Microb Ecol* 1991, **22**:1–13.

132. Gerritse J, Gottschal JC: **Two-membered mixed cultures of methanogenic and aerobic bacteria in O_2-limited chemostats.** *J Gen Microbiol* 1993, **139**:1853–1860.

133. Aburto A, Fahy A, Coulon F, Lethbridge G, Timmis KN, Ball AS, McGenity TJ: **Mixed aerobic and anaerobic microbial communities in benzene-contaminated groundwaters.** *J Appl Microbiol* 2009, **16**:317–328.

134. Chayabutra C, Ju L-K: **Degradation of *n*-hexadecane and its metabolites by *Pseudomonas aeruginosa* under microaerobic and anaerobic denitrifying conditions.** *Appl Environ Microbiol* 2000, **66**:493–498.

135. Vitte I, Duran R, Jézéquel R, Caumette P, Cravo-Laureau C: **Effect of oxic/anoxic switches on bacterial communities and PAH biodegradation in an oil-contaminated sludge.** *Environ Sci Pollut Res* 2011, **18**:1022–1032.

136. Rocchetti L, Beolchini F, Ciani M, Dell'Anno A: **Improvement of bioremediation performance for the degradation of petroleum hydrocarbons in contaminated sediments.** *Appl Environm Soil Sci* 2011, **2011**:319657.

137. Cravo-Laureau C, Hernandez-Raquet G, Vitte I, Jézéquel R, Bellet V, Godon J-J, Caumette P, Balaguer P, Duran R: **Role of environmental fluctuations and microbial diversity in degradation of hydrocarbons in contaminated sludge.** *Res Microbiol* 2011, **162**:888–895.

138. Field CB, Behrenfeld MJ, Randerson JT, Falkowski P: **Primary production of the biosphere: Integrating terrestrial and oceanic components.** *Science* 1998, **281**:237–240.

139. Biegala IC, Kennaway G, Alverca E, Lennon JF, Vaulot D, Simon N: **Identification of bacteria associated with dinoflagellates (Dinophyceae) *Alexandrium* spp. using tyramide signal amplification-fluorescent in situ hybridization and confocal microscopy.** *J Phycol* 2002, **38**:404–411.

140. Armbrust EV: **The life of diatoms in the world's oceans.** *Nature* 2009, **459**:185–192.

141. Crump BC, Armbrust EV, Baross JA: **Phylogenetic analysis of particle-attached and free-living bacterial communities in the Columbia River, its estuary, and the adjacent coastal ocean.** *Appl Environ Microbiol* 1999, **65**:3192–3204.

142. Cottrell M, Kirchman D: **Natural assemblages of marine proteobacteria and membersof the Cytophaga-Flavobacter cluster consuming low- and high-molecular-weight dissolved organic matter.** *Appl Environ Microbiol* 2000, **66**:1692–1697.

143. Allgaier M, Riebesell U, Vogt M, Thyrhaug R, Grossart HP: **Coupling of heterotrophic bacteria to phytoplankton bloom development at different pCO₂ levels: a mesocosm study.** *Biogeosciences* 2008, **5**:1007–1022.

144. Jasti S, Sieracki ME, Poulton NJ, Giewat MW, Rooney-Varga JN: **Phylogenetic diversity and specificity of bacteria closely associated with *Alexandrium* spp. and other phytoplankton.** *Appl Environ Microbiol* 2005, **71**:3483–3494.

145. Sapp M, Schwaderer AS, Wiltshire KH, Hoppe HG, Gerdts G, Wichels A: **Species-specific bacterial communities in the phycosphere of microalgae?** *Microb Ecol* 2007, **53**:683–699.

146. Hube AE, Heyduck-Söller B, Fischer U: **Phylogenetic classification of heterotrophic bacteria associated with filamentous marine cyanobacteria in culture.** *Syst Appl Microbiol* 2009, **32**:256–265.

147. Cole JJ: Interactions between bacteria and algae in aquatic ecosystems. *Annu Rev Ecol Evol Syst* 1982, **13**:291–314.

148. Rivas MO, Vargas P, Riquelme CE: Interactions of *Botryococcus braunii* cultures with bacterial biofilms. *Microb Ecol* 2010, **60**:628–635.

149. Bell WH, Lang JM, Mitchell R: Selective stimulation of marine bacteria by algal extracellular products. *Limnol Oceanogr* 1974, **19**:833–839.

150. Amin SA, Green DH, Hart MC, Küpper FC, Sunda WG, Carrano CJ: Photolysis of iron-siderophore chelates promotes bacterial-algal mutualism. *Proc Natl Acad Sci U S A* 2009, **106**:17071–17076.

151. Hopkinson BM, Roe K, Barbeau KA: Heme uptake by *Microscilla marina* and evidence for heme uptake systems in the genomes of diverse marine bacteria. *Appl Environ Microbiol* 2008, **74**:6263–6270.

152. Croft MT, Lawrence AD, Raux-Deery E, Warren MJ, Smith AG: Algae acquire vitamin B12 through a symbiotic relationship with bacteria. *Nature* 2005, **438**:90–93.

153. Mouget JL, Dakhama A, Lavoie MC, Delanoue J: Algal growth enhancement by bacteria - is consumption of photosynthetic oxygen involved. *FEMS Microbiol Ecol* 1995, **18**:35–43.

154. Huenken M, Harder J, Kirst GO: Epiphytic bacteria on the Antarctic ice diatom *Amphiprora kufferathii* Manguin cleave hydrogen peroxide produced during algal photosynthesis. *Plant Biology* 2008, **10**:519–526.

155. Villareal TA: Laboratory culture and preliminary characterization of the nitrogen-fixing *Rhizosolenia-Richelia* symbiosis. *Mar Ecol* 1990, **11**:117–132.

156. Kneip C, Lockhart P, Voß C Maier U-G: Nitrogen fixation in eukaryotes - new models for symbiosis. *BMC Evol Biol* 2007, **7**:55.

157. Moisander PH, Beinart RA, Voss M, Zehr JP: Diversity and abundance of diazotrophic microorganisms in the South China Sea during intermonsoon. *ISME J* 2008, **2**:954–967.

158. Matsuo Y, Imagawa H, Nishizawa M, Shizuri Y: Isolation of an algal morphogenesis inducer from a marine bacterium. *Science* 2005, **307**:1598.

159. Bruckner CG, Bahulikar R, Rahalkar M, Schink B, Kroth PG: Bacteria associated with benthic diatoms from Lake Constance: phylogeny and influences on diatom growth and secretion of extracellular polymeric substances. *Appl Environ Microbiol* 2008, **74**:7740–7749.

160. Gärdes A, Iversen MH, Grossart HP, Passow U, Ullrich MS: Diatom-associated bacteria are required for aggregation of *Thalassiosira weissflogii*. *ISME J* 2011, **5**:436–445.

161. Bolch CJS, Subramanian TA, Green DH: The toxic dinoflagellate *Gymnodinium catenatum* (dinophyceae) requires marine bacteria for growth. *J Phycol* 2011, **47**:1009–1022.

162. Bruckner CG, Rehm C, Grossart HP, Kroth PG: Growth and release of extracellular organic compounds by benthic diatoms depend on interactions with bacteria. *Environ Microbiol* 2011, **13**:1052–1063.

163. Kujawinski EB: The impact of microbial metabolism on marine dissolved organic matter. *Annu Rev Marine Sci* 2011, **3**:567–599.

164. González J, Figueiras FG, Aranguren-Gassis M, Crespo BG, Fernández E, Morán XAG, Nieto-Cid M: Effect of a simulated oil spill on natural assemblages of marine phytoplankton enclosed in microcosms. *Estuar Coastal Shelf Sci* 2009, **83**:265–276.

165. Al-Hasan RH, Khanafer M, Eliyas M, Radwan SS: Hydrocarbon accumulation by picocyanobacteria from the Arabian Gulf. *J Appl Microbiol* 2001, **91**:533–540.

166. Benthien M, Wieland A, de Oteyza T, Grimalt J, Kühl M: Oil-contamination effects on a hypersaline microbial mat community (Camargue, France) as studied with microsensors and geochemical analysis. *Ophelia* 2004, **58**:135–150.

167. Abed RM, Al-Thukair A, de Beer D: Bacterial diversity of a cyanobacterial mat degrading petroleum compounds at elevated salinities and temperatures. *FEMS Microbiol Ecol* 2006, **57**:290–301.

168. Cohen Y: Bioremediation of oil by marine microbial mats. *Int Microbiol* 2002, **5**:189–193.

169. van Bleijswijk J, Muyzer G: Genetic diversity of oxygenic phototrophs in microbial mats exposed to different levels of oil pollution. *Ophelia* 2004, **58**:157–164.

170. Païssé S, Coulon F, Goñi-Urriza M, Peperzak L, McGenity TJ, Duran R: Structure of bacterial communities along a hydrocarbon contamination gradient in a coastal sediment. *FEMS Microbiol Ecol* 2008, **66**:295–305.

171. Cerniglia CE, Van Baalen C, Gibson DT: Metabolism of naphthalene by the cyanobacterium *Oscillatoria* sp., strain JCM. *J Gen Microbiol* 1980, **116**:485–594.

172. Cerniglia CE, Gibson DT, Van Baalen C: Oxidation of naphthalene by cyanobacteria and microalgae. *J Gen Microbiol* 1980, **116**:495–500.

173. Semple KT, Cain RB, Schmidt S: Biodegradation of aromatic compounds by microalgae. *FEMS Microbiol Lett* 1999, **170**:291–300.

174. Todd SJ, Cain RB, Schmidt S: Biotransformation of naphthalene and diaryl ethers by green microalgae. *Biodegradation* 2002, **13**:229–238.

175. Al-Hasan RH, Sorkhoh NA, Al Bader D, Radwan SS: Utilization of hydrocarbons by cyanobacteria from microbial mats on oily coasts of the Gulf. *Appl Microbiol Biotechnol* 1994, **41**:615–619.

176. Lei AP, Wong YS, Tam NFY: Removal of pyrene by different microalgal species. *Water Sci Technol* 2002, **46**:195–201.

177. Abed RMM, Köster J: The direct role of aerobic heterotrophic bacteria associated with cyanobacteria in the degradation of oil compounds. *Int Biodeterior Biodegrad* 2005, **55**:29–37.

178. Raghukumar C, Vipparty V, David JJ, Chandramohan D: Degradation of crude oil by marine cyanobacteria. *Appl Microbiol Biotechnol* 2001, **57**:433–436.

179. de Oteyza TG, Grimalt JO, Diestra E, Sole A, Esteve I: Changes in the composition of polar and apolar crude oil fractions under the action of *Microcoleus* consortia. *Appl Microbiol Biotechnol* 2004, **66**:226–232.

180. Abed RMM, Safi NMD, Koster J, de Beer D, El-Nahhal Y, Rullkotter J, Garcia-Pichel F: Microbial diversity of a heavily polluted microbial mat and its community changes following degradation of petroleum compounds. *Appl Environ Microbiol* 2002, **68**:1674–1683.

181. Sánchez O, Diestra E, Esteve I, Mas J: Molecular characterization of an oil-degrading cyanobacterial consortium. *Microb Ecol* 2005, **50**:580–588.

182. Chaillan F, Gugger M, Sailot A, Coute A, Oudot J: Role of cyanobacteria in the biodegradation of crude oil by a tropical cyanobacterial mat. *Chemosphere* 2006, **62**:1574–1582.

183. Ghasemi Y, Rasoul-Amini S, Fotooh-Abadi E: The biotransformation, biodegradation, and bioremediation of organic compounds by microalgae. *J Phycol* 2011, **47**:969–980.

184. Chisti Y: Biodiesel from microalgae. *Biotechnology Advances* 2007, **25**:294–306.

185. Schirmer A, Rude MA, Li XZ, Popova E, del Cardayre SB: Microbial biosynthesis of alkanes. *Science* 2010, **329**:559–562.

186. Shaw SL, Gantt B, Meskhidze N: Production and emissions of marine isoprene and monoterpenes: a review. *Advances in Meteorology* 2010, doi:10.1155/2010/408696.

187. Exton DA, Steinke M, Suggett DJ, McGenity TJ: Spatial and temporal variability of biogenic isoprene emissions from a temperate estuary. *Global Biogeochem Cycles* 2012,**26**:GB2012. In press.

188. Acuña Alvarez L, Exton DA, Timmis KN, Suggett DJ, McGenity TJ: Characterization of marine isoprene-degrading communities. *Environ Microbiol* 2009, **11**:3280–3291.

189. Green DH, Llewellyn LE, Negri AP, Blackburn SI, Bolch CJS: Phylogenetic and functional diversity of the cultivable bacterial community associated with the paralytic shellfish poisoning dinoflagellate *Gymnodinium catenatum*. *FEMS Microbiol Ecol* 2004, **47**:345–357.

190. Radwan S, Mahmoud H, Khanafer M, Al-Habib A, Al-Hasan R: Identities of epilithic hydrocarbon-utilizing diazotrophic bacteria from the Arabian Gulf coasts, and their potential for oil bioremediation without nitrogen supplementation. *Microb Ecol* 2010, **60**:354–363.

191. Binark N, Guven KC, Gezgin T, Unlu S: Oil pollution of marine algae. *Bull Environ Contam Toxicol* 2000, **64**:866–872.

192. Kowalewska G: Phytoplankton-the main factor responsible for transport of polynuclear aromatic hydrocarbons from water to sediments in the Southern Baltic ecosystem. *ICES J Mar Sci* 1999, **56**:219–222.

193. Geng HF, Belas R: Molecular mechanisms underlying *Roseobacter*-phytoplankton symbioses. *Curr Opin Biotechnol* 2010, **21**:332–338.

194. Abed RMM: Interaction between cyanobacteria and aerobic heterotrophic bacteria in the degradation of hydrocarbons. *Int Biodeterior Biodegrad* 2010, **64**:58–64.

195. Warshawsky D, LaDow K, Schneider J: Enhanced degradation of benzo[a] pyrene by *Mycobacterium* sp. in conjunction with green alga. *Chemosphere* 2007, **69**:500–506.

196. Tang X, He LY, Tao XQ, Dang Z, Guo CL, Lu GN, Yi XY: Construction of an artificial microalgal-bacterial consortium that efficiently degrades crude oil. *J Hazard Mater* 2010, **181**:1158–1162.

197. Hofmann T, Hanlon ARM, Taylor JD, Ball AS, Osborn AM, Underwood GJC: Dynamics and compositional changes in extracellular

carbohydrates in estuarine sediments during degradation. *Mar Ecol Prog Ser* 2009, **379**:45–58.

198. Carmichael LM, Pfaender FK: **The effect of inorganic and organic supplements on the microbial degradation of phenanthrene and pyrene in soils.** *Biodegradation* 1997, **8**:1–13.

199. Wick LY, Pasche N, Bernasconi SM, Pelz O, Harms H: **Characterization of multiple substrate utilization by anthracene-degrading *Mycobacterium frederiksbergense* LB501T.** *Appl Environ Microbiol* 2003, **69**:6133–6142.

200. Teng Y, Luo YM, Ping LF, Zou DX, Li ZA, Christie P: **Effects of soil amendment with different carbon sources and other factors on the bioremediation of an aged PAH-contaminated soil.** *Biodegradation* 2010, **21**:167–178.

201. Sher D, Thompson JW, Kashtan N, Croal L, Chisholm CW: **Response of *Prochlorococcus* ecotypes to co-culture with diverse marine bacteria.** *ISME J* 2011, **5**:1125–1132.

202. Head IM, Swannell RP: **Bioremediation of petroleum hydrocarbon contaminants in marine habitats.** *Curr Opin Biotechnol* 1999, **10**:234–239.

203. Knowles R, Wishart C: **Nitrogen-fixation in Arctic marine sediments - effect of oil and hydrocarbon factions.** *Environ Pollut* 1977, **13**:133–149.

204. Musat F, Harder J, Widdel F: **Study of nitrogen fixation in microbial communities of oil-contaminated marine sediment microcosms.** *Environ Microbiol* 2006, **8**:1834–1843.

205. Thomson AD, Webb KL: **The effect of chronic oil pollution on salt-marsh nitrogen fixation (acetylene reduction).** *Estuaries* 1984, **7**:2–11.

206. Toccalino PL, Johnson RL, Boone DR: **Nitrogen limitation and nitrogen-fixation during alkane biodegradation in a sandy soil.** *Appl Environ Microbiol* 1993, **59**:2977–2983.

207. Taketani RG, dos Santos HF, van Elsas JD, Rosado AS: **Characterisation of the effect of a simulated hydrocarbon spill on diazotrophs in mangrove sediment mesocosm.** *Antonie Van Leeuwenhoek* 2009, **96**:343–354.

208. do Carmo FL, dos Santos HF, Martins EF, van Elsas JD, Rosado AS, Peixoto RS: **Bacterial structure and characterization of plant growth promoting and oil degrading bacteria from the rhizospheres of mangrove plants.** *J Microbiol* 2011, **49**:535–543.

209. Stoeck T, Edgcomb V: **Role of protists in microbial interactions with hydrocarbons.** In *Handbook of Hydrocarbon and Lipid Microbiology*. Edited by Timmis KN, McGenity TJ, van der Meer JR, de Lorenzo V. Berlin Heidelberg: Springer; 2010:2423–2434.

210. Gertler C, Näther DJ, Gerdts G, Malpass MC, Golyshin PN: **A mesocosm study of the changes in marine flagellate and ciliate communities in a crude oil bioremediation trial.** *Microb Ecol* 2010, **60**:180–191.

211. Dalby AP, Kormas KA, Christaki U, Karayanni H: **Cosmopolitan heterotrophic microeukaryotes are active bacterial grazers in experimental oil-polluted systems.** *Environ Microbiol* 2008, **10**:47–56.

212. Tso SF, Taghon GL: **Protozoan grazing increases mineralization of naphthalene in marine sediment.** *Microb Ecol* 2006, **51**:460–469.

213. Mattison RG, Harayama S: **The predatory soil flagellate *Heteromita globosa* stimulates toluene biodegradation by a *Pseudomonas* sp.** *FEMS Microbiol Lett* 2001, **194**:39–45.

214. Rogerson A, Berger J: **Enhancement of the microbial degradation of crude petroleum by the ciliate *Colpidium colpoda*.** *J Gen Appl Microbiol* 1983, **29**:41–50.

215. Näslund J, Nascimento FJA, Gunnarsson JS: **Meiofauna reduces bacterial mineralization of naphthalene in marine sediment.** *ISME J* 2010, **4**:1421–1430.

216. Carman KR, Bianchi TS, Kloep F: **Influence of grazing and nitrogen on benthic algal blooms in diesel fuel-contaminated saltmarsh sediments.** *Environ Sci Technol* 2000, **34**:107–111.

217. Fleeger JW, Carman KR, Nisbet RM: **Indirect effects of contaminants in aquatic ecosystems.** *Sci Total Environ* 2003, **317**:207–233.

218. Gertler C, Gerdts G, Timmis KN, Golyshin PN: **Microbial consortia in mesocosm bioremediation trial using oil sorbents, slow-release fertilizer and bioaugmentation.** *FEMS Microbiol Ecol* 2009, **69**:288–300.

219. Cappello S, Denaro R, Genovese M, Giuliano L, Yakimov MM: **Predominant growth of *Alcanivorax* during experiments on "oil spill bioremediation"in mesocosms.** *Microbiol Res* 2007, **162**:185–190.

220. Cochran PK, Kellogg CA, Paul JH: **Prophage induction of indigenous marine lysogenic bacteria by environmental pollutants.** *Mar Ecol Prog Ser* 1998, **164**:125–133.

221. Rosenberg E, Bittan-Banin G, Sharon G, Shon A, Hershko G, Levy I, Ron EZ: **The phage- driven microbial loop in petroleum bioremediation.** *Microb Biotechnol* 2010, **3**:467–472.

222. Lu Z-M, Deng Y, Van Nostrand JD, He Z-L, Voordeckers J, Zhou A-F, Lee YJ, Mason OU, Dubinsky EA, Chavarria KL, Tom LM, Fortney JL, Lamendella R, Jansson JK, D'haeseleer P, Hazen TC, Zhou J-Z: **Microbial gene functions enriched in the Deepwater Horizon deep-sea oil plume.** *ISME J* 2012, **6**:451–460.

223. Hazen TC, Dubinsky EA, Desantis TZ, Andersen GL, Piceno YM, Singh N, Jansson JK, Probst A, Borglin SE, Fortney JL, Stringfellow WT, Bill M, Conrad ME, Tom LM, Chavarria KL, Alusi TR, Lamendella R, Joyner DC, Spier C, Baelum J, Auer M, Zemla ML, Chakraborty R, Sonnenthal EL, D'haeseleer P, Holman HY, Osman S, Lu Z, Van Nostrand JD, Deng Y, Zhou J, Mason OU: **Deep-sea oil plume enriches indigenous oil- degrading bacteria.** *Science* 2010, **330**:204–208.

224. Herrick JB, Stuart-Keil KG, Ghiorse WC, Madsen EL: **Natural horizontal transfer of a naphthalene dioxygenase gene between bacteria native to a coal tar-contaminated field site.** *Appl Environ Microbiol* 1997, **63**:2330–2337.

225. Top EM, Springael D, Boon N: **Catabolic mobile genetic elements and their potential use in bioaugmentation of polluted soils and waters.** *FEMS Microbiol Ecol* 2002, **42**:199–208.

226. Cuny P, Miralles G, Cornet-Barthaux V, Acquaviva M, Stora G, Grossi V, Gilbert F: **Influence of bioturbation by the polychaete *Nereis diversicolor* on the structure of bacterial communities in oil contaminated coastal sediments.** *Mar Pollut Bull* 2007, **54**:452–459.

227. Gilbert F, Stora G, Desrosiers G, Deflandre B, Bertrand J, Poggiale J, Gagne J: **Alteration and release of aliphatic compounds by the polychaete *Nereis virens* (Sars) experimentally fed with hydrocarbons.** *J Exp Mar Biol Ecol* 2001, **256**:199–213.

228. Grossi V, Massias D, Stora G, Bertrand JC: **Burial, exportation and degradation of acyclic petroleum hydrocarbons following a simulated oil spill in bioturbated Mediterranean coastal sediments.** *Chemosphere* 2002, **48**:947–954.

229. Hickman ZA, Reid BJ: **Earthworm assisted bioremediation of organic contaminants.** *Environ Int* 2008, **34**:1072–1081.

230. International Maritime Organization: *Bioremediation in marine oil spills: guidance document for decision making and implementation of bioremediation in marine oil spills.* London: IMO Publishing; 2004.

231. Lin Q, Mendelssohn IA: **Determining tolerance limits for restoration and phytoremediation with *Spartina patens* in crude oil-contaminated sediment in greenhouse.** *Archiv Agrony Soil Sci* 2008, **54**:681–690.

232. Ramos JL, Duque E, van Dillewijn P, Daniels C, Krell T, Espinosa-Urgel M, Ramos-González M-I, Rodríguez S, Matilla M, Wittich R, Segura A: **Removal of hydrocarbons and other related chemicals via the rhizosphere of plants.** In *Handbook of Hydrocarbon and Lipid Microbiology*. Edited by Timmis KN, McGenity TJ, van der Meer JR, de Lorenzo V. Berlin Heidelberg: Springer; 2010:2575–2581.

233. Gallego JLR, Garcia-Martinez MJ, Llamas JF, Belloch C, Pelaez AI, Sanchez J: **Biodegradation of oil tank bottom sludge using microbial consortia.** *Biodegradation* 2007, **18**:269–281.

234. Jacques RJS, Okeke BC, Bento FM, Teixeira AS, Peralba MCR, Camargo FAO: **Microbial consortium bioaugmentation of polycyclic aromatic hydrocarbons in contaminated soil.** *Bioresour Technol* 2008, **99**:2637–2643.

235. Munoz R, Guieysse B: **Algal-bacterial processes for the treatment of hazardous contaminants: a review.** *Water Res* 2006, **40**:2799–2815.

236. Huang WE, Stoecker K, Griffiths R, Newbold L, Daims H, Whiteley AS, Wagner M: **Raman-FISH: combining stable-isotope Raman spectroscopy and fluorescence in situ hybridization for the single cell analysis of identity and function.** *Environ Microbiol* 2007, **9**:1878–1889.

237. Finzi-Hart JA, Pett-Ridge J, Weber PK, Popa R, Fallon SJ, Gunderson T, Hutcheon ID, Nealson KH, Capone DG: **Fixation and fate of C and N in the cyanobacterium *Trichodesmium* using nanometer-scale secondary ion mass spectrometry.** *Proc Natl Acad Sci USA* 2009, **106**:6345–6350.

238. Orphan VJ: **Methods for unveiling cryptic microbial partnerships in nature.** *Curr Opin Microbiol* 2009, **12**:231–237.

239. Klitgord N, Segre D: **Ecosystems biology of microbial metabolism.** *Curr Opin Biotechnol* 2011, **22**:541–546.

12

Benthic infaunal community structuring in an acidified tropical estuarine system

M Belal Hossain[1,2][*] and David J Marshall[1]

Abstract

Background: Recent studies suggest that increasing ocean acidification (OA) should have strong direct and indirect influences on marine invertebrates. While most theory and application for OA is based on relatively physically-stable oceanic ecological systems, less is known about the effects of acidification on nearshore and estuarine systems. Here, we investigated the structuring of a benthic infaunal community in a tropical estuarine system, along a steep salinity and pH gradient, arising largely from acid-sulphate groundwater inflows (Sungai Brunei Estuary, Borneo, July 2011- June 2012).

Results: Preliminary data indicate that sediment pore-water salinity (range: 8.07 - 29.6 psu) declined towards the mainland in correspondence with the above-sediment estuarine water salinity (range: 3.58 – 31.2 psu), whereas the pore-water pH (range: 6.47- 7.72) was generally lower and less variable than the estuarine water pH (range: 5.78- 8.3), along the estuary. Of the thirty six species (taxa) recorded, the polychaetes *Neanthes* sp., *Onuphis conchylega,* Nereididae sp. and the amphipod Corophiidae sp., were numerically dominant. Calcified microcrustaceans (e.g., Cyclopoida sp. and Corophiidae sp.) were abundant at all stations and there was no clear distinction in distribution pattern along the estuarine between calcified and non-calcified groups. Species richness increased seawards, though abundance (density) showed no distinct directional trend. Diversity indices were generally positively correlated (Spearman's rank correlation) with salinity and pH (p <0.05) and negatively with clay and organic matter, except for evenness values (p >0.05). Three faunistic assemblages were distinguished: (1) nereid-cyclopoid-sabellid, (2) corophiid-capitellid and (3) onuphid- nereid-capitellid. These respectively associated with lower salinity/pH and a muddy bottom, low salinity/pH and a sandy bottom, and high salinity/pH and a sandy bottom. However, CCA suggested that species distribution and community structuring is more strongly influenced by sediment particle characteristics than by the chemical properties of the water (pH and salinity).

Conclusions: Infaunal estuarine communities, which are typically adapted to survive relatively acidic conditions, may be **less exposed, less sensitive, and less vulnerable** than epibenthic or pelagic communities to further acidification of above-sediment waters. These data question the extent to which all marine infaunal communities, including oceanic communities, are likely to be affected by future global CO_2-driven acidification.

Keywords: Community structure, Infauna, Soft-bottom, Tropical estuary, Salinity, Acidification

Background

Many tropical and subtropical estuaries experience acidification (low pH) resulting from acid sulfate soil (ASS) inflows [1-5]. ASS perturbation occurs when pyrite is produced during bacterial breakdown of organic matter in the presence of sulfate and iron oxides in sediments [1]. Though the estuarine pH levels generally vary from 7.0 to 7.5 in the fresher sections, to between 8.0 and 8.6 in the more saline areas [6]; the runoff from the ASS can reduce the pH of adjacent estuaries to as low as 2 [7]. Many studies have demonstrated that aquatic organisms have trouble surviving if pH levels drop under 5 or rise above 9 [7-9]. Fluctuating pH may compromise optimally related life processes (such as metabolism and growth), but also potentially increase the solubility of calcium carbonates and the bioavailability of metals [8,10]. In the acidic condition, toxic metals in the estuarine sediment can be resuspended in the water columns which have indirect detrimental impacts on many

* Correspondence: mbhnstu@gmail.com
[1]Environmental and Life Sciences, Faculty of Science, Universiti Brunei Darussalam, BE1410 Jalan Tunkgu Link, Brunei Darussalam
[2]Department of Fisheries and Marine Science, Noakhali Science and Technology University, Sonapur 3814, Bangladesh

organisms. Field and short-term laboratory studies have shown that ASS runoff can cause significant shell dissolution and perforation of heavily shelled organisms, e.g. bivalves (oyster) and gastropods [1,10,11]. However, less information is available for non or weakly calcified organisms (crustaceans, polychaetes) and for community level responses in general for extraordinarily acidified tropical estuaries [3-5]. The timeframes for exposure to ASS groundwater runoff in estuarine systems are unclear, but exposure could last for decades or hundreds of years, allowing assessment of multi-generational impacts of low pH on biological systems [3].

In addition to ASS acidification, it is possible that estuarine biota may face increasing acidification stress through present and future elevations in atmospheric CO_2, a phenomenon affecting oceanic waters, known as 'ocean acidification' [10,12-15]. Because ocean acidification affects the balance of the carbonate systems, shelled organisms possessing calcium carbonate structures are expected to be especially threatened by this process. This brings to question how future OA might affect estuarine systems, and raises the importance of understanding of benthic community structuring in already acidified tropical estuarine systems. The few investigations that have looked at the effects of acidification on benthic fauna of estuaries have reported lower diversity and abundance in soft bottom communities in the acidified sites compared to reference sites [3,16,17]. However, these studies have mainly focused on the effects of ASS water inflows on epibenthic communities, without considering relationships between the pore-water pH/salinity and the faunas living within sediments. Soft-bottom infaunal organisms are expected to experience pH levels lower than those in the water column as a consequence of significant microbial and animal metabolism (including sulphide production) and poor ventilation within sediments.

Recent studies suggest that increasing acidification should have strong direct and indirect effects on marine invertebrates. These studies sometimes exclude details of the structural complexity, ecophysiological variability and genetic diversity characterizing natural communities [18]. It is however likely that communities which have evolved in acidified coastal and marine environments (such as acidified estuaries), possess species already well adapted for acidification exposure. Structural complexity should arise from taxonomic differences in physiological and behavioural adaptations, and tolerances, of low pH. As an example, crustaceans (crabs and prawns) are possibly better at tolerating reduced pH, given their ability to generally to regulate internal body fluids (with respect to most ions), compared to other taxa for which body fluids conform in ionic concentrations relative to the external environment (annelids and molluscs; see Wittman and Portner [19]). However, regulation comes at an energetic cost, and some groups are well adapted, behaviourally and physiologically, to isolate themselves from environmental perturbations. Furthermore, no studies found to show how benthic infaunal animals and communities (as opposed to epibenthic communities) of any marine system, which experience relatively stable, low pore-water pH's, are likely to respond to atmospheric CO_2-driven future acidification of marine waters.

The distributions of estuarine benthic faunas are well known to vary in relation to a variety of physicochemical gradients [20,21]. The main factors, however, that regulate benthic species composition and density are salinity, dissolved oxygen and substratum particle composition [20,22,23]. Among these variables, salinity fluctuations are usually overwhelming in their effect on the structure and functioning of estuarine systems [21,23-25]. Additionally, in the case of benthic infaunal communities, sediment granulometry has proven to be important [23], especially considering that estuarine sediments are extremely dynamic, comprising a variety of substrata from non-vegetated soft mud, fine and coarse sand, to vegetated saltmarsh, algal mats, seagrass beds and mangrove swamps [21,24].

Given the fact that physical stressors are likely to increase with reductions in salinity and pH of estuarine waters, we investigated whether communities in regions occupying low salinity/high acidity are characterized by relatively lowered abundance and diversity, in the case of benthic infauna communities of acidified estuarine systems. In particular, we determined variation in species abundance, species diversity and the structuring of these communities along the steep salinity and pH gradient of the Brunei estuarine system (BES, Brunei Darussalam, tropical South East Asia, Borneo). We further assessed the interaction strength for community attributes and the environmental parameters, including sediment properties.

Results

Pore-water characteristics

Overall the pore-water salinity varied between 8.07 and 29.6 psu, and pH between 6.47 and 7.72. Pore-water salinity was generally greater than the above-sediment water salinity, though these were well linearly related (pore water salinity $=7.62 + 0.72$ overlying water salinity; $r^2 = 0.875$ for means at each station, p $=0.062$) (Figure 1A). However, there was a poor relationship between pore-water and above-sediment estuarine water in the case of pH (pore-water pH $=5.49 + 0.228$ overlying water pH; $r^2 = 0.63$, p $=0.2$), due to in situ sediment biotic generation of fulvic and humic acids (through heterotrophic metabolism) at stations having relatively high pH (S5 for example) (Figure 1B). Nonetheless there was an overall tendency for pore-water salinity and pH to vary as predicted (low landwards and high seawards). The relationship between pore-water salinity and pH is: pH $=0.035$ Salinity $+6.42$ (Figure 1C).

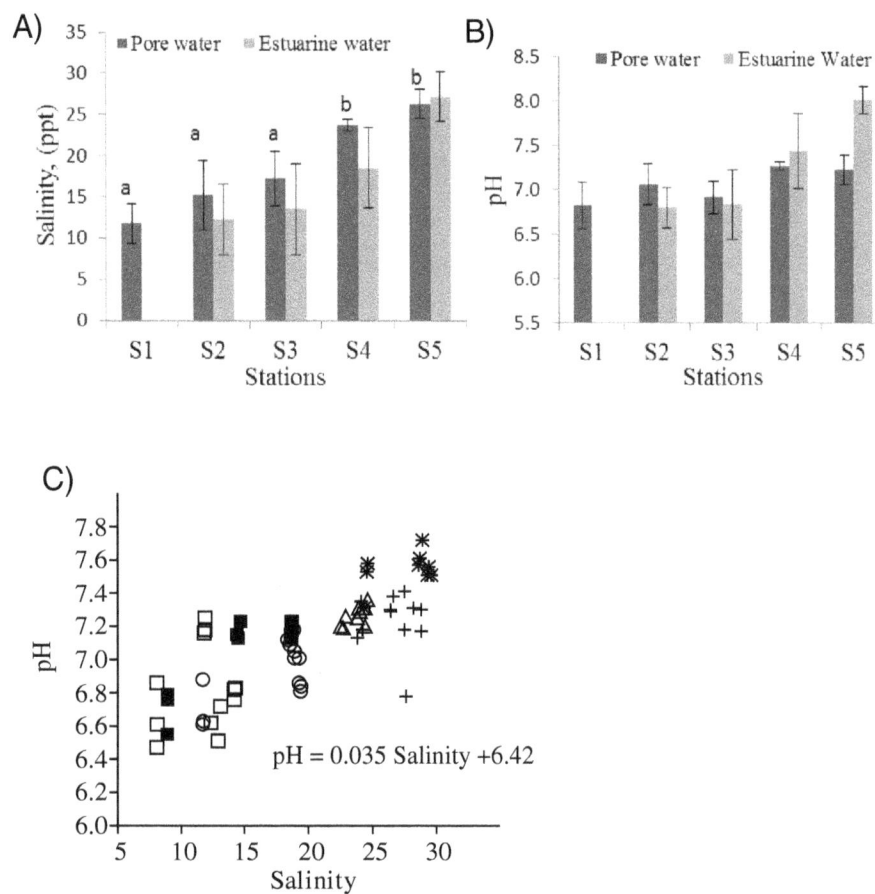

Figure 1 Sediment pore-water and above-sediment estuarine water properties at the five stations along the BES. Stations differed significantly in pore-water salinity (Wald =121.2; p <0.001) **[A]** and pH (Wald =14.2; p <0.006) **[B]**. 'a' and 'b' indicating significant difference at 5% level. The pore-water salinity and pH is correlated **[C]**. Estuarine water salinity and pH data were taken from Marshall et al. [11]; their study did not cover the site S1, hence the salinity and pH data were not available for S1.

Species composition

A total of 4174 individuals, belonging to 36 species from six taxonomic groups (Polychaeta, Copepoda, Amphipoda, Tanaidacea, Cumacea and Isopoda) were recorded in the Brunei estuarine system over the sampling period. The dominant group both in number of species (25 species) and individuals (75%) was the Polychaeta. Composition of species in different stations along the estuary clearly differed. The 10 most abundant species in the estuary, representing 93.3% of the collected benthic infauna, were *Neanthes* sp. (23.2%), *Onuphis conchylega* (9.8%), Nereididae sp.2 (9.7%), Corophiidae sp. (8.7%), Capitellidae sp.1 (8.3%), Cyclopoida sp. (7.9%), *Goniada* sp.(5.6%), Nereididae sp.3 (5.2%), *Prionospio* sp. (2.5%) and Amphipoda sp. 2 (2.44%). The dominant species also varied for different stations along the estuary: Nereididae sp.2 (43.8%) at S1, *Neanthes* sp. (35.0%) at S2, Corophiidae sp. (46.4%) at S3, *Onuphis conchylega* (18.4%) at S4 and Capitellidae sp.1 (21.1%) at S5.

Analyses based on the distinction of calcifiers and non-calcifiers, showed that calcifiers were highest (62.14%) in

the inner low pH station (S3) and the lowest (8.80%) in the downstream station (S4) (Figure 2). In contrast, non-calcifiers were the highest (91.20%) in the downstream high pH station (S4) and the lowest (37.86%) in the acidic station (S3). Although, there was no clear spatial trend in their % composition and mean abundance (Figure 2) along estuarine pH gradient, but species number of calcifiers and non-calcifiers tended to increase towards high pH downstream stations (Figure 2). All species were included for data analysis.

Variation of community parameters

Spatial variation in the number of species, density, species richness and diversity values for each station are presented in Figure 3. Except for evenness (λ^2 = 1.46, df =4, p =0.83), which was uniform among the stations, Kruskal-Wallis ANOVA indicated significant differences in species number (λ^2 = 40.86, df =4, p < 0.001), density (λ^2 = 14.72, df =4, p < 0.01), species richness (λ^2 = 38.18, df =4, p < 0.001) and diversity (λ^2 = 25.01, df =4, p < 0.001). The mean overall

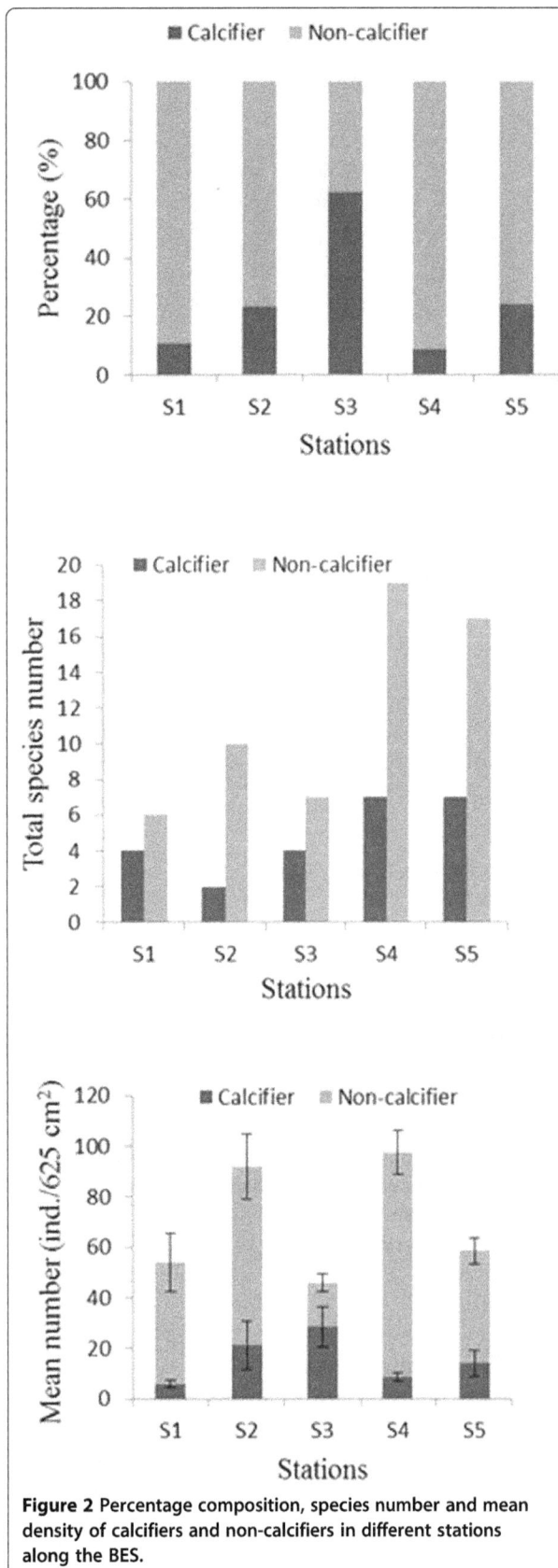

Figure 2 Percentage composition, species number and mean density of calcifiers and non-calcifiers in different stations along the BES.

species number and density per site ranged from 4 to 15 (mean 8) and 46 to 97 (mean 69) ind.625 cm^{-2} respectively. Species diversity and richness per site ranged from 1.01 to 3.15 (mean 1.72) and 0.98 to 2.03 (mean 1.38). All the measured diversity indices were the highest at station 4 and the lowest at station 3 (Figure 3). Species diversity and richness varied among sites, with a general trend of outer sites (e.g., 4 and 5) having significantly higher values than those in inner sites (e.g., 1, 2 and 3) ($P <0.05$ for all).

Variation of community structure

The MDS ordination grouped the infaunal assemblages into three main groups at 20% similarity level: the innermost stations S1 and S2, inner stations S3 and the outer stations S4 and S5 (Figure 4) and showed a horizontal zonation of assemblages, clearly separating outer stations from the rest of the inner stations. The low-stress value (r =0.15) in MDS showed that the community structure is well represented. Within each location, samples clustered mostly with those from the same sampling site. Samples between sites did not show a high level of similarity as indicated by the long terminal branches. The ANOSIM test confirmed the significant differences among stations (Global R = 0.977, p <0.001) (Table 1). R-statistic values for pair-wise comparisons provided by ANOSIM were used here to determine the dissimilarity between groups. Values close to 1 indicate very different composition, while values near zero show small difference. There was also a statistical evidence of temporal variability in infaunal assemblages between sampling periods (Global R = 0.548, p <0.001). SIMPER analysis of infaunal abundance data revealed that the innermost stations (S1 and S2) group was dominated by Nereididae sp.2, *Neanthes* sp. and Cylopoida sp., and the outer stations group (S4 and S5) by *Onuphis conchylega*, Capitellidae sp.1, Nereididae sp.3 and *Goniada* sp (Table 2). SIMPER revealed the average Bray-Curtis dissimilarity between locations ranged from 61.72 to 91.80%. The seaward station S4 had the highest dissimilarity values for infaunal species with most landward stations S1 (91.8%) and S2 (88.30%). The high abundance of *Onuphis conchylega* at S4 and absence of this species in S1 and S2 was a key cause of these high dissimilarities. S1 had an average dissimilarity of 82.9% with most seaward station S5, discriminating species being Nereididae sp.2 and *Neanthes* sp. The lowest dissimilarities were found between two inner stations (61.72%) and between two lower stations (64.76%), the discriminating species were Nereididae sp.3 and *Onuphis conchylega*. The dissimilarities between S1 and S3 was 84.89% and between S3 and S4 was 80.92%, and the discriminating species were Corophiidae sp. and *Onuphis conchylega*.

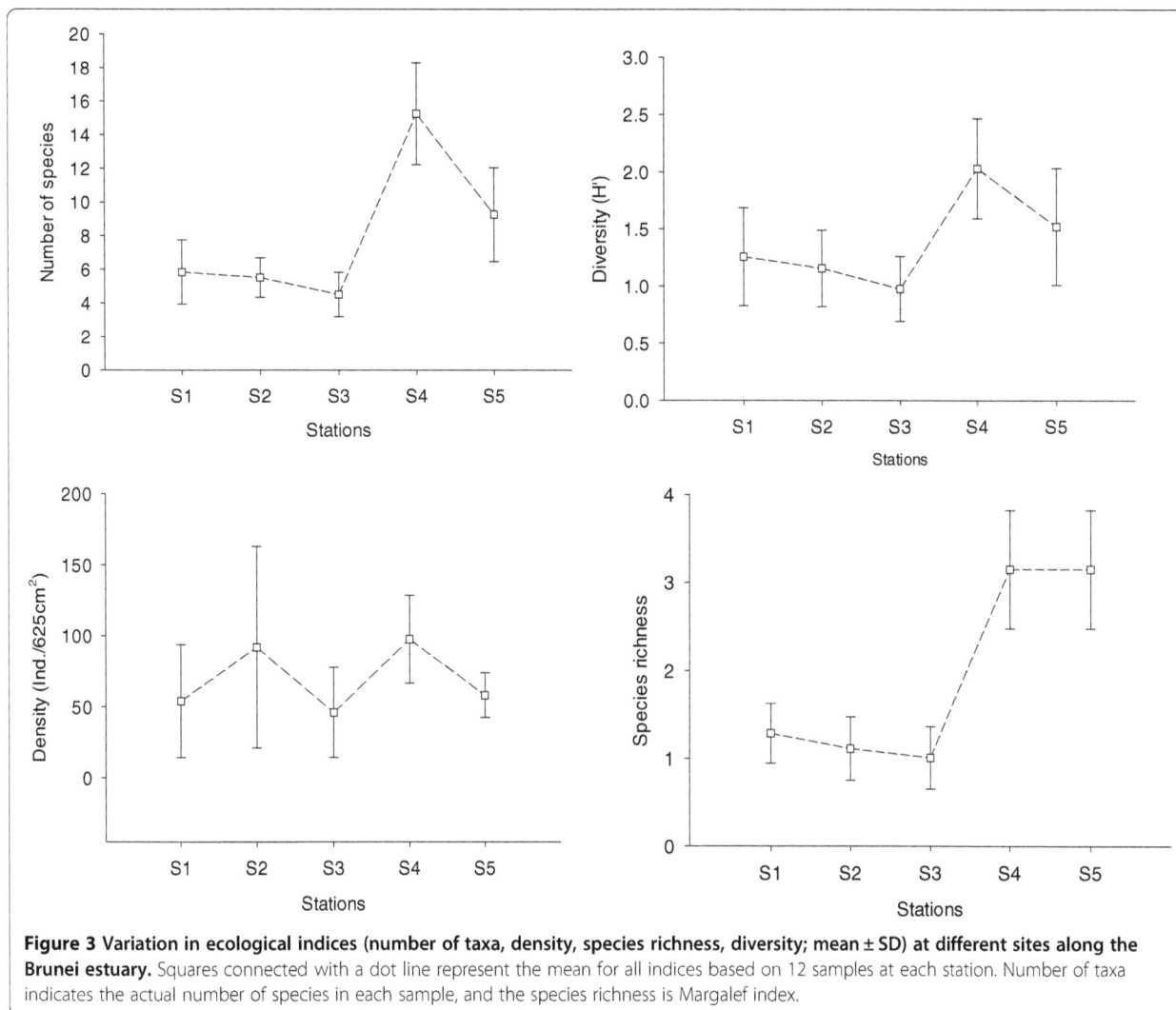

Figure 3 Variation in ecological indices (number of taxa, density, species richness, diversity; mean ± SD) at different sites along the Brunei estuary. Squares connected with a dot line represent the mean for all indices based on 12 samples at each station. Number of taxa indicates the actual number of species in each sample, and the species richness is Margalef index.

Correlation between community parameters and environmental variables

The Spearman's rank correlation analysis between the community parameters and the environmental variables indicated that all of the diversity indices were positively correlated with pore-water salinity and pH (p <0.05) and negatively correlated with clay and organic matter, with the exception of evenness values (p >0.05), which did not show any significant correlation with any of the environmental variables (Table 3). There was no significant correlation between community parameters and sand/silt (p >0.05).

Relationship between the infaunal assemblage and environmental factors

CCA for abundance of the 15 dominant species and six environmental variables produced an ordination plot in which the first two axes explain 71.31% of the variance in the species-environment relationships (Figure 5). The

first axis showed highest positive correlation with salinity (r =0.83), pH (r =0.82) and percentage of sand (r =0.78), and a negative correlation with percentage of clay (r = –0.84) and organic matter (r = –0.74). This axis mainly reflected salinity and % clay, which was closely related to the location of the stations. The second axis had the strongest negative correlation for a percentage silt (r = – 91). The CCA also revealed the relationships among 15 benthic infaunal species and environmental variables. *Onuphis conchylega*, Nereididae sp. 3, *Goniada* sp. Harpacticoida sp. Pilargidae sp., and Corophiidae sp. were placed on the right side of the plot. This indicates that the distribution of these taxa at the outer stations (i.e., high salinity/pH and sandy habitat) of the estuary. *Neanthes* sp., Nereididae sp.2, Cyclopoida sp. *Prionospio* sp, *Potamilla leptochaeta*, Spionidae sp. were found on the left side of the plot (lower salinity/pH and muddy habitat) suggesting these species are mainly distributed across the inner stations. Other dominant species

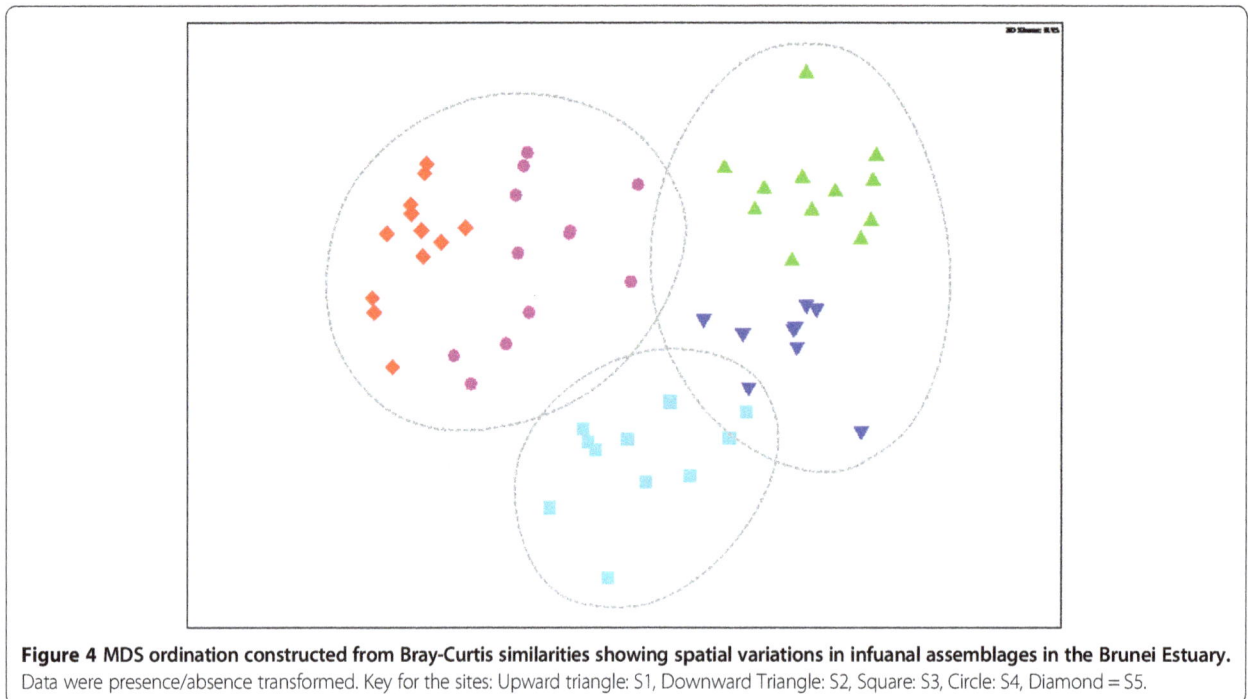

Figure 4 MDS ordination constructed from Bray-Curtis similarities showing spatial variations in infuanal assemblages in the Brunei Estuary.
Data were presence/absence transformed. Key for the sites: Upward triangle: S1, Downward Triangle: S2, Square: S3, Circle: S4, Diamond = S5.

Capitellidae sp.1, Amphipoda sp. 2 and *Sternaspis scutata*, which were widely distributed along the estuary, fell near the origin of the plot.

Discussion

Analysis of the sediment pore-water showed a decline of salinity and pH from sea to landwards, like the pattern reported for overlying estuarine water salinity and pH [11]. The salinity gradient in overlying water, due to the interaction of incoming sea water and outflowing of freshwater, along the estuary from inner to lower reaches is an established fact [22,23]. Ram and Zingde [27] reported that the interstitial water chlorinity is sensitive to the changes in the composition of the overlying water up to a certain depth decided by factors such as sediment type, porosity, the diffusion rate and bioturbation. However, pH variation in estuarine systems is strongly influenced by biological activity in addition to physical factors. The lowering of above sediment estuarine water pH at the landward stations of the Brunei estuary is influenced by the massive acidic freshwater outflow, but also biogenically through the production of carbonic acid from bacterial decomposition of organic matter [11]. High pH at the seaward stations results from the buffering effect of sea water [28]. Sasekumar [29] reported the acidic pH value of pore-water (varied from 6.3-7.1) from a Malayan mangrove shore and suggested that the acidity was caused by the activity of bacteria on oxidizable sulphur. The CO_2 arising from the decomposition of organic matter and animal respiration may also lower the pH values in the sediment pore-water [30,31].

Zhai et al. [31] reported a significant outgassing of CO_2 from Pearl River estuary which was associated with decomposition of organic pollutants by aerobic respiration. The higher pore-water salinity compared to overlying water observed in this study (Figure 1) may be attributed to evaporation of sediment surface water as the intertidal sediment at low tide was exposed to air and sunlight during sampling. However, the pore-water pH in the estuary was less variable than overlying water at the same station.

The infaunal species composition was found to vary along the BES salinity/pH gradient, though most species were euryhaline. The taxa, *Neanthes* sp., Nereididae sp., Capitellidae sp., *Potamilla leptochaeta*, *S. scutata*, Cyclopoida sp., *Leptochelia* sp., extended along the length of the estuary, even tolerating salinities of 8 ppt, but showed greatest densities in the upper estuary. The stenohaline marine component was represented by the species *Onuphis conchylega*, *Pholoe* sp., *Syllis* sp, *Pectinaria* sp. and Calanoida sp., dominated in the lower estuary. Similar patterns of species distribution are known to occur in estuaries around the world [21,22,24,28]. However, the true estuarine organisms (living below 30 ppt but not in the sea) are generally the most difficult to define [32]. *Onuphis conchylega* was most abundant in the middle reaches with abundance declining seaward and entirely absent at the upper low salinity stations, suggesting that it is a stenohaline species in this estuary.

Although there was no clear linear trend in community variables along the salinity/pH gradient as in the other estuaries [22,33], species richness, density, diversity and number

Table 1 Results of ANOSIM and pairwise tests for differences on infaunal structure between stations and sampling periods

Global test	R	p (%)
Between stations	0.977	0.1
Stations compared		
S1, S2	0.852	0.1
S1, S3	1	0.1
S1, S4	1	0.1
S1, S5	1	0.1
S2, S3	0.972	0.1
S2, S4	1	0.1
S2, S5	0.954	0.1
S3, S4	1	0.1
S3, S5	0.954	0.1
S4, S5	0.991	0.1

The pairwise R values give absolute measure of how separated the groups are. R >0.75: groups being well separated; R >0.5: groups overlapping but clearly different; R >0.25: groups barely separable at all. Bold types indicate separable communities.
Analyses performed on square-root transformed data.

of taxa in the benthic infauna were all positively related to salinity (p <0.05 for all). This is due to a reduction in stenohaline species at the inner stations [23,34] and lessened substrate diversity following a progression from sand at the lower reaches to mud at the upper reaches. The relationship between environmental variables and measured diversity indices explained this pattern as significant positive correlations were found for pore-water salinity, pH and % sand, while a negative correlation was found between % clay and organic matter (p <0.05) (Table 3). For infaunal macroinvertebrates, sediment grain size has frequently been reported as a crucial factor in determining the structure of many benthic communities [22,35-40].

Most studies suggest that calcifying animals should be more affected by low pH waters than non-calcifiers [3-5,15,41,42]. However, our abundance data showed no different trend in distribution of calcifiers and non-calcifiers along the pH gradient. The high abundance of few calcifiers in inner low pH stations (Cyclopoida sp. in station S2 and the amphipod Corophiidae sp. in S3) essentially differs from the predicted trend [15,43]. The wide-ranging abundance of infaunal calcifiers may relate to (1) the reduced gradient in the pore-water in the BES, (2) adaptations to endure lowered pHs, (3) biological interactions where the predators are less tolerant of slightly higher acidic conditions. The high number of tolerant species in the acidic areas may also be due to the absence of metal sensitive predators [16]. Furthermore, organisms in coastal and estuarine waters have been experiencing low pH for several hundreds of years and consequently have potentially had time to adapt to

low pH [3,42]. Several groups of invertebrates are physiologically resistant to low pH; for example, crabs can tolerate low pH through their physiological acid–base regulation [3,19,44]. Recent studies suggest that calcifiers are as sensitive to acidification as heavily calcified animals [3,15,41,42]. The findings of infaunal microcrustaceans (amphipods, isopods, cumaceans, tanaids, and copepods) are consistent with these previous findings suggesting a size-related difference in acidic vulnerability [5,15,42].

Many estuarine benthic studies [e.g., 22,33,45,46] have shown that community variables are principally correlated with water salinity, DO and sediment grain size, and this has been confirmed in the present study. Relatively few studies have explored the effect of pH on estuarine infaunal communities [16,17]. However, earlier work [16,17] emphasized the relationships between water column parameters and benthic infauna, rather than focusing on the sediment pore-water, with which these organisms directly interact. Therefore, this the first study investigating the effect of pore-water pH on infaunal community structuring in a tropical estuary. Consequently, the unavailability of data constrains comparison. Nonetheless, the community level responses observed here are consistent with observations on the effects of acidification for other aquatic (marine) systems, including (1) a decrease in total community diversity [43], (2) a high abundance of crustaceans [15], and (3) a shift in the community composition [47].

In conclusion, this study shows that sediment particle size probably overrides effects of the above sediment water, and that the sediment (pore) water pH did not vary as much as the above water pH (primarily due to biogenic acidification of the pore-water). The implication of the latter is that coastal infaunal communities may be less affected by water pH variations caused by various ways including elevation in atmospheric CO_2 than what has been described for epibenthic communities.

Materials and methods
Study site
The BES (Figure 6) constitutes the Inner Brunei Bay and three major river systems (Sungai Limbang, Sungai Temburong and Sungai Brunei) flanked by the South China Sea, Brunei and Malaysia in North-western Borneo (1° 00' 00" N and 114° 00' 00" E). It occupies an area of 1380 km². The equatorial tropical climate in the region ensures high rainfall and temperature conditions throughout the year, with regular freshwater inflow into the shallow (<5 m depth) and well-mixed system [26]. The shoreline is fringed predominantly (75%) by *Rhizophora* mangrove forests [48]. The tides are mainly diurnal with semi-diurnal on a few days, and the daily tidal amplitude ranges from about to 0.9 m to 2.0 m at the estuarine mouth. Although the agricultural and industrial sectors in the region are under-developed, the estuary is subject to significant

Table 2 SIMPER similarity analysis of infaunal species within stations along salinity-pH gradient

Species	Average abundance	Average similitude	Similitude/SD	Contribution %	Cumulative %
Station S1					
Average similarity: 52.65					
Nereididae sp.2	4.28	23.09	2.52	43.86	43.86
Prionospio sp	2.35	11.02	3.01	20.94	64.80
Cyclopoida sp.	1.68	5.73	0.77	10.89	75.69
Neanthes sp.	1.79	5.10	0.98	9.69	85.38
Potamilla leptochaeta	1.84	3.37	0.60	6.40	91.77
Station S2					
Average similarity: 54.96					
Neanthes sp.	5.74	19.25	1.48	35.03	35.03
Cyclopoida sp.	3.83	14.39	3.23	26.19	61.21
Nereididae sp.2	2.91	9.27	1.25	16.87	78.08
Capitellidae sp.1	1.93	7.65	1.81	13.91	91.99
Station S3					
Average similarity: 59.87					
Corophiidae sp.	4.71	27.80	2.82	46.44	46.44
Capitellidae sp.1	2.17	16.54	2.57	27.63	74.07
Neanthes sp.	2.69	11.42	1.08	19.07	93.14
Station S4					
Average similarity: 62.53					
Onuphis conchylega	4.91	11.52	1.88	18.42	18.42
Nereididae sp. 3	3.97	9.76	2.92	15.61	34.03
Capitellidae sp.1	3.01	7.97	5.03	12.75	46.79
Goniada sp.	3.14	5.73	1.15	9.17	55.95
Magelona sp.	1.66	4.19	4.39	6.70	62.66
Pilargidae sp.	1.70	3.96	1.87	6.34	69.00
Maldanidae sp.	1.38	3.18	1.72	5.09	74.09
Pholoe sp.	1.37	2.48	1.04	3.97	78.06
Nephtys sp.	1.11	2.17	1.01	3.47	81.53
Corophiidae sp.	1.01	1.70	0.82	2.71	84.24
Spionidae sp.	0.89	1.48	0.83	2.36	86.60
Harpacticoida sp.	1.06	1.40	0.65	2.24	88.84
Paranthuridae sp.	1.00	1.31	0.65	2.10	90.94
Station S5					
Average similarity: 46.79					
Capitellidae sp.1	2.55	9.88	3.54	21.11	21.11
Goniada sp.	2.20	7.91	1.65	16.91	38.02
Neanthes sp.	3.05	7.14	0.71	15.26	53.27
Pilargidae sp.	1.67	5.66	1.70	12.10	65.38
Onuphis conchylega	1.82	4.76	0.88	10.16	75.54
Harpacticoida sp.	1.28	3.21	1.01	6.86	82.40
Prionospio sp	0.94	3.04	1.06	6.51	88.91
Amphipoda sp. 2	1.66	2.11	0.45	4.51	93.41

Table 3 Spearman's rank correlation coefficients (r) between environmental and community variables estimated for all species

Variables	Number of species (S)	Density (D)	Richness (d)	Diversity (H')	Evenness (J')
Salinity	0.648****	0.279*	0.573****	0.446***	0.010 ns
pH	0.637***	0.315*	0.562****	0.492***	0.118 ns
Sand	0.114 ns	0.018 ns	0.150 ns	0.0565 ns	0.007 ns
Silt	−0.114 ns	−0.018 ns	−0.150 ns	−0.0565 ns	−0.007 ns
Clay	−0.451***	−0.107 ns	−0.450****	−0.349**	−0.096 ns
OM	−0.570****	−0.030 ns	−0.581****	−0.390**	−0.056 ns

*$p < 0.05$, **$p < 0.01$, ***$p < 0.001$, ****$p < 0.0001$.

solid waste pollution, including treated and untreated sewage and domestic waste from traditional water villages, urban developments (Bandar Seri Begawan, Kuilap and Gadong [11,49].

The estuary is naturally acidic, primarily due to eutrophication, heterotrophic metabolism and associated acid sulphate groundwater seeps [11]. Consequently, a steep cline in estuarine water salinity and pH extends across the system; low salinity/pH landwards and high salinity/pH seawards. Five study stations were established along the northern edge of the Inner Brunei Bay and Sungai Brunei estuary (Figure 6).

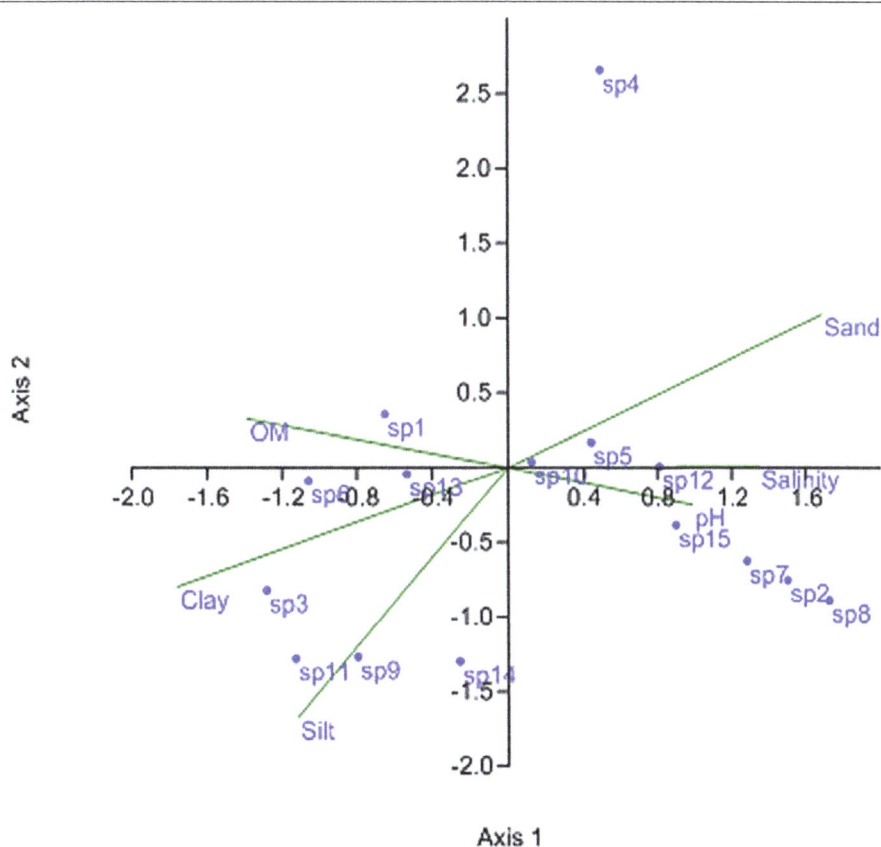

Figure 5 Canonical correspondence analysis (CCA) ordination diagrams for infaunal species abundance data (occurred >1% of total abundance). Species scores along the first and second axes in relation to environmental variables. Sediment parameters reported in Hossain et al. [26] were used. Environmental data were square root/ log transformed as necessary. The environmental variables are shown as line vectors, and the directions of which are obtained from the correlation of the variable to the axes. Species: sp1 = *Neanthes* sp; sp2 = *Onuphis conchylega*; sp3 = Nereididae sp.2; sp4 = Amphipoda sp.1; sp5 = Capitellidae sp.1; sp6 = Cyclopoida sp.; sp7 = *Goniada* sp.; sp8 = Nereididae sp. 3; sp9 = *Prionospio* sp.; sp10 = Amphipoda sp.1; sp11 = *Potamilla leptochaeta*, sp12 = Pilargidae sp.; sp13 = *Sternaspis scutata*; sp14 = Spionidae sp.; sp15 = Harpacticoida sp.

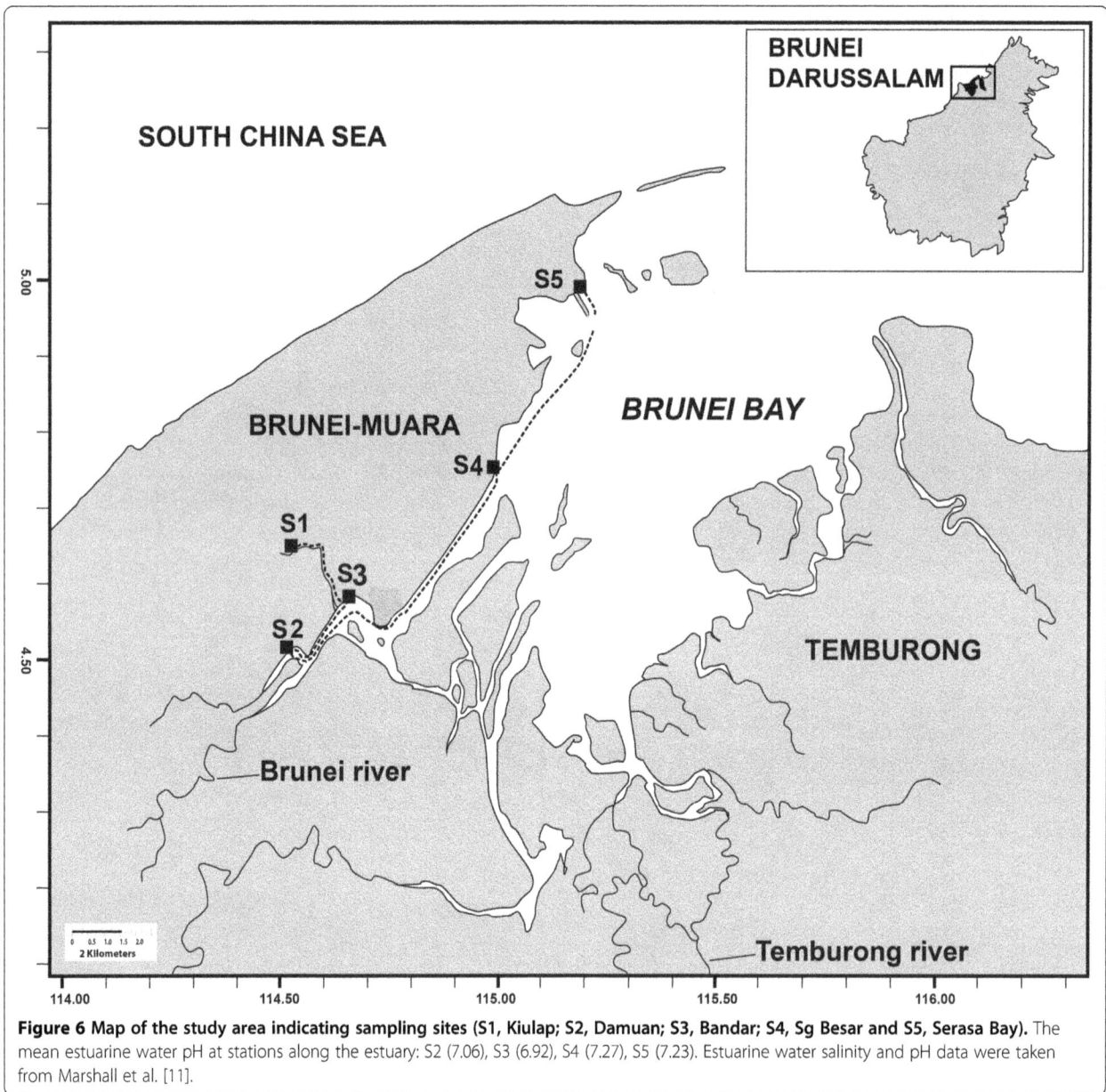

Figure 6 Map of the study area indicating sampling sites (S1, Kiulap; S2, Damuan; S3, Bandar; S4, Sg Besar and S5, Serasa Bay). The mean estuarine water pH at stations along the estuary: S2 (7.06), S3 (6.92), S4 (7.27), S5 (7.23). Estuarine water salinity and pH data were taken from Marshall et al. [11].

Pore-water salinity and pH determination

At each station, *in situ* pH and salinity of low-tide sediment pore-water was determined. A pit of 20 cm in depth was dug using spade, and pH and salinity of the water seep was measured using Hach HQ series portable salinity meters (CDC 401–01, USA) and Mettler Toledo pH probe (Type 1120, Germany) calibrated with Mettler Toledo SRM NIST precision buffer solution (using pH 4, 7, and 10 standards). In each field visit at least five stabilized readings were taken from each sampling spot. Details of methodology and results for sediment properties (% of sand, silt, clay and organic matter) were reported in the earlier study [26].

Infaunal sampling and determination

Macrobenthic infauna were sampled at five stations in Brunei estuary (Figure 6) on four occasions from July 2011 to June 2012. Stations 1 and 2, in the upper reaches of the estuary, had a muddy black bottom which was anoxic and with the smell of H_2S. Station 3, located in the upper reaches but with sandy mud bottom and station 4 and 5, at the lower end of the estuary, had a sandy mud bottom. Three replicate samples were taken at each station from mid-intertidal area on each date. One additional core was taken from each station for determination of sediment physical characteristics. Sediment samples were excavated from an area of 25 cm × 25 cm with a depth of

20 cm by using a spade during low tide. Excavated sediment samples were put into polyethylene bag and carried to the laboratory where sediment samples were washed through 0.5 mm mesh sieve with tap water. Detritus and organisms retained by the sieve were stained in dilute Rose Bengal and fixed in 5% formalin for one day. The formalin was later washed out and specimens were preserved in 70% ethanol. Individuals were separated from the sediment and detritus under a dissecting microscope and preserved and kept in small vials of ethanol for future identification. Animals were identified to the lowest known taxonomic level using available identification sources [e.g., 48-54]. Within each family, fauna were distinguished as 'morphospecies' [55]. The taxonomic status was checked and updated using the web portal WORMS (www.marine-species.org). Densities of taxa (number of individuals/ 625 cm^2 surface area) were determined by counting all organisms in samples. Only the dominant taxa (Polychaeta and Crustacea) were used in the analysis and we assumed interacted closely with the sediment pore-water. All the specimens are deposited in the Biology Department museum, Universiti Brunei Darussalam, Brunei.

Statistical analysis

Univariate statistical analyses were undertaken to determine for each station the number of species (taxa), average density (D), Margalef's species richness (d), Pielou's evenness (J'), Shannon-Weiner diversity (H', loge base). On the assumption that the data were not normally distributed, nonparametric Kruskal-Wallis ANOVA and Mann–Whitney pair-wise comparison tests were performed to assess significant differences in the diversity indices among sites. These analyses were run using PAST [56]. Multivariate statistics were used to investigate variations in the structure of the infaunal community throughout the study period. A Bray-Curtis similarity matrix was computed using abundances of fauna, from which cluster analysis and the non-metric multidimensional scaling (MDS) ordination plot was generated (based on presence/absence transformed data) to visualize the patterns in the spatial distribution using PRIMER V.6 [57,58]. The programme was also used to compute a two-way analysis of similarity (ANOSIM), to determine significant spatial variation in the faunal communities. Taxa making the highest contribution to the differences were detected using SIMPER (PRIMER).

Spearman's rank correlations were calculated to show the relationship between univariate measures and environmental parameters. Sediment parameters reported in Hossain et al. [26] were used in order to provide a basis for interpretation of the biological variables. Associations between assemblage patterns and environmental variables were quantified via canonical correspondence analysis (CCA), a nonlinear eigenvector ordination technique related to CA but which constrains the axes to be linear combinations of the measured environmental variables (run in PAST).

Competing interests

The authors declare that they have no competing interests.

Authors' contributions

This is Postgraduate research (Ph.D.) work of the first author. DJM supervised the work and contributed to this study significantly, and were involved at different times: sampling designing, data analyzing and improving the English language. Both authors read and approved the final manuscript.

Acknowledgments

MBH is indebted to UBD for providing GRS scholarship. We gratefully acknowledge the financial support provided through the UBD grant number UBD/GSR/S&T/16. The authors are grateful to Editor-in- Chief and anonymous referee for their constructive comments and suggestions which led to significant improvements to the manuscript.

References

1. Dove CM, Sammut J: Impacts of estuarine acidification on survival and growth of sydney rock oysters *Saccostrea glomerata* (Gould 1850). *J Shellfish Res* 2007, 26:519–527.
2. Salisbury J, Green M, Hunt C, Campbell J: Coastal acidification by rivers: a threat to shellfish? *Eos* 2008, 89:513–528.
3. Amaral V, Cabral HN, Bishop MJ: Resistance among wild invertebrate populations to recurrent estuarine acidification. *Estuar Coast Shelf Sci* 2011, 93:460–467.
4. Amaral V, Cabral HN, Bishop MJ: Effects of estuarine acidification on predator–prey interactions. *Mar Ecol Prog Ser* 2012, 445:117–127.
5. Amaral V, Cabral HN, Bishop MJ: Prior exposure influences the behavioural avoidance by an intertidal gastropod, *Bembicium auratum*, of acidified waters. *Estuar Coast Shelf Sci* 2014, 136:82–90.
6. U.S. Environmental Protection Agency (USEPA): pH and Alkalinity. In *Voluntary Estuary Monitoring Manual.* Washington DC: USEPA Office of Water; 2006:1–13. Available at http://water.epa.gov/type/oceb/nep/upload/ 2009_03_13_estuaries_monitor_chap11.pdf.
7. Sammut J, Melville MD, Callinan RB, Fraser GC: Estuarine acidification: impacts on aquatic biota of draining acid sulfate soils. *Aust Geogr Stud* 1995, 33:89–100.
8. Knutzen J: Effects of decreased pH on marine organisms. *Mar Pollut Bull* 1981, 12:25–29.
9. Amaral V, Cabral HN, Bishop MJ: Moderate acidification affects growth but not survival of 6-month-old oysters. *Aquat Ecol* 2012, 46:119–127.
10. Guinotte JM, Fabry VJ: Ocean acidification and its potential effects on marine ecosystems. *Ann N Y Acad Sci* 2008, 1134:320–342.
11. Marshall DJ, Santos JH, Leung KMY, Chak WH: Correlations between gastropod shell dissolution and water chemical properties in a tropical estuary. *Mar Environ Res* 2008, 66:422–429.
12. Caldeira K, Wickett ME: Oceanography: anthropogenic carbon and ocean pH. *Nature* 2003, 425:365.
13. Orr JC, Fabry VJ, Aumont O, Bopp L, Doney SC, Feely RA, Gnanadesikan A, Gruber N, Ishida A, Joos F, Key RM, Lindsay K, Maier-Reimer E, Matear R, Monfray P, Mouchet A, Najjar RG, Plattner GK, Rodgers KB, Sabine CL, Sarmiento JL, Schlitzer R, Slater RD, Totterdell IJ, Weirig MF, Yamanaka Y, Yool A: Anthropogenic ocean acidification over the twenty-first century and its impact on calcifying organisms. *Nature* 2005, 437(7059):681–686.
14. Feely RA, Simone RA, Newton J, Sabine CL, Warner M, Devol C, Krembs MC: The combined effects of ocean acidification, mixing, and respiration on pH and carbonate saturation in an urbanized estuary. *Estuar Coast Shelf Sci* 2010, 88:442–449.
15. Kroeker KJ, Micheli F, Gambi MC, Martz TR: Divergent ecosystem responses within a benthic marine community to Ocean acidification. *Proc Natl Acad Sci U S A* 2011, 108:14515–14520.
16. Roach AC: The effect of acid water inflow on the estuarine benthic and fish communities in the Richmond River, NSW Australia. *Australas J Ecotoxicol* 1997, 3:25–56.

17. Corfield J: The effects of acid sulphate run-off on a subtidal estuarine macrobenthic community in the Richmond River, NSW, Australia. *ICES J Mar Sci* 2000, **57**:1517–1523.

18. Lohbeck KT, Riebesell U, Reusch TB: Adaptive evolution of a key phytoplankton species to ocean acidification. *Nat Geosci* 2012, **5**(5):346–351.

19. Wittmann AC, Pörtner HO: Sensitivities of extant animal taxa to ocean acidification. *Nat Clim Chang* 2013, **3**:995–1001.

20. Teske PR, Wooldridge TH: What limits the distribution of subtidal macrobenthos in permanently open and temporarily open/closed South African estuaries? Salinity vs. sediment particle size. *Estuar Coast Shelf Sci* 2003, **57**:225–238.

21. Warwick RM: Comparative study of the structure of some tropical and temperate marine soft-bottom macrobenthic communities. *Mar Biol* 1987, **95**(4):641–649.

22. Ysebaert T, Herman PMJ, Meire P, Craeymeersch J, Verbeek H, Heip CHR: Large-scale spatial patterns in estuaries: estuarine macrobenthic communities in the Schelde estuary, NW Europe. *Estuar Coast Shelf Sci* 2003, **57**:335–355.

23. McLusky DS, Elliot M: *The Estuarine Ecosystem: Ecology, Threats, and Management.* 3rd edition. Oxford: Oxford University Press; 2006.

24. Barnes RSK: Distribution patterns of macrobenthic biodiversity in the intertidal seagrass beds of an estuarine system, and their conservation significance. *Biodivers Conserv* 2013, **22**:357–372.

25. McLusky DS, Hull SC, Elliott M: Variations in the intertidal and subtidal macrofauna and sediments along a salinity gradient in the Upper Forth estuary. *Neth J Aquat Ecol* 1993, **27**:101–109.

26. Hossain MB, Marshall DJ, Senapathi V: Sediment granulometry and organic matter content in the intertidal zone of the Sungai Brunei estuarine system, northwest coast of Borneo. *Carpathian J Earth Environ Sci* 2014, **9**(2):231–239.

27. Ram A, Zingde MD: Interstitial water chemistry and nutrients fluxes from tropical intertidal sediment. *Indian J Mar Sci* 2000, **29**(4):310–318.

28. Day JW, Hall CAS, Kemp WM, Yanez-Arancibia A: *Estuarine Ecology.* New York: J. Wiley and Sons; 1989:558.

29. Sasekumar A: Distribution of macrofauna on a Malayan mangrove shore. *J Anim Ecol* 1974, **43**(1):51–69.

30. Zhai W, Dai M, Guo X: Carbonate system and CO_2 degassing fluxes in the inner estuary of Changjiang (Yangtze) River. *Chin Mar Chem* 2007, **107**(3):342–356.

31. Zhai W, Dai M, Cai WJ, Wang Y, Wang Z: High partial pressure of CO_2 and its maintaining mechanism in a subtropical estuary: the Pearl River estuary. *Chin Mar Chem* 2005, **93**(1):21–32.

32. Gray JS, Elliott M: *Ecology of Marine Sediments: science to management.* 2nd edition. Oxford: Oxford University Press; 2009:225.

33. Selleslagh J, Lobry J, N'Zigou AR, Bachelet G, Blanchet H, Chaalali, Sautour AB, Boet P: Seasonal succession of estuarine fish, shrimps, macrozoobenthos and plankton: physico-chemical and trophic influence. The Gironde Estuary as a Case Study. *Estuar Coast Shelf Sci* 2012, **112**:243–254.

34. Telesh I, Schubert H, Skarlato S: Life in the salinity gradient: Discovering mechanisms behind a new biodiversity pattern. *Estuar Coast Shelf Sci* 2013, in Press.

35. Gray JS: Animal-sediment relationships. *Oceanogr Mar Biol Ann Rev* 1974, **12**(223):61.

36. Anderson MJ, Ford RB, Honeywill C, Feary D: Quantitative measures of sedimentation in an estuarine system and its relationship with intertidal soft-sediment infauna. *Mar Ecol Prog Ser* 2004, **272**:33–48.

37. Giménez L, Dimitriadis C, Carranza A, Borthagaray AI, Rodríguez M: Unravelling the complex structure of a benthic community: a multiscale-multianalytical approach to an estuarine sandflat. *Estuar Coast Shelf Sci* 2006, **68**:462–472.

38. Norkko A, Thrush SF, Hewitt JE, Cummings VJ, Norkko J, Ellis JI, MacDonald I: Smothering of estuarine sandflats by terrigenous clay: the role of wind-wave disturbance and bioturbation in site-dependent macrofaunal recovery. *Mar Ecol Prog Ser* 2002, **234**:23–42.

39. Ysebaert T, Herman PM: Spatial and temporal variation in benthic macrofauna and relationships with environmental variables in an estuarine, intertidal soft-sediment environment. *Mar Ecol Prog Ser* 2002, **244**:105–124.

40. Giménez L, Venturini N, Kandratavicius N, Hutton M, Lanfranconi A, Rodríguez M, Muniz P: Macrofaunal patterns and animal–sediment relationships in Uruguayan estuaries and coastal lagoons (Atlantic coast of South America). *J Sea Res* 2014, **87**:46–55.

41. Hendriks IE, Duarte CM, Alvarez M: Vulnerability of marine biodiversity to ocean acidification: a meta-analysis. *Estuar Coast Shelf Sci* 2010, **86**:157–164.

42. Hofmann GE, Barry JP, Edmunds PJ, Gates RD, Hutchins DA, Klinger T, Sewell MA: The effect of ocean acidification on calcifying organisms in marine ecosystems: an organism-to-ecosystem perspective. *Annu Rev Ecol Evol Syst* 2010, **41**:127–147.

43. Hall-Spencer JM, Rodolfo-Metalpa R, Martin S, Ransome E, Fine M, Turner SM, Buia MC: Volcanic carbon dioxide vents show ecosystem effects of ocean acidification. *Nature* 2008, **454**(7200):96–99.

44. Pörtner HO: Ecosystem effects of ocean acidification in times of ocean warming: a physiologist's view. *Mar Ecol Prog Ser* 2008, **373**:203–217.

45. Hampel H, Elliott M, Cattrijsse A: Macrofaunal communities in the habitats of intertidal marshes along the salinity gradient of the Schelde estuary. *Estuar Coast Shelf Sci* 2009, **84**:45–53.

46. Hossain MB: *Macrobenthic Community Structure from a Tropical Estuary.* Germany: LAP publishing company; 2011:84.

47. Traister EM, McDowell WD, Krám P, Fottová D, Kolaříková K: Persistent effects of acidification on stream ecosystem structure and function. *Freshw Sci* 2013, **32**:586–596.

48. Chua T-E, Chou LM, Sadorra MSM: *The coastal environmental profile of Brunei Darussalam: resource assessment and management issues.* Brunei Darussalam: ICLARM Technical Report 18, Fisheries Department, Ministry of Development; 1987:193.

49. Fauvel P: *The Fauna of India, including Pakistan, Ceylon, Burma and Malaya. Annelida Polychaeta.* Allahabad: The Indian Press; 1953:507.

50. Day JH: A monograph of the Polychaeta of Southern Africa. *Lond Br Nat Hist Mus* 1967, **1**:458.

51. Fauchald K: The polychaete worms, definitions and keys to the orders, families and genera. *Nat Hist Mus Los Angeles Cty: Los Angeles CA (USA) Sci Ser* 1977, **28**:1–188.

52. Day JH: A monograph of the polychaeta of Southern Africa. *Lond Br Nat Hist Mus* 1967, **2**:420.

53. Hutchings P, Reid A: The Nereididae (Polychaeta) from Australia - Leonnates, Platynereis and Solomononereis. *Rec Aust Mus* 1991, **43**:47–62.

54. Barnes RD, Ruppert EE: *Invertebrate Zoology.* USA: Brooks; 1994:928.

55. Oliver I, Beattie AJ: A possible method for the rapid assessment of biodiversity. *Conserv Biol* 1993, **7**:562–568.

56. Hammer Ø, Harper DAT, Ryan PD: PAST: paleontological statistics software package for education and data analysis. *Palaeontol Electron* 2004, **4**, art. 4.

57. Clarke KR, Warwick RM: *Change in Marine Communities: An Approach to Statistical Analysis and Interpretation.* Plymouth: Plymouth Marine Laboratory; 2001:172.

58. Clarke KR, Gorley RN: *PRIMER v.6: User Manual/Tutorial.* Plymouth: PRIMER-E; 2006.

13

The Manx shearwater (*Puffinus puffinus*) as a candidate sentinel of Atlantic Ocean health

Maíra Duarte Cardoso[1*], Jailson Fulgencio de Moura[2], Davi C Tavares[3], Rodrigo A Gonçalves[4], Fernanda I Colabuono[5], Emily M Roges[6], Roberta Laine de Souza[6], Dalia Dos Prazeres Rodrigues[6], Rosalinda C Montone[5] and Salvatore Siciliano[3]

Abstract

Introduction: Seabirds have been historically used to monitor environmental contamination. The aim of the present study was to test the suitability of a species belonging to the Procellariiformes group, the Manx shearwater, *Puffinus puffinus*, as a sentinel of environmental health, by determining contaminant levels (trace metals and organochlorine compounds) from carcass tissues and by isolating *Vibrio* spp. and *Aeromonas* spp. from live specimens. To this end, 35 *Puffinus puffinus* carcasses wrecked on the north-central coast of the state of Rio de Janeiro, Brazil, and two carcasses recovered in Aracruz, on the coast of the state of Espírito Santo, Brazil, were sampled, and fragments of muscle and hepatic tissues were collected for contaminant analyses. Swabs from eleven birds found alive at the north-central coast of Rio de Janeiro were collected for isolation of the aforementioned bacteria.

Results: The average concentration in dry weight (dw) of the trace metals were: mercury 7.19 mg kg^{-1}(liver) and 1.23 mg kg^{-1} (muscle); selenium 34.66 mg kg^{-1} (liver) and 7.98 mg kg^{-1} (muscle); cadmium 22.33 mg kg^{-1} (liver) and 1.11 mg kg^{-1} (muscle); and lead, 0.1 mg kg^{-1} (liver) and 0.16 mg kg^{-1} (muscle). Organochlorine compounds were detected in all specimens, and hexachlorbiphenyls, heptachlorbiphenyls and DDTs presented the highest levels. Regarding microbiological contamination, bacteria from the *Vibrio* genus were isolated from 91% of the analyzed specimens. *Vibrio harveyi* was the predominant species. Bacteria from the *Aeromonas* genus were isolated from 18% of the specimens. *Aeromonas sobria* was the only identified species.

Conclusions: The results indicate that *Puffinus puffinus* seems to be a competent ocean health sentinel. Therefore, the monitoring of contaminant levels and the isolation of public health interest bacteria should proceed in order to consolidate this species importance as a sentinel.

Keywords: *Puffinus puffinus*, Brazil, Sentinel, Metal, Organochlorines, *Vibrio*, *Aeromonas*

Introduction

Oceans cover approximately 70% of the earth surface [1] and about 60% of the human population lives in coastal areas. Many of these populations depend on the ocean for their subsistence [2].

As environmental degradation accelerates, science has increasingly focused on the influence of the environment on human health. Environmental degradation has direct impacts on life quality and health conditions of the human

population [3]. Consequently, ocean processes, which are influenced by human activity, have important public health implications [4,5].

Some substances, such as persistent organic pollutants (POPs), polycyclic aromatic hydrocarbon (PAHs) and toxic metals show negative impacts on the health of humans and other animals and also on the oceans themselves [5-8]. The same is true with regard to pathogenic microorganisms, especially those autochthonous of marine environments, like bacteria from the *Vibrio* and *Aeromonas* genera [5,7].

As these substances reach the marine environment, they impact the biota in a negative manner [9]. Metals are naturally present in marine environments, and they only

* Correspondence: mairadc@gmail.com

[1]Programa de Pós-Graduação em Saúde Pública e Meio Ambiente, ENSP/Fiocruz, Rua Leopoldo Bulhões, 1480, Manguinhos, Rio de Janeiro 21041-210, RJ, Brasil

Full list of author information is available at the end of the article

become toxic when their concentration levels are increased beyond a certain point [10]. Many organochlorine compounds are synthetic and reach the environment mainly by anthropic action [9]. These two classes of pollutants are persistent in the marine environment and are capable of bioaccumulation and biomagnification in the marine food web [10].

There is a need for coastal countries to develop ocean monitoring strategies [11] and one way of doing this is through the use of sentinel species [12]. These species are capable of accumulating pollutants in their organisms without significant adverse effects and are used to measure the amount of bioavailable pollutants [13]. Seabirds have been historically used as sentinels because they are well-known, conspicuous, ubiquitous, abundant, large, long-living, well-liked by people, and, most importantly, they are top predators in the food chain. This is important, since the determination of pollutants that are capable of bioaccumulation and biomagnification is most adequate in higher level organisms [14-16].

In this context, the present study suggests the use of the Manx shearwater (*Puffinus puffinus*) as a sentinel of Atlantic Ocean health, as this species fulfills the requisites of a good sentinel species [17]. The oldest specimen ever recorded was over fifty years old [18] and a 31-year-old specimen was recovered from the Brazilian coast in 2009 (GEMM-Lagos, unpublished data). Manx shearwater colonies are located in the North Atlantic Ocean, mostly in the United Kingdom [17]. During the northern winter, the Manx shearwater migrates to the South Atlantic Ocean, with the Brazilian coast as its main destination [17,19]. The objection that a migratory species should not be used for this purpose, as they are not specific to that particular environment, does not stand when the goal is to gather data at a large scale [14].

Thus, the aim of the present study was to test the usefulness of *P. puffinus* as a sentinel of environmental health, by determining contaminant levels (trace metals and organochlorine compounds) from carcass tissues and by isolating *Vibrio* spp. and *Aeromonas* spp. from live specimens.

Results and discussion
Biometric data
The length of the carcasses collected in this study was of 32.88 ± 0.49 (mean ± standard deviation), ranging from

30 to 36 cm. The wingspan was of 72.24 ± 0.52 (mean ± standard deviation), ranging from 69 to 76 cm.

Metal analyses
Element concentrations are presented in Table 1, as mean ± standard deviation (SD) and range (min-max) on a dry weight basis. Tables 2 and 3 show the values obtained and the recovery levels of the certified reference material analyses. The method limits of quantification for muscle tissue were 0.02 mg.kg^{-1} for Hg; 0.072 mg.kg^{-1} for Se; 0.071 mg.kg^{-1} for Cd; and 0.097 mg.kg^{-1} for Pb. The limits for hepatic tissue were 0.02 mg.kg^{-1} for Hg; 0.59 mg.kg^{-1} for Se; 0.28 mg.kg^{-1} for Cd; and 0.009 mg.kg^{-1} for Pb.

The results of this study are in accordance with worldwide literature. Mean Hg concentrations in *P. puffinus* hepatic tissue were slightly higher than those reported by Dale *et al.* [20] but somewhat lower than those reported by Osborn *et al.* [21]. The means in muscle tissue were somewhat higher than the results by Osborn *et al.* [21]. These comparisons suggest that mercury concentrations in this species have not varied much from the 1970s to the present day.

Mean hepatic Cd concentrations in this study are higher than those described by Osborn *et al.* [21] and Garcia [22], but this is mainly due to outliers. If the median (16.25 mg.kg^{-1}) had been used instead of the means, the results would have been very close to those reported by the aforementioned studies. In muscle tissue, Cd concentrations were lower than those described by Osborn *et al.* [21] and Garcia [22].

Comparing the results of this study with others involving different *Puffinus* genus species with similar habits, lower metal concentrations are usually reported in species that breed in the southern hemisphere (*P. gravis*, *P. assimilis* and *P. griseus*) [23-26]. It is possible that the North Atlantic Ocean, where *P. puffinus* colonies are located, shows higher metal contamination rates, due to earlier and more intense industrialization, but the differences observed could also be due to different diets within the species [23].

On the other hand, Cd concentrations reported by Muirhead & Furness [23] in *P. gravis* and *P. assimilis* are almost twice those observed in the present study. Garcia [22] also reported higher concentrations in *P. gravis* tissues. The concentration of this metal tends to be higher in species whose diet consists of a high amount of cephalopods [23], since these animals are important transfer

Table 1 Metal concentrations tissues of in *Puffinus puffinus* wrecked at study area

Tissue	[Hg] (dw) (mg.kg^{-1})	[Se] (dw) (mg.kg^{-1})	[Cd] (dw) (mg.kg^{-1})	[Pb] (dw) (mg.kg^{-1})
Hepatic (n = 20)	7.19 ± 3.37 (1.16 – 14.22)	34.66 ± 20.14 (10.56 – 75.20)	22.33 ± 25.46 (2.31 – 113.01)	0.1 ± 0.06 (0.036 – 0.28)
Muscle (n = 37)	1.23 ± 0.53 (0.47 – 2.31)	7.98 ± 3.68 (3.17 – 19.01)	1.11 ± 1.72 (<LQ* – 8.94)	0.16 ± 0.09 (<LQ* – 0.43)

*< LQ = below the limits of quantification.
Detailed legend: Metal concentrations tissues of in *Puffinus puffinus* wrecked at study area, from 2005 to 2011.

Table 2 Analyses and recovery of the certified reference materials (DORM-2)

Metal	Reference value DORM-2 (mg.kg^{-1})	Mean of obtained values (mg.kg^{-1})	Recovery (%)	n
Cd	0.043 ± 0.008	0.04	97%	3
Hg	4.64 ± 0.26	4.83	104%	3
Pb	0.065 ± 0.007	0.07	104%	3
Se	1.4 ± 0.09	1.39	99%	3

vectors of cadmium in the food chain [27,28]. Cephalopods are important prey in *P. assimilis*, *P. gravis* and *P. puffinus* diets [17,23,29], so it is possible that the species studied by Muirhead & Furness [23] and Garcia [22] fed on cephalopods that accumulate higher cadmium concentrations, or simply fed on a higher amount of cephalopods, as compared to the *P. puffinus* specimens analyzed in this study. Differences in the diet of this species can be seen in the study by Petry *et al.* [29]. Another hypothesis is that the specimens of the previously cited studies were older than those used in the present study, since cadmium tends to accumulate over the years [30].

Pb concentrations reported in studies involving *P. gravis* and *P. griseus* were very low, in accordance to the present study [25,26]. This is due to the biodilution that the element seems to undergo in the marine food chain [31,32].

It is important to highlight that most cephalopod species and all the fish species cited as Manx shearwater food sources in several studies, such as *Ammodytes tobianus*, *Mallotus villosus*, *Paralonchurus brasiliensis*, *Clupea* genus juveniles and juvenile squid from the Ommastrephidae family [17,29,33], are also consumed by humans, indicating that humans can be exposed to these contaminants through their diet.

Concentrations of the same metal in hepatic and muscle tissues

Spearman's rank correlation coefficient was used in order to verify the strength of the association between the concentration of an element in the liver and the concentration of the same element in the muscle. The classification of the strength of the association followed the scale described by Bryman & Cramer [34].

Moderate positive correlations were found between hepatic and muscle Hg ($\rho = 0.50$) and between hepatic and muscle Se ($\rho = 0.48$). A strong positive correlation was found between hepatic and muscle Cd ($\rho = 0.81$), and a weak positive correlation, with no statistical significance, was found between hepatic and muscle Pb ($\rho = 0.29$).

The differences in element concentrations in both tissues can be seen in Figure 1, where all elements, except for Pb, show the same pattern, with higher concentrations in hepatic tissue when compared to muscle.

In birds, 90% of the existing cadmium is accumulated in the liver and kidneys, the latter being the main site of cadmium toxicity, since kidneys do not resist the toxic effects of cadmium as well as the liver. For this reason, hepatic tissue is considered the best locus to monitor cadmium exposition. It is also the best choice for monitoring mercury and selenium exposure [35]. This is possibly the reason why, in the present study, Cd levels in the liver were always higher than in muscle.

In this study, Pb behaved differently from the other analyzed metals, but it is important to note that Pb levels were very low and the differences between muscle and liver concentrations were very small. In birds, this element accumulates primarily in the bones and among the soft tissues, with the primary accumulation site being the kidneys [35,36]. It is known that lead can occupy the binding sites of calcium [37], so this is possibly the reason why, in the present study, the element was found at its highest concentration in muscle tissues.

Interelemental relationships

The correlations between the elements were tested by the Spearman's rank correlation coefficient. Very weak positive correlations, with no statistical significance, were found between Hg and Se ($\rho = 0.05$) and Se and Cd ($\rho = 0.04$), both in muscle tissue; and also between Cd and Pb ($\rho = 0.16$) and Hg and Pb ($\rho = 0.10$), both in hepatic tissue. Very weak negative correlations, with no statistical significance, were found between Hg and Cd ($\rho = -0.04$) and Hg and Pb ($\rho = -0.06$), both in muscle tissue. Weak positive correlations, with no statistical significance, were

Table 3 Analyses and recovery of the certified reference materials (DOLT-3)

Metal	Reference value DORM-2 (mg.kg-1)	Mean of obtained values (mg.kg-1)	Recovery (%)	n
Cd	19.4 ± 0.6	17.45	90%	3
Hg	3.37 ± 0.14	3.17	94%	3
Pb	0.319 ± 0.045	0.29	92%	3
Se	7.06 ± 0.48	6.98	99%	3

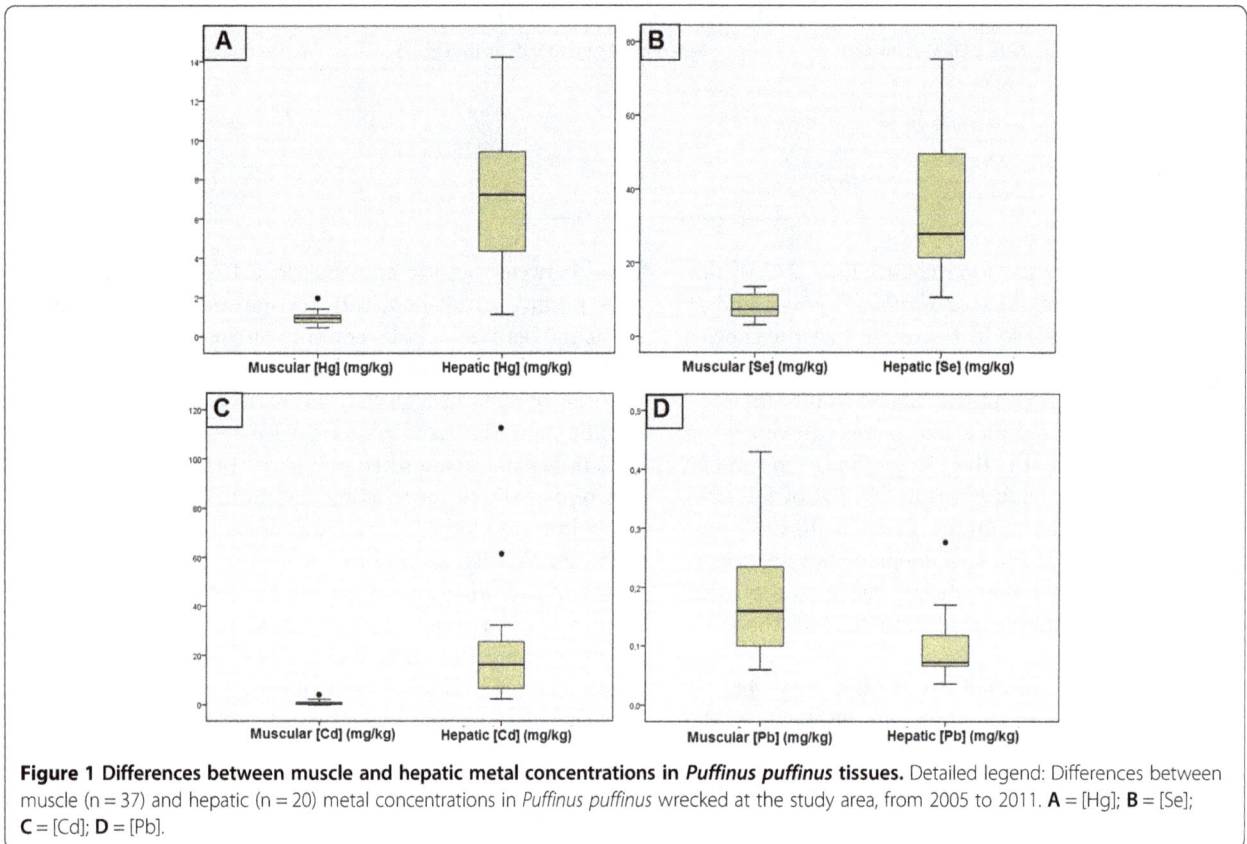

Figure 1 Differences between muscle and hepatic metal concentrations in *Puffinus puffinus* tissues. Detailed legend: Differences between muscle (n = 37) and hepatic (n = 20) metal concentrations in *Puffinus puffinus* wrecked at the study area, from 2005 to 2011. **A** = [Hg]; **B** = [Se]; **C** = [Cd]; **D** = [Pb].

found between Se and Pb ($\rho = 0.20$) and Cd and Pb ($\rho = 0.20$), both in muscle tissue; and the same was observed between Se and Pb in hepatic tissue ($\rho = 0.20$).

However, some statistically significant correlations were found, such as the moderate positive correlations found between Hg and Se ($\rho = 0.46$), Hg and Cd ($\rho = 0.52$) and Se and Cd ($\rho = 0.50$), all in hepatic tissue.

Selenium and mercury tend to co-accumulate in bird livers, where selenium acts in reducing mercury toxicity [38-40], although this detoxifying process is not well elucidated. Some possible mechanisms involved in this are the redistribution of mercury throughout the organism in the presence of selenium; competition between the two elements for binding sites; formation of complexes between these elements; conversion of toxic forms of mercury into less toxic forms, such as the demethylation of methyl mercury by selenium; and prevention of oxidative stress caused by mercury [38,41,42].

In marine mammals, a molar ratio of 1:1 between Hg and Se is found when Se is being used to detoxify Hg and vice-versa [43]. In birds, this ratio is almost never found [38,43,44], as in the present study, where a 1:5 ratio was found. This is possibly the reason why the correlation between Hg and Se found here is moderate, and not strong or very strong, as has been reported in several studies involving marine mammals [45-47], since this protection

mechanism seems to be inherent to marine mammals [43]. Other studies involving aquatic birds have found a larger quantity of selenium relative to mercury, precluding the 1:1 molar ratio, as in the present study [42,44,48].

Current literature has also suggested a correlation between Se and Cd in marine birds [30]. Besides detoxifying mercury, selenium can also detoxify cadmium in these organisms, probably by altering the availability of this metal [30].

Selenium seems to be highly effective against damage caused by cadmium and also against methyl mercury and inorganic mercury toxicity, and seems to have very little effect against lead toxicity [41]. This is possibly the reason why, in the present study, Se, Cd and Hg showed correlations, while Pb did not.

Organochlorine compounds analyses

Organochlorine compound concentrations are presented in Table 4, as a sum of each group and range (min-max) in ng.g^{-1}, on a dry weight basis.

In this study, organochlorines were found in all specimens. Among the OCPs, the DDT group predominated, especially pp' DDE. Among the PCBs, there was a predominance of hexachlorbiphenyls and heptachlorbiphenyls.

The standard deviation for each compound was extremely high, which indicates great individual variation

Table 4 Concentrations of organochlorine compounds in tissues of *Puffinus puffinus* wrecked at study area

Compounds	Results (dw) (ng.g^{-1}) (n = 13)
ΣHCHs[a]	<LQ*
HCB	16.7 (5.97 – 34.3)
ΣChlordanes[b]	9.59 (<LQ* – 20.8)
ΣDrins[c]	31.39 (<LQ* – 65.2)
ΣDDTs[d]	193.42 (<LQ* – 626.0)
Endosulfan II	<LQ*
Metoxychlor	<LQ*
Mirex	8.94 (<LQ* – 36.9)
ΣPCBs[e]	729.16 (<LQ* – 632.0)

[a] = sum of means of α-, β-, γ- and δ-isomer; [b] = sum of means of γ-, α –chlordane, oxychlordane, heptachlor and heptachlor epoxide A and B; [c] = sum of means of aldrin, endrin, isodrin and dieldrin; [d] = sum of means of pp' DDT, op' DDT, pp' DDE, op' DDE, pp' DDD and op' DDD; [e] = sum of means of 51 congeners; *< LQ = below limits of quantification. Detailed legend: Concentrations of organochlorine compounds in tissues of *Puffinus puffinus* wrecked at study area, from 2005 to 2011.

in the contaminant burden. This variation is expected in long-living birds with delayed sexual maturity, even among specimens at the same maturity stage. This can be related to distribution, migration, diet or age of the individuals [49]. The fact that *P. puffinus* is a migratory species must also be taken into account, which makes it likely that different individuals have different diets depending on the places visited throughout the migration period, reflecting the contamination of breeding and migration sites [50].

In the DDT group, pp' DDE was predominant and found in greater quantities among all the OCPs, in accordance with other studies [49,51]. This compound is usually found in the tissues of top predators due to its stability, bioconcentration and bioaccumulation [52]. In birds, DDT is metabolized to DDD and DDE, and DDD to DDE and DDMU, therefore it is normal that DDT and DDD concentrations decrease while DDE increases [53].

HCHs were not detected in any sample, probably due to the rapid metabolization and elimination that these compounds undergo in birds [54].

Oxychlordane is a metabolite belonging to the chlordanes group. It is persistent and presents difficult biotransformation in seabirds. It is therefore accumulated in greater amount in these organisms than other compounds of the group [49,55]. However, in the present study, this compound was only detected in five specimens and in very low concentrations.

From the drins group, only dieldrin was detected. This can be due to the rapid metabolization of aldrin into dieldrin and to the storage of this form in animals [56]. Isodrin is an aldrin isomer and endrin may not have been detected due to its rapid metabolization and excretion [57].

Endosulfan and methoxychlor are also metabolized and excreted fairly quickly [58,59], which may explain their absence in the analyzed tissues. In contrast, HCB and mirex, compounds of great persistence in the environment and biota [60,61], were present in the specimens analyzed in this study.

With regard to PCBs, there was a predominance of hexachlorbiphenyls and heptachlorbiphenyls (Figure 2). This is due to the fact that birds tend to metabolize and excrete low molecular weight PCBs congeners and accumulate high weight congeners that present a higher degree of halogenation [62].

In another study involving *P. puffinus*, in addition to other Procellariiformes, conducted at Rio Grande do Sul, Brazil, by Colabuono *et al.* [49], a similar pattern of contamination was found in tissues: a predominance of pentachlorobiphenyls, hexachlorobiphenyls and heptachlorobiphenyls among PCBs and a predominance of DDTs, especially DDE, among the OCPs. A predominance of DDE has also been reported by Bourne & Bogan [51], in a study involving seabirds from the North Atlantic, including the Manx shearwater. However, in contrast to the present study and to Colabuono *et al.* [49], uniformity in organochlorine levels was detected in the analyzed bird tissues.

In another study involving this species, higher levels of pp' DDE and lower levels of PCBs were reported in tissues of specimens from the Mediterranean and from the Black Sea [63]. Another study conducted in Brazil detected all of the analyzed PCBs congeners and, in contrast to the present study, pentachlorobiphenyls were predominant and levels of hexa- and heptacholobiphenyls were low [64].

Microbiological analyses

Bacteria from the *Vibrio* genus were detected in 90% of the analyzed specimens (n = 11), with *V. harveyi* being the most frequent species, followed by *Vibrio mediterranei*

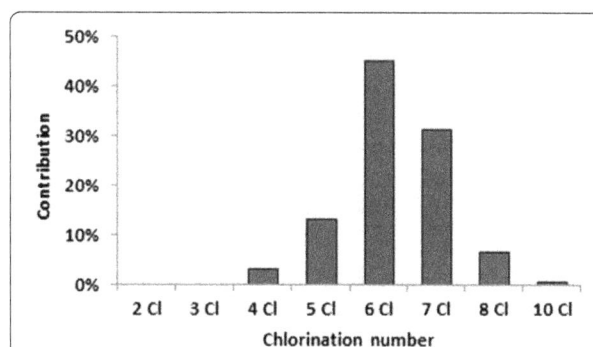

Figure 2 Mean relative contribution of PCB homologs in *Puffinus puffinus* muscle. Detailed legend: Mean relative contribution of PCB homologs from two to ten chlorine atoms found in *Puffinus puffinus* muscle.

and *V. parahaemolyticus*, each one present in 19% of the cases. Other species were also isolated: *V. fluvialis* (7%), *V. fisheri* (7%), *V. cincinnatiensis* (7%), *V. orientalis* (4%), *V. cholerae* non-01/non-0139 (4%), *V. gazogenes* (4%) and *V. alginolyticus* (4%). In 3% of the cases, classification by species was not possible.

Bacteria from the *Aeromonas* genus were isolated in only 18% of the specimens (n = 11) and in most cases (67%), classification by species was not possible. In the remaining cases (33%), only one species was isolated: *Aeromonas sobria*.

Vibrio anguillarum and *V. tapetis* are the bacteria most frequently related to diseases in aquatic animals [65]. None of these was isolated in the present study. Some species are known to cause diseases in both animals and humans, being potential zoonoses agents, such as *V. alginolyticus*, *V. harveyi*, *V. cholerae*, *V. fluvialis*, *V. furnissii*, V. mimicus, *V. metschnikovii*, *V. parahaemolyticus* and *V. vulnificus* [65]. Attention should be paid to the fact that five of these species were isolated in the present study, considering that an important infection route for animals and humans is through contaminated seafood [66,67].

Some coincident species were isolated in studies involving aquatic birds in the USA [68,69], Brazil [70], Japan [66] and England [71].

Migratory birds, such as the Manx shearwater, are important dispersers of micro-organisms. Additionally, when migrating, these birds tend to meet at certain locations, which facilitates interindividual and interspecies transmission, even more so since migration stress contributes in decreasing resistance to infection [72]. An important pathogen that can be carried by migratory seabirds is *Vibrio cholerae* [67,72], which can allow for cholera outbreaks in regions distant from endemic areas [72]. In this study, as in the study of Lee *et al.* [71], and as was the case of most birds studied by Ogg *et al.* [68], the *Vibrio cholerae* isolated was not from the O1 or O139 serogroups, which are responsible for cholera disease [73].

In some *Vibrio* species, such as *V. vulnificus*, *V. parahaemolyticus* and *V. cholera*, survival is related to water temperature, with most infections taking place during summer, through the consumption of contaminated seafood or through contact of contaminated water with wounds [65,66,71]. With this in mind, some studies [66,71] were able to isolate these species in aquatic birds even during winter, when the frequency of isolation in water was low, which demonstrates that these bacteria can multiply in birds even when environmental conditions are not favorable and that the survival of this species in the gastrointestinal tract by just a few days is already sufficient to disperse those microorganisms throughout large distances [71].

The study area (north-central coast of Rio de Janeiro) is under influence of the Cabo Frio upwelling, which causes a decrease in water temperatures during spring and summer, due to an outcropping of cold, deep and nutrient-rich waters (South Atlantic Central Water, SACW) [74]. From September to April, the surface water temperature rarely exceeds 18°C, and in deeper waters, it is often below 15°C [74,75]. Coincidentally, this is the period when the Manx shearwater is present in the Brazilian coast. Therefore, the species may be an important *Vibrio* spp. carrier when the water temperature is low.

With regard to *Aeromonas* spp., some species also depend on warmer waters to survive, as most isolations and cases of gastroenteritis caused by this genus also occur during summer [76-78]. This fact suggests that the hypothesis that waterfowl are sponsors for these microorganisms during periods of unfavorable environmental temperature can also be applied here, and the period of the influence of the upwelling on the study area should be taken into account.

In a study regarding the source of diarrhea infections caused by *Aeromonas* spp., Moyer [77] reported that some patients may have been contaminated by the ingestion of bivalves and by fishing and swimming in untreated waters. In addition, one of the patients worked at a fish market. These forms of infection indicate the importance of aquatic animals in the transmission of diseases caused by *Aeromonas* spp., and aquatic birds are among those most related as carrying different species from the genus in their gastrointestinal tract [79].

In two studies conducted in Canada, Lévesque *et al.* [80,81] isolated high concentrations of *Aeromonas* spp. from ring-billed gulls feces. These studies emphasize the importance of monitoring these bacteria, as birds can contaminate recreational waters through feces, which may lead to contamination of humans by means of contact between wounds with water or through the ingestion of contaminated seafood [80,81].

Conclusions

The present study leads to the conclusion that Manx shearwater seems to be an effective sentinel of Atlantic Ocean health, since different types of chemical contaminants could be detected in their tissues and common species of bacteria from aquatic environments could be isolated from swabs collected from this species.

In the present study, metal levels were in accordance with other studies involving this species. The same occurred concerning the pattern of contamination of organochlorine compounds. The results reflect the environmental contamination in breeding sites and throughout the migration route.

As described, *P. puffinus* feeds on fish species and cephalopods which are also consumed by humans, which highlights the contribution of this study to the public health field. It is worth emphasizing the importance of beach monitoring activities in an effort to prevent contamination.

It is recommended that contaminant levels and the frequency of micro-organisms isolation continue to be monitored in this species and in others, in order to evaluate a possible increase in environmental degradation. It is also important to continue studying the species in order to consolidate the Manx shearwater as a sentinel species.

Methods

For the purposes of this study, between 2005 and 2011, thirty-five carcasses of wrecked Manx shearwater were collected during beach monitoring at the north-central coast of the state of Rio de Janeiro (from Saquarema – south, 22°55'12"S; 42°30'37"W – to São Francisco do Itabapoana – north, 21°18'07"S; 40°57'4"W) and two carcasses were collected at Aracruz (19°49'13"S; 40°16'24"W), in the state of Espírito Santo, Brazil. The carcasses were measured, necropsied and fragments of hepatic and muscle tissues were collected and stored at –20°C for contaminant analyses. For the metal analyses, 37 muscle samples and 20 liver samples were used. For the organochlorine compounds analyses, 13 muscle samples were used. Most carcasses were found in 2010, a year of severe mortality for this species on the Southeastern coast of Brazil.

Furthermore, from 2009 to 2012, cloaca, oral, ocular and tracheal swabs were collected from eleven specimens found alive on beaches of the north-central coast of Rio de Janeiro (from Saquarema to São Francisco do Itabapoana) for the analysis of bacteria from the *Vibrio* and *Aeromonas* genera.

Analytical methods for metal analyses

The determination of cadmium (Cd), lead (Pb) and selenium (Se) levels was performed by inductively coupled plasma mass spectrometry (ICP-MS). The determination of mercury (Hg) levels was performed by cold vapor atomic absorption spectrometry (CV-AAS).

Samples were defrosted and homogenized with a food microprocessor (HC31 Black & Decker). For Cd, Pb and Se analyses, approximately 0.5 g of each muscle sample, in triplicate, and approximately 0.1 g of each liver sample in duplicate were digested with 5 mL and 1 mL of nitric acid (HNO_3) (Vetec), respectively, in a heating block (Quimis) (80°C), until total dissolution. Three procedural blanks were prepared with each sample batch in the heating block. Ultra-pure water (obtained from Master All water purificator, Gehaka) was added until a final volume of 50 mL was reached. Metal concentrations

were determined using an ICP-MS 7500 Series (Agilent Technologies).

For the Hg analyses, approximately 0.5 g of each muscle sample and approximately 0.1 g of each liver sample, both in duplicate, were digested with 5 mL of a sulphuric-nitric acid mixture ($HNO_3/H_2SO_4/V_2O_5$) (Vetec) in a heating block (80°C), until total dissolution. The procedural blanks were performed in the same manner applied to the previous analyses. The samples and the blanks were then cooled and 5 mL of $KMnO_4$ 5% (Vetec) were added. The purpose of this process was to ensure that mercury remained in the sample until reading was performed. At this moment, 1 mL of hydroxylamine (Vetec) was added and then ultra-pure water, until a final volume of 50 mL was reached. Finally, concentrations were determined using a 3300 spectrometer (Perkin Elmer).

The accuracy of the analytical methods was ensured by the use of Merck certified material (Hg, Cd, Pb and Se Titrisol Standard Solutions) and the quality of the methods used was ensured by the use of National Research Council of Canada certified reference materials (DORM-2 - Dogfish Muscle Certified Reference Material for Trace Metals; and DOLT-3 - Dogfish Liver Certified Reference Material for Trace Metals), analyzed in parallel, in triplicate, with average recovery ranging from 90% to 104%.

Samples were analyzed as wet weight and, subsequently, aliquots of all samples were weighed and dried in a 315 SE oven (Fanem) until constant weight. This procedure made it possible to obtain humidity factors, which were applied to the results in order to convert them into dry weight.

The limits of quantification (LQ) were provided by equipment software, using the formula: $LQ = [10*(SDbr)]/S$, where SDbr is the standard deviation from 10 blank readings, and S is the inclination of the calibration curve. This formula provides the instrumental LQ. In order to calculate the final LQ, this formula was multiplied by the dilution factor.

Analytical methods for organochlorine compounds analyses

The organochlorine pesticides (OCPs) analyzed in this study were: hexachlorocyclohexanes (HCH) (α-, β- , γ- and δ-isomer), hexachlorobenzene (HCB), heptachlor, heptachlor epoxide A and B, chlordanes (oxychlordane, α- and γ- chlordane), drins (aldrin, isodrin, dieldrin, endrin), dichlorodiphenyltrichloroethane (op' DDT, pp' DDT) and its metabolites, dichlorodiphenyldichloroethylene (op' DDE, pp' DDE) and dichlorodiphenyldichloroethane (op' DDD, pp' DDD), endosulfan II, methoxychlor and mirex. The analyzed polychlorinated biphenyls (PCBs) congeners were the following IUPAC numbers: 8, 28, 31, 33, 44, 49, 52, 56/

60, 66, 70, 74, 77, 81, 87, 95, 97, 99, 101, 105, 110, 114, 118, 123, 126, 128, 132, 138, 141, 149, 151, 153, 156, 157, 158, 167, 169, 170, 174, 177, 180, 183, 187, 189, 194, 195, 203, 206 and 209.

The analytical procedure followed the protocol described by MacLeod et al. (1986) [82], with minor modifications, as described by Colabuono et al. (2012) [49].

At first, 2,2',4,5',6-pentachlorobiphenyl (PCB 103) and 2,2',3,3',4,5,5',6-octachlorobiphenyl (PCB 198) were added to all samples, blanks and reference material as surrogates for OCPs and PCBs. Then, approximately 2.5 g of each lyophilized muscle sample were extracted in a Soxhlet apparatus for 8 h using 80 mL of n-hexane and methylene chloride (1:1, v/v). The determination of extractable lipids was made by gravimetric analyses. The extracts were cleaned-up through the use of column chromatography with 8 g of silica and 16 g of alumina, both 5% water deactivated, eluted with 80 mL of n-hexane and methylene chloride (1:1, v/v). The fraction was purified once again, in order to remove lipid excess, now through high-performance liquid chromatography (HPLC), using methylene chloride as eluent with a flow of 5 mL min^{-1}. The extract was concentrated to a volume of 0.9 mL in hexane. The internal standard 2,4,5,6-tetrachlorometaxylene (TCMX) was added before the gas chromatographic analysis was undertaken and a procedural blank was included in the set of samples.

Identification and quantification analyses of the organochlorine pesticides were performed with a 6890 N gas chromatograph with an electron capture detector (GC-ECD) (Agilent Technologies), using a 30 m × 0.25 mm i.d. capillary column coated with 5% phenyl-substituted dimethylpolysiloxane phase (0.5 μm film thickness). Automatic splitless injections of 2 μL were applied and the total purge rate was adjusted to 50 mL min^{-1}. The carrier gas used was hydrogen (constant pressure of 40 kPa at 100°C), and the makeup gas was nitrogen, at a rate of 60 mL min^{-1}. Injector and detector temperatures were 280°C and 320°C, respectively. Oven temperature was programmed as follows: 70°C for 1 min, raised at 40°C.min^{-1} until 170°C, then raised at 1.5°C.min^{-1} until 230°C (held for 1 min), and at 20°C.min^{-1} until 300°C with a final hold of 5 min.

The quantitative PCBs analyses were performed by a 5973 N gas chromatograph coupled to a mass spectrometer (GC–MS) (Agilent Technologies), in a selected ion mode (SIM 70 eV), using a 30 m × 0.25 mm i.d. capillary column coated with 5% phenyl-substituted dimethylpolysiloxane phase (0.25 μm film thickness). Injections were made with 1 μL in automatic splitless mode. The carrier gas was helium (constant flow of 1.1 mL min^{-1}). The interface, source and quadrupole temperatures were 280°C, 300°C and 200°C, respectively. Oven temperature was

programmed as follows: 75°C for 3 min, raised at 15°C. min^{-1} until 150°C, then raised at 2.0°C.min^{-1}, until 260°C and at 20°C.min^{-1}, until 300°C with a final hold of 10 min.

The analytical methodology was validated through the use of a standard reference (SRM 1945 – organics in whale blubber), from the National Institute of Standards and Technology, USA, for quality assurance and quality control. This material was analyzed in parallel, in duplicate, with an average analyte recovery inside the range accepted by the NS&T [83]. The same occurred in the analyte recovery in spiked blanks and matrices (67–115%). Analytes in laboratory blanks were subtracted from the samples. Analyte quantification was performed using a nine-level analytical curve following the internal standard procedure. All solvents were residue-analyzed grade from JT Baker. Standard solutions were from AccuStandard. Surrogate recoveries were acceptable and presented mean ± standard deviation = 97 ± 7. Method limits of quantification (LQ) ranged from 1.02 ng.g^{-1} to 8.5 ng.g^{-1} dry weight (dw).

Analytical methods for microbiological analyses

Cloacal, oral, ocular and tracheal swabs were carefully collected in order to avoid external contamination and accommodated in Cary-Blair media for transportation.

Samples were enriched with Alkaline Peptone Water (APW) containing 1% sodium chloride (NaCl) (37°C/18-24 hours). Samples were then streaked onto Thiossulfate Citrate Bile Salts Sucrose Agar (TCBS) and onto Glutamate Starch Phenol Red Agar (GSP) and incubated to 37°C over night (model 31483, Thelco). Suspected colonies were transferred to Kligler Iron Agar, Lysine Iron Agar and Nutrient Agar with 1% NaCl. Posteriorly, biochemical tests were performed in order to identify species from the Vibrionaceae and Aeromonadaceae families, according to Noguerola & Blanch (2008) [84] and Janda & Abbott (2010) [85], respectively.

Statistical analyses

Statistical analyses were performed using the SPSS Statistics 17.0 software (IBM). Results from metal, organochlorine and bacteriological analysis were analyzed separately. Basic descriptive statistics was conducted. Data were tested for normal distribution using the Shapiro-Wilk's test. Since most data were not normally distributed, nonparametric tests were used. Spearman's rank correlation coefficient was used to verify the strength of the association between the concentration of the same metal in hepatic and muscular tissue and to verify the strength of the association in interelement relationship. A p-value of less than 0.05 was used to indicate statistical significance. For bacteriological analyses, the frequencies of isolation of Vibrio spp. and Aeromonas spp. were calculated, as well as the frequency of isolation of each species.

Competing interests
The authors declare that they have no competing interests.

Authors' contributions
MDC conceived of the study, carried out metal and statistical analyses and drafted the manuscript; JFM and DCT collected samples in the field, carried out necropsies and helped in several steps, including metal and statistical analyses and in the drafting of the manuscript; RAG helped in the metal analyses; FIC and RCM carried out the organochlorine compound analyses; EMR, RLS and DPR carried out the microbiological analyses; SS conceived the study, participated in its design and coordination and helped draft the manuscript. All authors read and approved the final manuscript.

Acknowledgements
To CAPES (Coordination of Improvement of Higher Education Personnel) for the financial support; to Reinaldo Calixto de Campos (in memoriam) for allowing the use of the Laboratory of Atomic Absorption (LAATOM) at PUC-Rio; and to all employees of all the laboratories involved in this work.

Author details
[1]Programa de Pós-Graduação em Saúde Pública e Meio Ambiente, ENSP/ Fiocruz, Rua Leopoldo Bulhões, 1480, Manguinhos, Rio de Janeiro 21041-210, RJ, Brasil. [2]Systems Ecology, Leibniz Center for Tropical Marine Ecology (ZMT), Fahrenheitstrasse 6, 28359 Bremen, Germany. [3]Departamento de Endemias Samuel Pessoa – DENSP & Grupo de Estudos de Mamíferos Marinhos da Região dos Lagos – GEMM-Lagos, Escola Nacional de Saúde Pública /FICORUZ, Rua Leopoldo Bulhões, 1.480, 6° andar, Sala 611, Manguinhos, Rio de Janeiro 21041-210, RJ, Brasil. [4]Departamento de Química, Pontifícia Universidade Católica do Rio de Janeiro, Rua Marquês de São Vicente, 225, Gávea, Rio de Janeiro 22453-900, RJ, Brasil. [5]Universidade de São Paulo, Instituto Oceanográfico, Praça do Oceanográfico 191, Cidade Universitária, São Paulo 05508-120, SP, Brasil. [6]Instituto Oswaldo Cruz/FIOCRUZ, Laboratório de Referência Nacional de Enteroinfecções Bacterianas, Av. Brasil, 4365, Manguinhos, Rio de Janeiro 21040-360, RJ, Brasil.

References
1. Odum EP, Barret GW: **Ecologia Regional: Principais Tipos de Ecossistemas e Biomas.** In *Fundamentos de Ecologia.* 5th edition. Edited by Odum EP, Barret GW. São Paulo: Cengage Learning; 2008:412–458.
2. Joint Group of Experts on the Scientific Aspects of Marine Environmental Protection (GESAMP): *Protecting the Oceans from Land-Based Activities.* New York: 2001.
3. Augusto LGS, Câmara VM, Carneiro FF, Câncio J, Gouveia N: **Saúde e Ambiente: uma reflexão da Associação Brasileira de Pós-Graduação em Saúde Coletiva – ABRASCO.** *Rev Bras Epidemiol* 2003, **6**(2):87–94.
4. Laws EA, Fleming LE, Stegeman JJ: **Centers for Oceans and Human Health: contributions to an emerging discipline.** *Environl Health* 2008, **7**:1–5.
5. Kite-Powell HL, Fleming LE, Backer LC, Faustman EM, Hoagland P, Tsuchiya A, Younglove LR, Wilcox BA, Gast RJ: **Linking the oceans to public health: current efforts and future directions.** *Environ Health* 2008, **7**:1–15.
6. Knap A, Dewailly E, Furgal C, Galvin J, Baden D, Bowen RE, Depledge M, Duguay L, Fleming LE, Ford T, Moser F, Owen R, Suk WA, Unluata U: **Indicators of ocean health and human health: developing a research and monitoring framework.** *Environ Health Perspect* 2002, **110**(9):839–845.
7. Fleming LE, Broad K, Clement A, Dewailly E, Elmir S, Knap A, Pomponi SA, Smith S, Solo-Gabriele H, Walsh P: **Oceans and human health: emerging public health risks in the marine environmental.** *Mar Pollut Bull* 2006, **53**:545–560.
8. Moura JF, Cardoso M, Belo MSSP, Hacon S, Siciliano S: **A interface da saúde pública com a saúde dos oceanos: produção de doenças, impactos socioeconômicos e relações benéficas.** *Ciênc Saúde Colet* 2008, **1144**:1–14.
9. Niencheski LF, Fillman G: **Contaminantes: Metais, Hidrocarbonetos e Organoclorados.** In *Avaliação Ambiental de Estuários Brasileiros: Aspectos Metodológicos.* 1st edition. Edited by Lana PC, Bianchini A, Ribeiro C, Niencheski LF, Fillman G, Santos CSG. Rio de Janeiro: Museu Nacional; 2006:63–118.
10. Tait RV, Dipper FA: **Human impact on the marine environment.** In *Elements of Marine Ecology.* 4th edition. Edited by Tait RV, Dipper FA. Oxford: Butterworth-Heinemann; 1998:395–435.
11. Strain PM, Macdonald RW: **Design and implementation of a program to monitor ocean health.** *Ocean Coast Manage* 2002, **45**:325–355.
12. Stewart JR, Gast RJ, Fujioka RS, Solo-Gabriele HM, Meschke JS, Amaral-Zettler LA, Castillo E, Polz MF, Collier TK, Strom MS, Sinigalliano CD, Moeller PDR, Holland AF: **The coastal environment and human health: microbial indicators, pathogens, sentinels and reservoirs.** *Environ Health* 2008, **7**:1–14.
13. Beeby A: **What do sentinels stand for?** *Environ Pollut* 2001, **112**:285–298.
14. Furness RW, Camphuysen CJ: **Seabirds as monitors of the marine environment.** *ICES J Mar Sci* 1997, **54**:726–737.
15. Burger J, Gochfeld M: **Marine birds as sentinels of environmental pollution.** *Ecohealth* 2004, **1**:263–274.
16. Siciliano S, Alves VC, Hacon S: **Aves e mamíferos marinhos como sentinelas ecológicas da saúde ambiental: uma revisão do conhecimento brasileiro.** *Cad Saúde Colet* 2005, **13**(4):927–946.
17. Thompson KR: *The ecology of Manx shearwater Puffinus puffinus on Rhum, West Scotland,* PhD thesis. University of Glasgow, Faculty of Science; 1987.
18. British Trust for Ornithology (BTO): **Manx Shearwater** *Puffinus puffinus* **(Brünnich, 1764).** http://blx1.bto.org/birdfacts/results/bob460.htm.
19. Guilford T, Meade J, Willis J, Phillips RA, Boyle D, Roberts S, Collett M, Freeman R, Perrins CM: **Migration and stopover in a small pelagic seabird, the Manx shearwater** *Puffinus puffinus:* **insights from machine learning.** *Proc R Soc B* 2009, **276**:1215–1223.
20. Dale IM, Baxter MS, Bogan JA, Bourne WRP: **Mercury in seabirds.** *Mar Pollut Bull* 1973, **4**(5):77–79.
21. Osborn D, Harris MP, Nicholson JK: **Comparative tissue distribution of mercury, cadmium and zinc in three species of pelagic seabirds.** *Comp Biochem Phys C* 1979, **64**:61–67.
22. Garcia JG: **Avaliação da concentração de metais traço em tecidos de petréis (Aves: Procellariidae) encontrados nas áreas central e sul das praias do Rio Grande do Sul.** In *Monograph.* Fundação Universidade do Rio Grande, Oceanology Department; 2008.
23. Muirhead SJ, Furness RW: **Heavy metal concentrations in the tissues of seabirds from Gough Island, South Atlantic Ocean.** *Mar Pollut Bull* 1988, **19**(6):278–283.
24. Thompson DR, Furness RW: **The chemical form of mercury stored in South Atlantic Seabirds.** *Environ Polut* 1989, **60**:305–317.
25. Barbieri E, Garcia CAB, Passos EA, Aragão KAS, Alves JPH: **Heavy metal concentration in tissues of Puffinus gravis sampled on the Brazilian coast.** *Rev Bras Ornitol* 2007, **15**(1):69–72.
26. Bekhit AE-D, Al-Amer S, Gooneratne R, Mason SL, Osman KA, Clucas L: **Concentrations of trace elementals and organochlorines in Mutton bird (Puffinus griseus).** *Ecotox Environ Safe* 2011, **74**:1742–1746.
27. Bustamante P, Caurant F, Fowler SW, Miramand P: **Cephalopods as a vector for the transfer of cadmium to top marine predators in the north-east Atlantic Ocean.** *Sci Total Environ* 1998, **220**(1):71–80.
28. Dorneles PR, Lailson-Brito J, Santos RA, Costa PAS, Malm O, Azevedo AF, Torres JPM: **Cephalopods and cetaceans as indicators of offshore bioavailability of cadmium off Central South Brazil Bight.** *Environ Pollut* 2007, **148**(1):352–359.
29. Petry MV, Fonseca VSS, Krüger-Garcia L, Piuco RC, Brummelhaus J: **Shearwater diet during migration along the coast of Rio Grande do Sul, Brazil.** *Mar Biol* 2008, **154**:613–621.
30. Norheim G: **Levels and interactions of heavy metals in Sea Birds from Svalbard and the Antarctic.** *Environ Pollut* 1987, **47**:83–94.
31. Chen CY, Stemberger RS, Klaue B, Blum JD, Pickhardt PC, Folt CL: **Accumulation of heavy metals in food web components across a gradient of lakes.** *Limnol Oceanogr* 2000, **45**(7):1525–1536.
32. Campbell LM, Norstrom RJ, Hobson KA, Muir DCG, Backus S, Fisk AT: **Mercury and other trace elements in a pelagic Arctic marine food web (Northwater Polynya, Baffin Bay).** *Sci Total Environ* 2005, **351**:247–263.
33. Bundy A, Lilly RG, Shelton PA: *A mass balance model of the Newfoundland-Labrador Shelf.* Dartmouth: Canadian Technical Report of Fisheries and Aquatic Sciences 2310; 2000.
34. Bryman A, Cramer D: *Quantitative data analysis with IBM SPSS Statistics 17, 18 and 19: A guide for social scientists.* Hove: Psychology Press; 2011.
35. Scheuhammer AM: **The chronic toxicity of aluminium, cadmium, mercury, and lead in birds: a review.** *Environ Pollut* 1987, **46**:263–295.

36. Custer TW, Franson JC, Pattee OH: **Tissue lead distribution and hematologic effects in american kestrels (*Falco sparverius* L.) fed biologically incorporated lead.** *J Wildl Dis* 1984, **20**(1):39–43.

37. Barton JC, Conrad ME, Harrison L, Nuby S: **Effects of calcium on the absorption and retention of lead.** *J Lab Clin Med* 1978, **91**(3):366–376.

38. Cuvin-Aralar MLA, Furness RW: **Mercury and selenium interaction: a review.** *Ecotox Environ Safe* 1991, **21**:348–364.

39. Scheuhammer AM, Wong AHK, Bond D: **Mercury and selenium accumulation in common loons (*Gavia immer*) and common mergansers (*Mergus merganser*) from Eastern Canada.** *Environ Toxicol Chem* 1998, **17**(2):197–201.

40. Eagles-Smith CA, Jackerman JT, Yee J, Adelsbach TL: **Mercury demethylation in waterbird livers: dose–response thresholds and differences among species.** *Environ Toxicol Chem* 2009, **28**(3):568–577.

41. Ridlington JW, Whanger PD: **Interactions of selenium and antioxidants with mercury, cadmium and silver.** *Fund Appl Toxicol* 1981, **1**:368–375.

42. Ji X, Hu W, Cheng J, Yuan T, Xu F, Qu L, Wang W: **Oxidative stress on domestic ducks (Shaoxing duck) chronically exposed in a Mercury-Selenium coexisting mining area in China.** *Ecotox Environ Safe* 2006, **64**:171–177.

43. Koeman JH, Van De Vem WSM, Goeij JJM, Tjioe PS, Van Haaften JL: **Mercury and selenium in marine mammals and birds.** *Sci Total Environ* 1975, **3**:279–287.

44. Goede AA, Wolterbeek HT: **Have high selenium concentrations in wading birds their origin in mercury?** *Sci Total Environ* 1994, **144**:247–253.

45. Caurant F, Amiard JC, Amiard-Triquet C, Sauriau PG: **Ecological and biological factors controlling the concentrations of trace elements (As, Cd, Cu, Hg, Se, Zn) in delphinids *Globicephala melas* from the North Atlantic Ocean.** *Mar Ecol Prog Ser* 1994, **103**:207–219.

46. Kunito T, Nakamura S, Ikemoto T, Anan Y, Kubota R, Tanabe S, Rosas FCW, Fillmann G, Readman JW: **Concentration and subcellular distribution of trace elements in liver of small cetaceans incidentally caught along the Brazilian coast.** *Mar Pollut Bull* 2004, **49**(7–8):574–587.

47. Lemos LS: **Avaliação das concentrações de Cd, Cu, Hg, Mn, Se e Zn em pequenos cetáceos da costa norte do estado do Rio de Janeiro, Brasil.** In *Master thesis*. Fundação Oswaldo Cruz, Escola Nacional de Saúde Pública; 2012.

48. Dietz R, Riget F, Born EW: **An assessment of selenium to mercury in Greenland marine animals.** *Sci Total Environ* 2000, **245**:15–24.

49. Colabuono FI, Taniguchi S, Montone RC: **Organochlorine contaminants in albatrosses and petrels during migration in South Atlantic Ocean.** *Chemosphere* 2012, **86**:701–708.

50. Tanaka H, Ogi H, Tanabe S, Tatsukawa R, Oka N: **Bioaccumulation and metabolism of PCBs and DDE in short-tailed shearwater *Puffinus tenuirostris* during its transequatorial migration and in the wintering and breeding grounds.** *Mem Nat Inst Polar Res* 1986, **40**:434–442.

51. Bourne WRP, Bogan JA: **Polychlorinated biphenyls in North Atlantic seabirds.** *Mar Pollut Bull* 1972, **3**(11):171–175.

52. Jones KC, Voogt P: **Persistent organic pollutants (POPs): state of the science.** *Environ Pollut* 1999, **100**(1–3):209–221.

53. Bailey S, Bunyan PJ, Rennison BD, Taylor A: **The metabolism of 1,1-di(p-chlorophenyl)-2,2,2-trichloroethane and 1,1-di(p-chlorophenyl)-2,2-dichloroethane in the pigeon.** *Toxicol Appl Pharmacol* 1969, **14**:13–22.

54. Moisey J, Fisk AT, Hobson KA, Norstrom RJ: **Hexachlorocyclohexane (HCH) Isomers and Chiral Signatures of α-HCH in the Arctic Marine Food Web of the Northwater Polynya.** *Environ Sci Technol* 2001, **35**:1920–1927.

55. Guruge KS, Watanabe M, Tanaka H, Tanabe S: **Accumulation status of persistent organochlorines in albatrosses from the North Pacific and the Southern Ocean.** *Environ Pollut* 2001, **114**(3):389–398.

56. Bann JM, Decino TJ, Earle NW, Sun Y-P: **The fate of Aldrin and Dieldrin in the animal body.** *J Agric Food Chem* 1956, **4**(11):937–941.

57. Agency for Toxic Substances and Disease Registry (ATSDR): *Toxicological Profile for Endrin.* Atlanta 1996.

58. Kapoor IP, Metcalf RL, Nystrom RF, Sangha GK: **Comparative metabolism of methoxychlor, methiochlor, and DDT in mouse, insects, and in a model ecosystem.** *J Agric Food Chem* 1970, **18**(6):1145–1152.

59. Agency for Toxic Substances and Disease Registry (ATSDR): *Toxicological Profile for Endosulfan.* Atlanta 2000.

60. Agency for Toxic Substances and Disease Registry (ATSDR): *Toxicological Profile for Mirex and Cholordecone.* Atlanta 1995.

61. Agency for Toxic Substances and Disease Registry (ATSDR): *Toxicological Profile for Hexachlorobenzene.* Atlanta 2002.

62. Maervoet J, Chu SG, De Vos S, Covaci A, Voorspoels S, De Schrijver R, Schepens P: **Accumulation and tissue distribution of selected polychlorinated biphenyl congeners in chickens.** *Chemosphere* 2004, **57**:61–66.

63. Walker CH: **Persistent pollutants in fish-eating sea birds – bioaccumulation, metabolism and effects.** *Aquat Toxicol* 1990, **17**:293–324.

64. Ferreira AP: **Measurement of chlorinated dioxins, furans and PCBs in *Ardea alba* Great Egret, and *Puffinus puffinus* Manx Shearwater: Ilha Grande Bay, Rio de Janeiro, Brazil.** *J Integrat Coast Zone Manage* 2012, **12**(1):7–15.

65. Austin B: **Vibrios as causal agents of zoonoses.** *Vet Microbiol* 2010, **140**:310–317.

66. Miyasaka J, Yahiro S, Arahira Y, Tokunaga H, Katsuki K, Hara-Kudo Y: **Isolation of *Vibrio parahaemolyticus* and *Vibrio vulnificus* from wild aquatic birds in Japan.** *Epidemiol Infect* 2006, **134**(4):780–785.

67. Halpern M, Senderovich Y, Izhaki I: **Waterfowl – the missing Link in epidemic and pandemic cholera dissemination.** *PLoS Pathog* 2008, **4**(10):1–3.

68. Ogg JE, Ryder RA, Smith HL Jr: **Isolation of *Vibrio cholerae* from aquatic birds in Colorado and Utah.** *Appl Environ Microbiol* 1989, **55**(1):96–99.

69. Buck JD: **Isolation of *Candida albicans* and Halophilic *Vibrio* spp. from Aquatic Birds in Connecticut and Florida.** *Appl Environ Microbiol* 1990, **56**(3):826–828.

70. Roges EM, Souza RL, Santos AFM, Siciliano S, Ott PH, Moreno IB, Pereira CS, Reis EMF, Lazaro NS, Rodrigues DP: **Distribution of *Vibrio* sp. in marine mammals, seabirds and turtles beached or accidentaly captured in fishing nets in coastal regions of Brazil.** In *Vibrios In The Environment 2010.* Mississipi 2010:S30–S32. Abstract.

71. Lee JV, Bashford DJ, Donovan TJ, Furniss AL, West PA: **The incidence of *Vibrio cholerae* in water, animals and birds in Kent, England.** *J Appl Bacteriol* 1982, **52**:281–291.

72. Hubálek Z: **An annotated checklist of pathogenic microorganisms associated with migratory birds.** *J Wildl Dis* 2004, **40**(4):639–659.

73. Rabbani GH, Greenough WB III: **Food as a vehicle of transmission of cholera.** *J Diarrhoeal Dis Res* 1999, **17**(1):1–9.

74. Valentin JL: **Analyse des paramètres hydrobiologiques dans la remontée de Cabo Frio (Brésil).** *Mar Biol* 1984, **82**:259–276.

75. Leite GS, Dourado MS, Candella RN: **Estudo preliminar da climatologia da ressurgência em Arraial do Cabo,RJ.** In *XI ENAPET; Florianópolis.* Edited by XI ENAPET Amostra de Atividades Petianas. 2006:1–11.

76. Burke V, Robinson J, Gracey M, Peterson D, Partridge K: **Isolation of aeromonas hydrophila from a metropolitan water supply: seasonal correlation with clinical isolates.** *Appl Environ Microbiol* 1984, **48**(2):361–366.

77. Moyer NP: **Clinical significance of Aeromonas species isolated from patients with diarrhea.** *J Clin Microbiol* 1987, **25**(11):2044–2048.

78. Pereira CS, Amorim SD, Santos AFM, Reis CMF, Theophilo GND, Rodrigues DP: **Caracterização de *Aeromonas* spp isoladas de neonatos hospitalizados.** *Rev Soc Bras Med Trop* 2008, **41**(2):179–182.

79. Staples P: ***Aeromonas, Plesiomonas* and *Vibrio* bacteria isolated from animals in New Zeland.** *Surveillance* 2000, **27**(1):3–4.

80. Lévesque B, Brousseau P, Simard P, Dewailly E, Meisels M, Ramsay D, Joly J: **Impact of the ring-billed gull (*Larus delawarensis*) on the microbiological quality of recreational water.** *Appl Environ Microbiol* 1993, **59**(4):1228–1230.

81. Lévesque B, Brousseau P, Bernier F, Dewailly E, Joly J: **Study of the bacterial content of the ring-billed gull droppings in relation to recreational water quality.** *Water Res* 2000, **34**(4):1089–1096.

82. Macleod WD, Brown DW, Friedman AJ, Burrows DG, Maynes O, Pearce RW, Wigren CA, Bogar RG: *Standard Analytical Procedures of the NOAA National Analytical Facility, 1985–1986. Extractable Toxic Organic Compounds.* Springfield: U.S. Department of Commerce, NOAA Technical Memorandum NMFS F/NWC-92; 1986.

83. Wade TL, Cantillo AY: *Use of Standards and Reference Materials in the Measurement of Chlorinated Hydrocarbon Residues*, Chemistry Workbook. Silver Spring: U.S. Department of Commerce, NOAA Technical Memorandum NOS ORCA 77; 1994.

84. Noguerola I, Blanch AR: **Identification of *Vibrio* spp. with a set of dichotomous keys.** *J Appl Microbiol* 2008, **105**(1):175–185.

85. Janda JM, Abbot SL: **The genus aeromonas: taxonomy, pathogenicity, and infection.** *Clin Microbiol Rev* 2010, **23**(1):35–73.

14

High throughput screening of CO_2-tolerating microalgae using GasPak bags

Zheng Liu, Fan Zhang and Feng Chen[*]

Abstract

Background: Microalgae are diverse in terms of their speciation and function. More than 35,000 algal strains have been described, and thousands of algal cultures are maintained in different culture collection centers. The ability of CO_2 uptake by microalgae varies dramatically among algal species. It becomes challenging to select suitable algal candidates that can proliferate under high CO_2 concentration from a large collection of algal cultures.

Results: Here, we described a high throughput screening method to rapidly identify high CO_2 affinity microalgae. The system integrates a CO_2 mixer, GasPak bags and microplates. Microalgae on the microplates will be cultivated in GasPak bags charged with different CO_2 concentrations. Using this method, we identified 17 algal strains whose growth rates were not influenced when the concentration of CO_2 was increased from 2 to 20% (v/v). Most CO_2 tolerant strains identified in this study were closely related to the species *Scenedesmus* and *Chlorococcum*. One of *Scenedesmus* strains (E7A) has been successfully tested in in the scale up photo bioreactors (500 L) bubbled with flue gas which contains 10-12% CO_2.

Conclusion: Our high throughput CO_2 testing system provides a rapid and reliable way for identifying microalgal candidate strains that can grow under high CO_2 condition from a large pool of culture collection species. This high throughput system can also be modified for selecting algal strains that can tolerate other gases, such as NOx, SOx, or flue gas.

Keywords: CO_2 sequestration, Microalgae, High through-put selection

Background

Increasing atmospheric greenhouse gas emission by human activities has been regarded as a major challenge of global sustainability. CO_2 is a primary greenhouse gas, which makes up approximately 83.6% of the total greenhouse gas emission [1].

Increasing level of CO_2 causes global warming and the subsequent environmental issues such as the rising sea level and snow or ice melting [2,3].

Biological fixation carried out by photosynthetic plants and microalgae has attracted increasing attention as an environmentally friendly CO_2 mitigation strategy [4,5]. Photosynthesis renews oxygen in the atmosphere while fixing CO_2 into potentially useful biomass. Microalgae are emerging as a promising biological fixation system; each acre of microalgae is able to fix three to five times more CO_2 than the same area of terrestrial plants [6].

Meanwhile, microalgae are also able to remove nitrogen, phosphorus, and heavy metals from wastewater, and algal biomass can be converted into useful products, such as biofuels, nutraceutical products, animal feed and fodder for aquaculture [7-9].

Exhaust gases from power plants attribute to ca. 40% of the U.S. annual CO_2 emission in 2010 [10]. Earlier studies have reported that microalgae can be used to sequester CO_2 in power plant flue gases [11-13]. The concentration of CO_2 in power plant exhausts varies from 10-15% depending on the source of fuels [10]. Therefore, the ideal microalgal candidates for sequestering CO_2 in flue gases should be able to grow under CO_2 concentration above 10%. It is known that different species of microalgae can tolerate different levels of CO_2. For examples, it has been reported that *Chlorella sp.* and *Euglena gracilis* can tolerate up to 40% CO_2 [14], *Chlorococcum littorale* could endure 60% CO_2 [15], *Scenedesmus sp.* could grow under 80% CO_2 [14], and *Cyandium caldarium* were successfully grown under 100% CO_2 [16]. However, it is difficult to

* Correspondence: chenf@umces.edu
Institute of Marine and Environmental Technology, University of Maryland Center for Environmental Science, 701 E Pratt St, Baltimore, MD 21202, USA

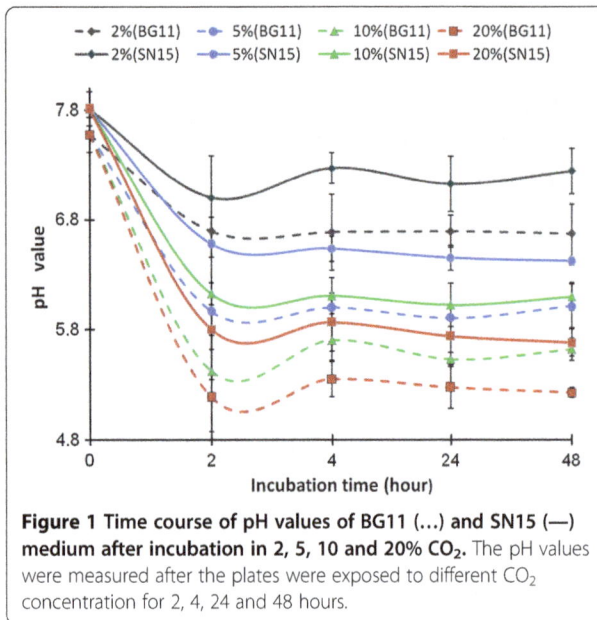

Figure 1 Time course of pH values of BG11 (...) and SN15 (—) medium after incubation in 2, 5, 10 and 20% CO$_2$. The pH values were measured after the plates were exposed to different CO$_2$ concentration for 2, 4, 24 and 48 hours.

evaluate and compare the actual growth rates of these algae under high levels of CO$_2$ because some strains just tolerate but do not grow under high CO$_2$ condition. Moreover, a comparison within the same species in different studies can also be challenging due to the different experimental setup. It has been reported that *Scenedesmus obliquus* can tolerate up to 18% CO$_2$, but the optimal growth was observed with 6% CO$_2$ [17]. It is also reported

that *S. obliquus* grew successfully under 70% CO$_2$, however, the highest growth rate occurred below 10% CO$_2$ [18]. Nevertheless, in a separate study, the optimal growth of *S. obliquus* at 15% CO$_2$ was observed [19]. The inconsistency of these results may be caused by the difference in the experimental setup. Many earlier studies conducted the CO$_2$ tolerance tests using flasks bubbled with certain levels of CO$_2$. In this case, the actual concentration of CO$_2$ that algae are exposed to is hard to monitor because a certain amount of CO$_2$ can be lost to air due to bubbling [20]. Difference in light, temperature, culture media and containments, bubbling rates, and other factors may all contribute to variable CO$_2$ tolerance within the same species [21,22].

Microalgae are diverse in the natural environment. It has been estimated that about 200,000-800,000 algal species exist in nature, of which about 35,000 species have been described [23,24]. Thousands of algal strains have been isolated, characterized and maintained in different laboratories and culture collection centers. To select the candidates that have a high CO$_2$ affinity from this large pool of algal collections can be very time-consuming and technical challenging, particularly when the growth of algae need to be measured. A high throughput method for evaluating the capability of these algal strains for CO$_2$ tolerance could facilitate the research in algal sequestration of CO$_2$ pollutions.

In this study, we designed a high throughput system to select microalgae based on their CO$_2$ tolerating capability.

Figure 2 Top view of algal cultures in microplates after exposing to different concentrations of CO$_2$ (2, 5, 10 and 20% v/v respectively), (a) strains grown in BG11 medium and (b) strains grown in SN15 medium.

The system we described here includes a CO_2-air mixture device that provide a desired CO_2 level, and a DB GasPak™ EZ bags to hold the CO_2 gas. Microalgal cultures can be dispensed into a 24, 48, or 96-well plate that will be incubated inside the GasPak™ bag. This high throughput system can also be charged with flue gas. It provides a high-throughput, uniform and repeatable method for CO_2 tolerant strains selection and comparison. Using this system, we identified 17 strains of microalgae from our culture collection that can grow under 20% CO_2 condition.

Results and discussion

The pH value of the medium is used as a direct indicator of the ambient CO_2. To test the stability of GasPak bags for holding desired CO_2 concentrations during the incubation time, we monitored pH in the wells of microplates over a two-day period. We monitored pH for 2 days because the GasPak bags will be opened for OD reading every 48 hours and recharged with CO_2. The pH values in the medium dropped sharply and reached equilibriums within 2 hours after the bags were filled with CO_2 (Figure 1). The value of pH reached different equilibriums depending on the percentage of CO_2 charged to the bags, suggesting that the system is sensitive to a small difference in CO_2 input. In all the treatments, CO_2 levels remained relatively constant in 48 hrs, suggesting that the GasPak bag can provide a stable environment to test the effect of different concentrations of CO_2 on the algal growth. When charged with same amount of CO_2, the marine medium (SN15) was able to maintain higher pH values compared to the freshwater medium (BG11). For example, under 20% CO_2, pH in the BG11 medium is 5.2, while pH in the SN15 medium is 5.7. It has been known that seawater is a good

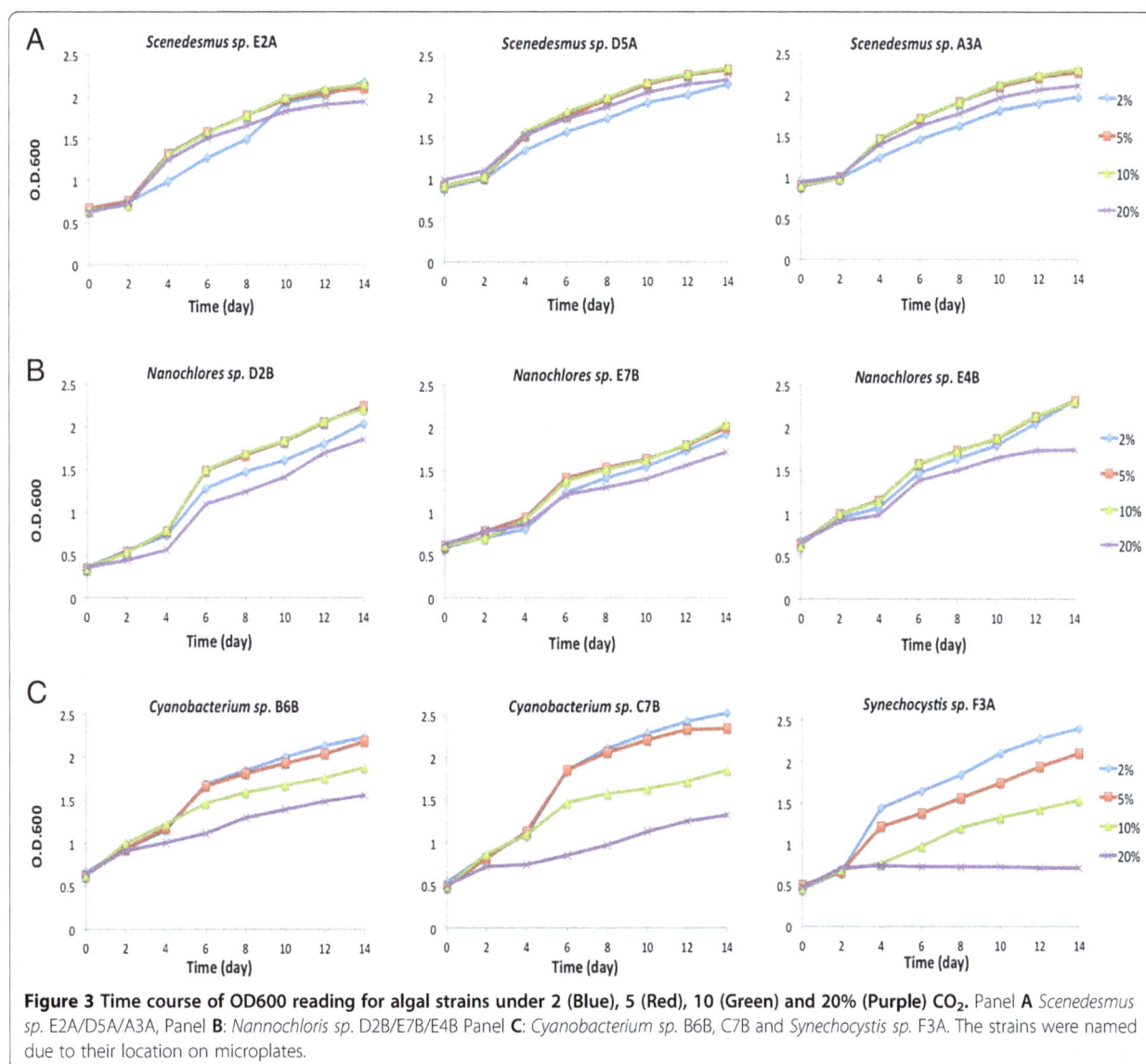

Figure 3 Time course of OD600 reading for algal strains under 2 (Blue), 5 (Red), 10 (Green) and 20% (Purple) CO_2. Panel **A** *Scenedesmus sp.* E2A/D5A/A3A, Panel **B**: *Nannochloris sp.* D2B/E7B/E4B Panel **C**: *Cyanobacterium sp.* B6B, C7B and *Synechocystis sp.* F3A. The strains were named due to their location on microplates.

buffering system, and the solubility of CO_2 decreases when the salinity increases [25].

The GasPak bag system provides a uniform environment for testing many algal strains at the same time. After 14 days, distinct growth performance of different algal strains can be visualized (Figure 2). At the lower CO_2 level (2%), nearly all the algal strains grew well and showed healthy green or blue-green color at day 14. The inhibition of growth was visible on many algal strains when the CO_2 level increased to 10 or 20%.

Growing multiple algal strains in a 48-well or 96-well plate allows a quick measurement and direct comparison of algal cell density using a plate reader. Using the system we developed here, growth performance of all tested strains at different CO_2 concentrations were monitored. The growth curves allowed us to compare and evaluate the tolerance capacity of selected algae under different CO_2 conditions. For example, three *Scenedesmus* strains showed a rapid growth under 4 different CO_2 concentrations (2, 5, 10 and 20%) (Figure 3, panel A). The growths of these algal strains were not affected by increasing CO_2 level (up to 20%), suggesting that they may tolerate

even higher level of CO_2. In contrast, the growths of three cyanobacterial strains were inhibited when the CO_2 level was increased to 10 or 20% (Figure 3, panel C). The degree of growth inhibition increased with increasing concentration of CO_2 suggesting that this system provides sufficient sensitivity for distinguishing algal strains capable of tolerating different levels of CO_2.

Within all the 96 strains tested, 17 strains were able to maintain similar growth rates with CO_2 concentration ranging from 2 to 20%, and these algal strains were considered to be high CO_2 tolerant strains (Figure 2). The algal strains that only grew under 5% CO_2 concentration were considered as CO_2 sensitive strains and may not be suitable for the CO_2 mitigation purpose. In general, the seawater strains tend to show better performance under elevated CO_2 stress compared with freshwater strains. One explanation would be that seawater medium (SN15) is a better buffering system than freshwater medium (BG11), therefore smaller decrease in pH in the seawater medium poses less acidification stress to microalgae when both media are exposed to the same ambient CO_2.

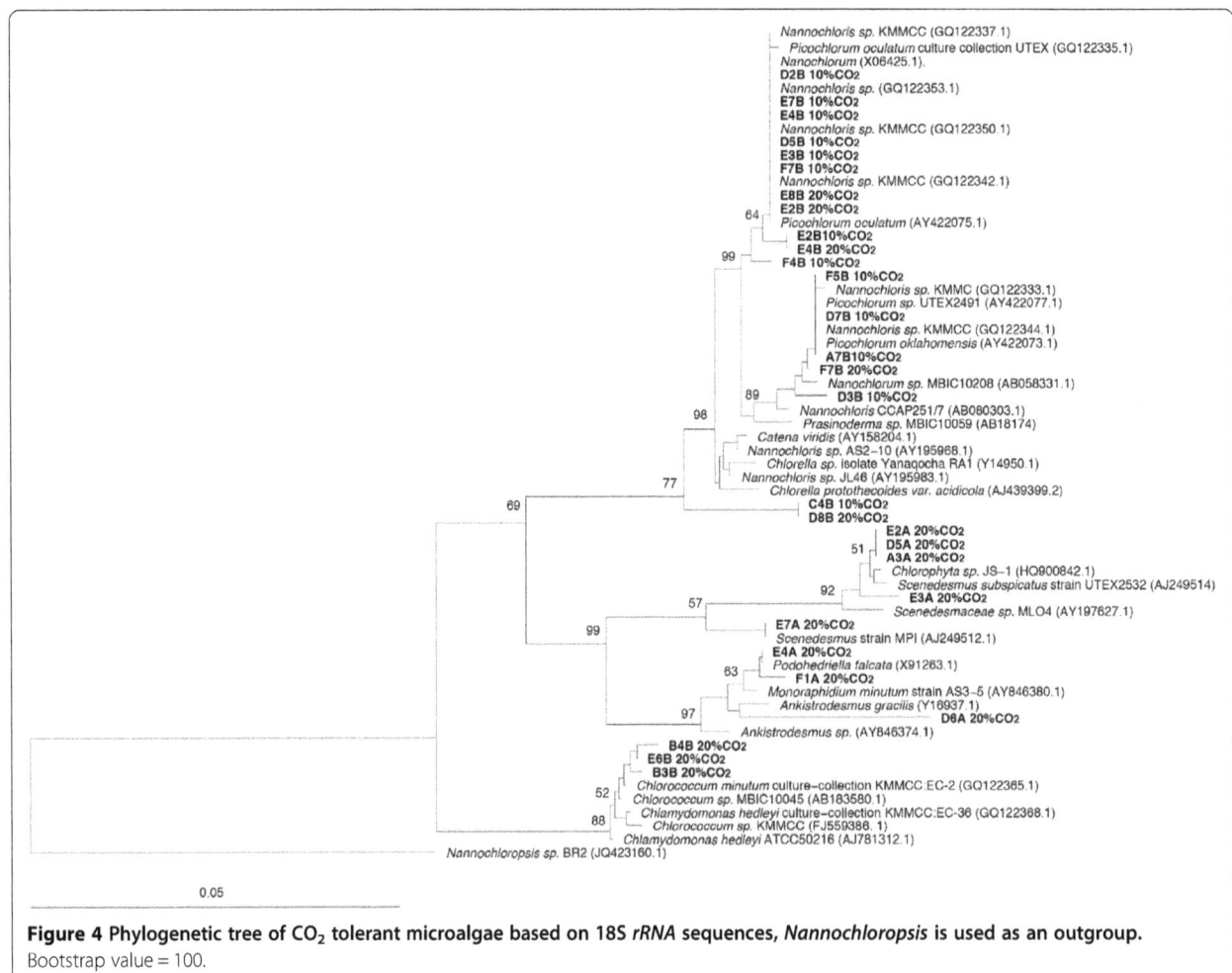

Figure 4 Phylogenetic tree of CO_2 tolerant microalgae based on 18S *rRNA* sequences, *Nannochloropsis* is used as an outgroup.
Bootstrap value = 100.

The algal strains that can tolerate 10 and 20% CO_2 were identified by sequencing the partial 18S *rRNA* gene or 16S *rRNA* gene. In this study, 5 strains of *Scenedesmus* and 3 strains of *Chlorococcum* can tolerate 20% CO_2, and 10 strains of *Nannochloris* can tolerate 10% CO_2. The majority of CO_2-tolerating strains are closely related to *Scenedesmus sp., Nannochloris sp.* and *Chlorococcum sp.* (Figure 4). Three closely related *Scenedesmus* strains (E2A, D5A, and A3A) showed little effect on their growth when exposed to 2, 5, 10 and 20% CO_2 (Figure 3, panel A). Growths of three *Nannochloris* strains (D28, E7B, and E4B) were slightly inhibited at 20% CO_2, but were similar at 2, 5, and 10% CO_2 (Figure 3, panel B). These results suggest that many microalgae in genera *Scenedemus* and *Nannochloris* can grow under high levels of CO_2. Other studies have also reported that many algal species from these two genera can tolerate high concentration of CO_2 [5,18,21]. In contrast, tested cyanobacterial strains (B6B, C7B, and F3A) grew poorly under high concentrations of CO_2 (Figure 3, panel C). Among the limited number of algal strains we tested, it appears that algal species from *Scenedesmus, Nannochloris* and *Chlorococcum* are good potential candidates for sequestering CO_2 in power plant flue gas.

The majority of the 20% CO_2 tolerant strains formed several separate branches, represented by genera *Scenedesmus, Chlorococcum* and *Ankistrodesmus* (Figure 4), suggesting that certain groups or genotypes of algae tend to perform better under high CO_2 level compare with other algal groups. When more algal strains from diverse taxa were tested for CO_2 tolerance, the phylogenetic information may provide a useful link to the potential of CO_2 tolerance of algal strains in the future.

One of *Scenedesmus* strains (E7A) has been tested in large photo-bioreactors (500 L) charged with flue gas (10-12% CO_2), and it was able to maintain vigorous growth and consume the vast majority of influx CO_2 (data not published). This test suggests that the algal strains selected using our high throughput system may be suitable for large scale cultivation.

The GasPak system we demonstrated here is designed for high throughput selection of CO_2 tolerant algal strains. Ideally, it would be useful to integrate a CO_2 sensor into the system so that the actual concentration of CO_2 in the GasPak chamber can be monitored. It is possible that the CO_2 concentration in the chamber could decrease significantly as algae continue to grow over a longer period. Given the fact that algae showed consistent growth trends under different CO_2 concentrations, we believe that the GasPak system is able to maintain desirable CO_2 levels during the 2-week experiment.

Conclusion

We introduced a high throughput system that can be used to quickly select microalgae or other microorganisms that can grow under different concentrations of CO_2 or other type of gases. Our system provides an adjustable gas input and yields reproducible growth measurements. The growth performance of hundreds of algal strains can be compared at the uniform and sustainable condition using this system. In addition, the system can be used for high throughput screening for algal strains that can tolerate other gases, such as NOx, SOx, or flue gas.

Methods

Organisms and culture conditions

The algal strains used in this study were isolated from waters collected from different parts of the Chesapeake Bay including the Baltimore Inner Harbor and the Back River (Baltimore, Maryland), using agar plates made of BG11 [26] as a freshwater medium and SN medium [27] as a seawater medium. Single algal colonies were picked and transferred to 96-well plates, and scaled up to large culture flasks. The algal cultures were illuminated continually using the plant light (Agro-Lite R20, 50 W, PHILIPS) at $25\mu E/ m^2/s$. This light level was carefully selected for growing algae in the small volume of 48 well Costar plates (Coring, NY, USA). Considering the low starting algal density and the amount of photon received

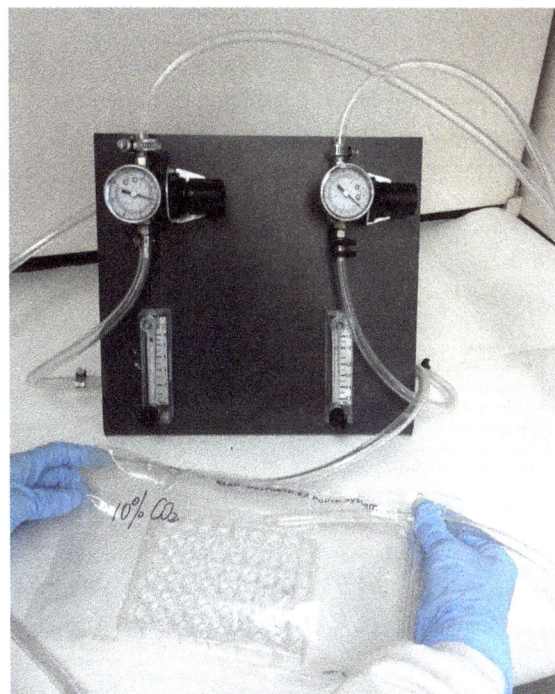

Figure 5 System used for charging specific concentration of CO_2 to the GasPak bags. For instance, to generate 10% (v/v) CO_2, the flow of pure CO_2 was set at the speed of 1 L/min using the controller on the right hand side and the flow of air was set at a velocity of 9 L/min using the controller on the left hand site. The CO_2/air mixture was then used to inflate the GasPark bag as shown.

by individual cells, the light intensity in this setting is within the appropriate range.

The CO_2 mixing and incubation system

To set up different concentrations of CO_2, pure CO_2 and air were blended using a device with two gas flow meters (Figure 5).

CO_2 equilibrium experiment

In order to test the stability of CO_2 concentration inside GasPak™ bag (Becton Dickinson, NJ, USA), one milliliter of BG11 or SN15 medium was added to individual wells on a 48-well microtiter plate. Four identical plates were prepared and placed into 4 GasPak™ bags which were charged with 2, 5, 10 and 20% CO_2, respectively. The pH values in the culture media were measured at 2, 4, 24 and 48 hours, using an Accumet Basic pH meter (Fisher Scientific). Only culture media (no algal inoculation) were used in this test, and the same experiment was repeated 3 times.

High throughput CO_2 tolerant strain screening

A 48-well-microtiter plate (without lid) that contains multiple algal strains was placed inside a GasPak™ bag and the bag was aerated with desired concentration of CO_2. The sealed bags were incubated with light at $25 \mu E/ m^2/s$.

In this experiment, different algal strains were dispensed into 48-well plates and charged with 2, 5, 10 and 20% of CO_2 respectively. The growth of algae was monitored by cell density (OD600) every other day using a multi-mode microplate reader (Molecular devices, SpectraMax M5). After reading, the culture plates were placed back into the bags, and the system was re-charged with desired concentration of CO_2.

Identification of algal strains

Genomic DNA of selected strains were extracted and 18S or 16S ribosomal RNA gene was amplified using the universal primers for eukaryotes and prokaryotes, respectively [28]. Phylogenetic trees were constructed based on partial 18S or 16S *rRNA* gene sequences using ARB Neighbor-joining algorithms with 100 bootstrap [29]. Comparison was carried out between selected strains from this work and high CO_2 tolerant species reported from other studies.

Competing interests
The authors have declared that no competing interests exist.

Authors' contribution
Experimental design: FC, LZ. Performing the experiments: LZ FZ. Data analysis: FC LZ FZ. Manuscript preparation: LZ FC FZ. All authors read and approved the final manuscript.

Acknowledgements
We thank Robert Mroz and Jack French for providing the unpublished data on the scale up cultivation of *Scenedesmus* sp. We also thank Odi Zmora and Mathew Moore for helping the design of the CO_2 mixer. The funding from Maryland Industrial Partnership supports this work.

References

1. EPA: *Inventory of U.S. Greenhouse Gas Emission and Sinks 1990–2010*. Washington, DC: Environmental Protection Agency; 2012.
2. Bernstein L, Bosch P, Canziani O, Chen Z, Christ R, Davidson O, Hare W, Huq S, Karoly D, Kattsov V, *et al*: *Climate Change 2007: Synthesis Report: An Assessment of the Intergovernmental Panel on Climate Change*. Geneva, Switzerland: IPCC; 2008:104.
3. Omer AM: **Energy, environment and sustainable development.** *Renew Sustain Energy Rev* 2008, **12**(9):2265–2300.
4. Stewart C, Hessami M-A: **A study of methods of carbon dioxide capture and sequestration––the sustainability of a photosynthetic bioreactor approach.** *Energy Convers Manag* 2005, **46**:403–420.
5. Wang B, Li Y, Wu N, Lan C: **CO2 bio-mitigation using microalgae.** *Appl Microbiol Biotechnol* 2008, **79**(5):707–718.
6. Hase R, Oikawa H, Sasao C, Morita M, Watanabe Y: **Photosynthetic production of microalgal biomass in a raceway system under greenhouse conditions in Sendai city.** *J Biosci Bioeng* 2000, **89**:157–163.
7. Chiu S-Y, Kao C-Y, Huang T-T, Lin C-J, Ong S-C, Chen C-D, Chang J-S, Lin C-S: **Microalgal biomass production and on-site bioremediation of carbon dioxide, nitrogen oxide and sulfur dioxide from flue gas using Chlorella sp. cultures.** *Bioresour Technol* 2011, **102**(19):9135–9142.
8. Brennan L, Owende P: **Biofuels from microalgae—a review of technologies for production, processing, and extractions of biofuels and co-products.** *Renew Sustain Energy Rev* 2010, **14**:557–577.
9. Chen C-Y, Yeh K-L, Aisyah R, Lee D-J, Chang J-S: **Cultivation, photobioreactor design and harvesting of microalgae for biodiesel production: a critical review.** *Bioresour Technol* 2011, **102**(1):71–81.
10. EEA: *Air pollution from electricity-generating large combustion plants*. Copenhagen: European Environment Agency; 2008.
11. Douskova I, Doucha J, Livansky K, Machat J, Novak P, Umysova D, Zachleder V, Vitova M: **Simultaneous flue gas bioremediation and reduction of microalgal biomass production costs.** *Appl Microbiol Biotechnol* 2009, **82**(1):179–185.
12. Kadam KL: **Power plant flue gas as a source of CO2 for microalgae cultivation: economic impact of different process options.** *Energy Convers Manag* 1997, **38**:S505–S510.
13. Maeda K, Owada M, Kimura N, Omata K, Karube I: **CO2 fixation from the flue gas on coal-fired thermal power plant by microalgae.** *Energy Convers Manag* 1995, **36**(6–9):717–720.
14. Hanagata N, Takeuchi T, Fukuju Y, Barnes DJ, Karube I: **Tolerance of microalgae to high CO2 and high temperature.** *Phytochemistry* 1992, **31**:3345–3348.
15. Kodama M, Ikemoto H, Miyachi S: **A new species of highly CO2-tolerant fast-growing marine microalga suitable for high-density culture.** *J Mar Biotechnol* 1993, **1**(1):21–25.
16. Seckbach J, Baker FA, Shugarman PM: *Algae thrive under pure CO2*; 1970.
17. de Morais M, Costa J: **Carbon dioxide fixation by Chlorella kessleri, C. vulgaris, Scenedesmus obliquus and Spirulina sp. cultivated in flasks and vertical tubular photobioreactors.** *Biotechnol Lett* 2007, **29**:1349–1352.
18. Ho S-H, Chen W-M, Chang J-S: **Scenedesmus obliquus CNW-N as a potential candidate for CO2 mitigation and biodiesel production.** *Bioresour Technol* 2010, **101**:8725–8730.
19. Kaewkannetra P, Enmak P, Chiu T: **The effect of CO2 and salinity on the cultivation of Scenedesmus obliquus for biodiesel production.** *Biotechnol Bioproc E* 2012, **17**(3):591–597.
20. Rodriguez-Maroto JM, Jimenez C, Aguilera J, Niell FX: **Air bubbling results in carbon loss during microalgal cultivation in bicarbonate-enriched media: experimental data and process modeling.** *Aquac Eng* 2005, **32**(3):493–508.
21. Kumar K, Dasgupta CN, Nayak B, Lindblad P, Das D: **Development of suitable photobioreactors for CO2 sequestration addressing global**

warming using green algae and cyanobacteria. *Bioresour Technol* 2011, **102**(8):4945–4953.

22. Singh RN, Sharma S: **Development of suitable photobioreactor for algae production – A review.** *Renew Sustain Energy Rev* 2012, **16**(4):2347–2353.
23. Díez B, Pedrós-Alió C, Massana R: **Study of genetic diversity of eukaryotic picoplankton in different oceanic regions by small-subunit rRNA gene cloning and sequencing.** *Appl Environ Microbiol* 2001, **67**(7):2932–2941.
24. Ebenezer V, Medlin L, Ki J-S: **Molecular detection, quantification, and diversity evaluation of microalgae.** *Marine Biotechnol* 2012, **14**:129–142.
25. Colt J: **Computation of dissolved gas concentrations in water as functions of temperature, salinity, and pressure.** *Am Fish Soc Spec Publ* 1984, **14**:110–111.
26. Stanier RY, Kunisawa R, Mandel M, Cohen-Bazire G: **Purification and properties of unicellular blue-green algae (order Chroococcales).** *Bacteriol Rev* 1971, **35**:171–205.
27. Waterbury JB, Watson SW, Valois FW, Franks DG: **Biological and ecological characterization of the marine unicellular cyanobacterium Synechococcus.** *Can Bull Fish Aquat Sci* 1986, **214**:71–120.
28. Rasoul-Amini S, Ghasemi Y, Morowvat MH, Mohagheghzadeh A: **PCR amplification of 18S rRNA, single cell protein production and fatty acid evaluation of some naturally isolated microalgae.** *Food Chem* 2009, **116**(1):129–136.
29. Ludwig W, Strunk O, Westram R, Richter L, Meier H, Yadhukumar, Buchner A, Lai T, Steppi S, Jobb G, *et al*: **ARB: a software environment for sequence data.** *Nucleic Acids Res* 2004, **32**:1363–1371.

15

Assessing environmental impacts of offshore wind farms: lessons learned and recommendations for the future

Helen Bailey[1*], Kate L Brookes[2] and Paul M Thompson[3]

Abstract

Offshore wind power provides a valuable source of renewable energy that can help reduce carbon emissions. Technological advances are allowing higher capacity turbines to be installed and in deeper water, but there is still much that is unknown about the effects on the environment. Here we describe the lessons learned based on the recent literature and our experience with assessing impacts of offshore wind developments on marine mammals and seabirds, and make recommendations for future monitoring and assessment as interest in offshore wind energy grows around the world. The four key lessons learned that we discuss are: 1) Identifying the area over which biological effects may occur to inform baseline data collection and determining the connectivity between key populations and proposed wind energy sites, 2) The need to put impacts into a population level context to determine whether they are biologically significant, 3) Measuring responses to wind farm construction and operation to determine disturbance effects and avoidance responses, and 4) Learn from other industries to inform risk assessments and the effectiveness of mitigation measures. As the number and size of offshore wind developments increases, there will be a growing need to consider the population level consequences and cumulative impacts of these activities on marine species. Strategically targeted data collection and modeling aimed at answering questions for the consenting process will also allow regulators to make decisions based on the best available information, and achieve a balance between climate change targets and environmental legislation.

Keywords: Marine mammals, Seabirds, Wind turbine, Underwater noise, Collision risk, Human impacts, Cumulative impact assessment, Population consequences

Introduction

Efforts to reduce carbon emissions and increase production from renewable energy sources have led to rapid growth in offshore wind energy generation, particularly in northern European waters [1,2]. The first commercial scale offshore wind farm, Horns Rev 1 (160 MW with 80 turbines of 2 MW), became operational in 2002. The average capacity of turbines and size of offshore wind farms have been increasing since then, and they are being installed in deeper waters further from the coast. By the end of 2013, operational wind farms were in an average water depth of 16 m and 29 km from shore in Europe [3] (Figure 1). With technological advances in the future [4]

there is likely to be a continued increase in the size of offshore wind projects [3], but there are still uncertainties about the effects on the environment [5]. The novelty of the technology and construction processes make it difficult to identify all of the stressors on marine species and to estimate the effect of these activities [6].

The major environmental concerns related to offshore wind developments are increased noise levels, risk of collisions, changes to benthic and pelagic habitats, alterations to food webs, and pollution from increased vessel traffic or release of contaminants from seabed sediments. There are several reviews of the potential impacts of offshore wind energy on marine species e.g. [5-7]. As well as potential adverse impacts, there are possible environmental benefits. For example, wind turbine foundations may act as artificial reefs, providing a surface to which animals attach. Consequently there can be increases in the number of

* Correspondence: hbailey@umces.edu
[1]Chesapeake Biological Laboratory, University of Maryland Center for Environmental Science, 146 Williams Street, Solomons, MD 20688, USA
Full list of author information is available at the end of the article

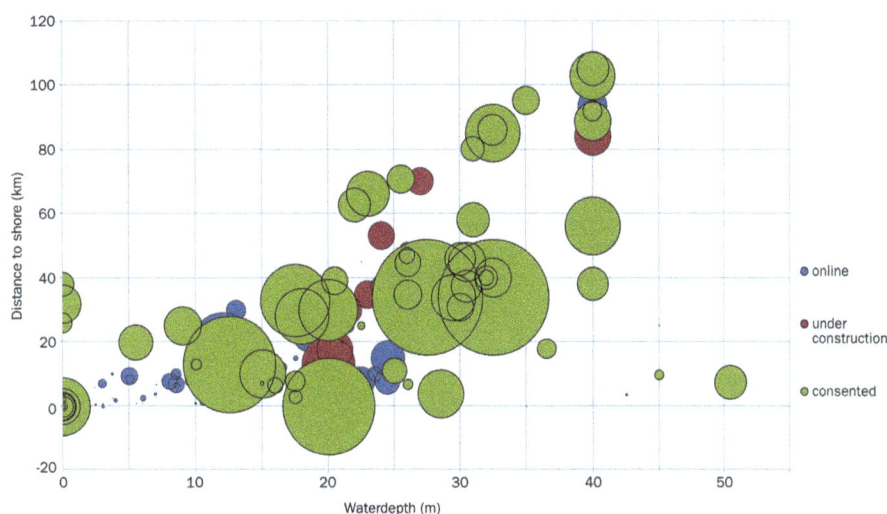

Figure 1 Average water depth and distance to shore of offshore wind farms (reproduced from ref. 3, source EWEA). Operational (online), under construction and consented wind farms in Europe up to the end of 2013 are occurring at increasing water depths and distances from shore. The circle size represents the total power capacity of the wind farm.

shellfish, and the animals that feed on them, including fish and marine mammals [8-11]. A second possible benefit is the sheltering effect. A safety buffer zone surrounding the wind turbines may become a de-facto marine reserve, as the exclusion of boats within this zone would reduce disturbance from shipping. Exclusion of some or all types of fishing could also result in local increases in prey abundance for top predators, whilst reducing the risk of bycatch in fishing gear [9]. Further research is required to understand the ability of wind turbines to attract marine species and the effect of excluding fisheries. Finally, there may also be opportunities in the future to combine offshore wind farms with open ocean aquaculture [12].

Over 2,000 wind turbines are installed in 69 offshore wind farms across Europe, with the greatest installed capacity currently in the U.K. (Figure 2) [3]. As the number of offshore wind farms has increased, approaches for environmental monitoring and assessment have improved over time. However, there are still few studies that have measured the responses of marine species to offshore wind farm construction and operation, and none have yet assessed longer term impacts at the population level. In Europe, legislation requires consideration of cumulative impacts, defined as impacts that result from incremental changes caused by other past, present or foreseeable actions together with the project [13]. However, approaches for cumulative impact assessments currently vary in terms of their transparency, efficiency and complexity, and this is an active area of research development [14]. In addition to assessing and measuring impacts, it is also necessary to develop decision support tools that will assist regulators with

determining whether a proposed development can be legally consented.

In this paper, we first briefly review the potential impacts of offshore wind developments on marine species. We then identify the key lessons that have been learned from our own studies and others in Europe, primarily focusing on marine mammals and seabirds. Much of the environmental research that has been conducted in relation to offshore wind energy has concerned the impact of sound exposure for marine mammals and the risk of collisions with turbines for seabirds. We identify where knowledge gaps exist that could help to improve current models and impact assessments. Finally, we discuss emerging technologies and make recommendations for future research to support regulators, developers and researchers involved in proposed developments, particularly in countries where the implementation of offshore wind energy is still in its early stages.

Impact pathways

The potential effects of offshore wind farm construction and operation will differ among species, depending on their likelihood of interaction with the structures and cables, sensitivities, and avoidance responses. Studies have generally focused on marine mammals and seabirds because of stakeholder concerns and legal protection for these species and their habitats. The construction phase is likely to have the greatest impact on marine mammals and the activities of greatest concern are pile driving and increased vessel traffic [15]. Pile driving is currently the most common method used to secure the turbine foundation to the seafloor, although other foundation types are being developed [4]. The loud sounds emitted during pile driving

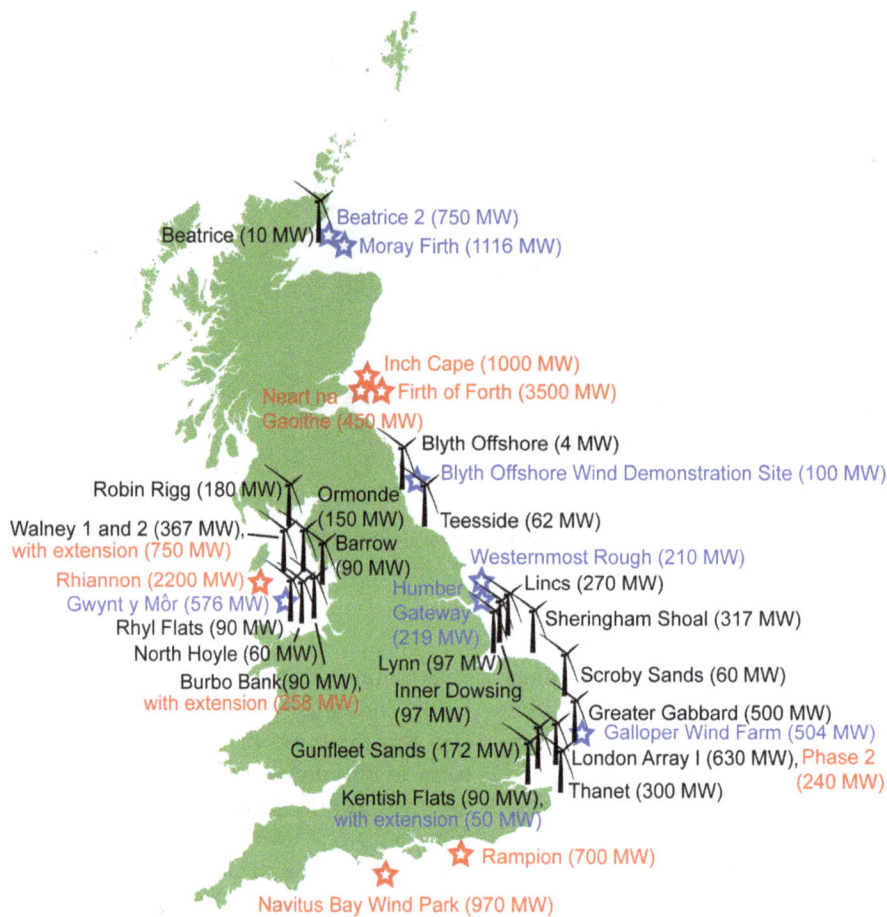

Figure 2 Offshore wind farms around the U.K., July 2014. This includes wind farms in operation (black wind turbines), those consented and under development (blue stars), and in the proposal and planning stages (red stars).

could potentially cause hearing damage, masking of calls or spatial displacement as animals move out of the area to avoid the noise [16,17]. Fish could similarly be affected by these sounds [17-20]. There is also a risk to marine mammals, sea turtles and fish of collision and disturbance from vessel movements associated with surveying and installation activities.

During operation of the wind turbines, underwater sound levels are unlikely to reach dangerous levels or mask acoustic communication of marine mammals [21,22]. However, this phase of the development is of greatest concern for seabirds. Mortality can be caused by collision with the moving turbine blades, and avoidance responses may result in displacement from key habitat or increase energetic costs [23,24]. This may affect birds migrating through the area as well as those that breed or forage in the vicinity.

During operation, cables transmitting the produced electricity will also emit electromagnetic fields. This could affect the movements and navigation of species that are sensitive to electro- or magnetic fields, which

includes fish species, particularly elasmobranchs and some teleost fish and decapod crustaceans, and sea turtles [25-27]. Commercial fish species may potentially be positively affected if fishing is prohibited in the vicinity of the wind farm, although this could result in a displacement of fisheries effort and consequent change in catches and bycatch.

The specific species of greatest concern will differ among regions depending on their occurrence and protection status. For example, assessments of impacts upon marine mammals in Europe have generally focused on small cetaceans (particularly harbor porpoises (*Phocoena phocoena*)) and pinnipeds (primarily harbor seals (*Phoca vitulina*)). These species are common in such areas and the EU Habitats Directive (92/43/EEC) requires governments to establish Special Areas of Conservation for their protection. However, in other locations, marine mammal species listed under the Endangered Species Act, such as the North Atlantic right whale (*Eubalaena glacialis*), blue whale (*Balaenoptera musculus*), humpback whale (*Megaptera novaeangliae*), and fin whale (*Balaenoptera physalus*), may be of greater concern. Based on their call frequencies, these large whales are considered to be sensitive to the low frequency sounds produced during pile driving [16,28,29].

There is also a paucity of information on the effects of human-generated sound on fish [18,20,30,31]. Evidence of injury from pile driving sounds in a laboratory simulated environment has been reported for several fish species [32-34]. Recovery tended to occur within 10 days of exposure and is unlikely to have affected the survival of the exposed animals. Common sole larvae (*Solea solea*) also survived high levels of pile-driving sound in controlled exposure experiments [35]. However, a behavioral response was triggered in cod (*Gadus morhua*) and sole by playbacks of pile driving sounds in the field and was initiated at a much lower received sound level [36]. This could consequently result in a large zone of behavioral response. The sounds produced by offshore wind farms may also mask fish communication and orientation signals [30]. These responses need to be investigated further to determine their potential effect on foraging, breeding and migration, and require the ability to record the movements of fish as well as the measurement of sound pressure levels and particle motion since fish are sensitive to both [18,37]. Some fish species are short-lived and highly fecund reducing the likelihood of any longer-term population level effects from wind farm noise and disturbance. However, this is not true of all fish species and there are endangered species, such as the Atlantic sturgeon (*Acipenser oxyrinchus*), those listed as vulnerable, such as the basking shark (*Cetorhinus maximus*), and potential impacts to fisheries which may need to be taken into consideration.

Much of the early work investigating impacts upon bird populations at European sites has focused on species of migratory or wintering waterfowl [23,38]. There is much less known about potential collision risk or displacement for the broader suite of seabird species that occur in many of the areas currently being considered for large scale wind farm developments. Migrating bats have also been found to occur offshore [39,40], although relatively little research on their offshore distribution, collision risk and potential displacement by offshore wind farms has been done compared to that for wind farms on land [39,41].

Other taxonomic groups such as sea turtles are rare visitors to coastal European waters, and have not been considered at high risk from the effects of offshore wind farms. However, in other areas, for example along the North American coast, there may be sea turtle nesting or breeding grounds in the vicinity of proposed sites [42]. It has recently been determined that the hearing sensitivity of leatherback turtles overlaps with the frequencies and source levels produced by many anthropogenic sounds, including pile-driving [43]. This highlights the need for a better understanding of the potential physiological and behavioral impacts on sea turtles.

Lessons learned

Environmental research for offshore wind energy has evolved over time in Europe as a better understanding of the type of information and analysis that best informs decisions about the siting of offshore wind facilities has been developed. Other countries interested in offshore wind energy, such as the U.S.A. (Figure 3), may therefore benefit from the European experience and hindsight to maximize the potential success of their projects [44]. Based on our experiences relating to marine mammals and seabirds, the key lessons learned that we have identified are:

1. Define the area of potential effect
   - Identifying the area over which biological effects may occur to inform baseline data collection.
   - Determining the connectivity between key populations and proposed wind energy sites.
2. Identify the scale and significance of population level impacts
   - The need to define populations, identify which populations occur within the wind energy site and the area of potential effect, and their current status.
   - The requirement for demographic data and information on vital rates to link individual responses to population level consequences.
   - Research to test and validate modeling assumptions and parameters.
3. Validate models through measuring responses in the field
   - Use of a gradient design to determine the extent of spatial displacement as a result of offshore wind energy development and how this may change over time.
   - Utilization of techniques with the power to detect changes.
   - Coordination of human activities and monitoring in the vicinity of wind energy sites.
4. Learn from other industries to inform risk assessments and the effectiveness of mitigation measures.
   - Onshore wind energy, seismic surveys and floating oil platforms.

We will discuss each of these lessons learned in more detail and then provide recommendations for further research to fill identified knowledge gaps, test existing

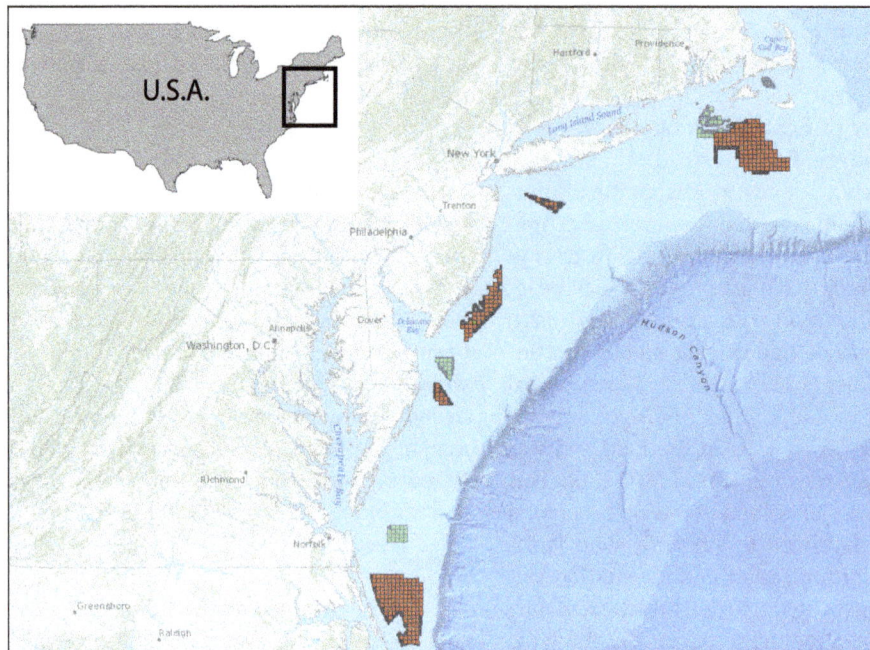

Figure 3 Potential wind energy areas in the Mid-Atlantic off the U.S.A. (source BOEM). There are no commercial wind farms currently on the Atlantic outer continental shelf off the U.S.A., but the Bureau of Ocean Energy Management (BOEM) has designated wind energy areas, which are in the planning (orange squares) and leased (green squares) stages (July 2014).

models and improve future environmental impact assessments (EIA).

The area of potential effect

To evaluate the impact of a proposed activity on marine species, it is necessary to have sufficient baseline information on distribution, abundance and their trends within the area of potential effect. This is particularly challenging for many marine species since some stressors, such as underwater sound, can travel long distances, and these species are often highly mobile and/or migratory. Consequently, the area of potential effect can extend far beyond the immediate vicinity of the proposed development. For example, the sound produced during pile driving may travel tens of kilometers underwater, which could cause behavioral disturbance to marine mammals and fish over a large area [17]. Early baseline studies in relation to marine mammals and the impact upon them from underwater pile driving noise were designed on relatively uncertain estimates of the area of potential effect, and some of the control sites used were subsequently identified as being impacted [45,46].

Pile-driving sounds were recorded to determine the received levels at a range of distances during the construction of the Beatrice Demonstrator Wind Farm project in the Moray Firth, Scotland [47] (Figure 2). This wind farm consisted of two 5 MW turbines in water 42 m deep with 1.8 m diameter tubular steel piles to secure the jacket foundation structure to the seafloor. Pile-driving sounds

were recorded at 0.1 to 70 km away and the peak to peak sound pressure levels ranged from 122–205 dB re 1 μPa. Based on the measurements within 1 km, the source level was back-calculated and estimated to be 226 dB re 1 μPa at 1 m [47]. Each pile required thousands of hammer blows and was struck about once every second.

Since pile-driving involves multiple strikes, it is considered a multiple pulse sound. The sound exposure level (SEL) is a measure of the energy of a sound and depends on both the pressure level and duration [28]. This can be summed over multiple strikes to give the cumulative SEL [19,28]. The cumulative energy level over the full pile-driving duration gives a measure of the dose of exposure, assuming no recovery of hearing between repeated strikes, and is necessary for assessing cumulative impacts.

There were, and still are, insufficient data available to develop noise exposure criteria for behavioral responses of marine mammals to multiple-pulsed noise such as pile driving [28]. Evidence of behavioral disturbance from sounds arising from pile-driving has been obtained through simulated, playback, and live conditions, and indicates that the zone of responsiveness for harbor porpoises may extend to 20 km or more [45,46,48-52]. However, response distances will vary depending on the activity being undertaken by the animal when it is exposed to the sound, the sound source level, sound propagation, and ambient noise levels [16,28,53]. Limited understanding of the role of these different environmental, physical and biological factors currently constrain assessments of the potential scale of impact at particular development sites.

Collecting baseline information for such large areas of potential effect presents a number of challenges. In cases where there is little or no existing information on the species of concern, such as their distribution and abundance in more offshore areas, it can be difficult to determine appropriate designs for impact studies [49]. The logistical difficulties of working offshore, along with financial limitations, may additionally restrict the number of sampling sites and replicates. It has been recommended that at least two years of baseline data are necessary for a sufficient description of species occurrences [54]. However, whilst this may provide information on seasonal variability, longer time-series of data are ideally required to capture inter-annual variability in order to identify the effects of construction activities over natural variation (which may be high) [54,55]. Given that data collection over these large spatial and temporal scales will be so difficult to achieve, it is crucial that studies are targeted to focus upon those data which are critical for supporting decision making. Baseline data and modeling should be targeted to answer specific questions relating to the consenting process, meaning that monitoring and site characterization requirements may differ under different legislative systems. For example, consent may require an understanding of the connectivity of a proposed wind energy site with key protected populations. In Scottish waters, this requirement has focused research efforts on areas within and surrounding proposed wind energy sites, and between these sites and EU designated Special Areas of Conservation (SACs) for harbor seals and for bottlenose dolphins [56-58].

For birds, the operational phase of wind farms is likely to present the biggest risk. Vulnerability and mortality at onshore wind turbines has been identified as being related to a combination of site-specific, species-specific and seasonal factors [59]. The development of collision risk models for seabirds requires information on their spatial distribution and flight heights to determine the likelihood of co-occurrence with the wind turbine blades, and their avoidance response to estimate the mortality risk [60]. However, much of this relies on expert-based estimates because there are very few empirical data on flight heights for different seabird species [24]. Although there have been estimates of flight heights during ship-based surveys where they are classified into altitude categories there can be large inter-observer differences [61]. One recent approach to address this data gap is to model flight height distributions based on compilations of survey data [62,63].

While information on site-specific flight heights of bird species is lacking, there is even less information on avoidance responses to large offshore wind farms by birds. The few studies examining avoidance behavior involved tracking eider ducks (*Somateria mollissima*) and geese by radar.

These studies documented a substantial avoidance response by these migrating birds, which reduced the collision risk [23,64]. There is a need for empirical data on both broad and fine-scale avoidance responses to improve the reliability of predictions from collision risk models [65]. There should also be a focus not only on estimates of mortality, but also of the energetic consequences of avoidance and displacement behaviors [66], and their impacts on survival and fecundity. The cumulative impacts of different disturbance activities (such as ship and helicopter traffic) and multiple wind energy sites within the migration pathway or home range of a population should also be considered [67].

Population level impacts

The regulatory requirements for assessing the impacts of a proposed activity and determining whether it is biologically significant will vary among countries. However, in general this process will require populations to be defined, identifying which of these populations occur within the area of potential effect, and understanding their current status to determine whether the impact will be significant. The complexity of approaches, models and simulation tools to support these assessments has greatly increased over time [24,58,68]. However, there are still many knowledge gaps concerning behavioral responses, particularly on the consequences of any behavioral change on vital rates. For example, there is a growing understanding that anthropogenic noise, such as pile-driving, may affect the behavior of marine mammals and lead to spatial displacement [69]. However, there have been no empirical studies linking the consequences of this behavioral response to longer term population change. Similarly, there are concerns that the presence of wind farms may displace seabirds from preferred foraging areas [24], but there is limited understanding either of the extent of such effects or of the individual and population consequences of displacement, should it occur.

For other management issues, such as bycatch, estimates of Potential Biological Removal have provided management limits for human-caused mortality in mammals [70,71], but this approach cannot be used for assessing non-lethal impacts. To address this, a framework was developed called "Population Consequences of Acoustic Disturbance" (PCAD) [29,72]. The aim of this approach is to link behavioral responses by individuals and their vital rates to determine the consequences for the population. In addition to the spatially-explicit information on distribution and abundance typically collected for impact assessments, this approach also requires knowledge of dose–response relationships to link behavioral responses and demographic parameters. Since a general characterization of the dose–response relationship between received noise levels and changes in vital rates does not exist for marine mammals, expert judgment has been used to link individual impacts to changes in survival or reproductive rates [73].

Given the uncertainties involved, the population level assessments required from developers by U.K. regulators have been very conservative, and are expected to overestimate the impacts to populations. Nevertheless, the application of such an approach to a harbor seal population suggested that the population trends were largely driven by the baseline dynamics of the population and, even in a worst-case scenario of impacts, only a short term reduction in numbers would be expected to occur [58]. The long-term dynamics appeared relatively robust to uncertainty in key assumptions, but there is still a strong reliance on expert judgment and many assumptions are made. Focused studies around subsequent developments are now required to test these modeling assumptions and frameworks to ensure they are robust and, if appropriate, made less conservative in the future.

There is also a strong reliance on expert judgment in seabird collision risk models and sensitivity indices [24,60]. Avoidance rates are applied to collision risk models, but for many species they are not based on empirical data. Work is ongoing to provide estimates of these, but has been hampered by a lack of suitable techniques. The importance of this human-induced mortality on seabirds may depend on the current status of the population, with conservation concerns potentially being greater for populations that are currently in decline. For example, black-legged kittiwakes (*Rissa tridactyla*) have declined by more than 50% since 1990 in the North Sea [74]. The cause of this decline has mainly been attributed to poor breeding success as a result of reduced recruitment of their prey species, the lesser sandeel (*Ammodytes marinus*), linked to warm winters and the presence of a local sandeel fishery [74]. Black-legged kittiwakes generally fly below the minimum height of any turbine's rotor blades, but there were approximately 15.7% of flights that occurred within a generic collision risk height band defined as 20–150 m above sea level [62] and their avoidance response is unknown. The potential additional mortality that offshore wind farms could induce for this declining population makes this species of particular concern in the environmental assessment and consenting process.

Seabirds are considered at their most vulnerable when wind energy sites are proposed near their breeding colonies. During the breeding season, they make regular trips between their nest and foraging grounds. This could reduce the collision risk for wind farms proposed further offshore, but there is generally less known about the distribution and habitat use of seabirds in these areas outside of the breeding season, and their connectivity with any protected areas. As wind farms move further offshore such knowledge gaps will need to addressed.

Measuring responses

A BACI (before-after-control-impact) design was initially recommended to assess the responses of marine mammals to wind farm construction and operation [54]. However, this type of design is limited in its ability to characterize spatial variability, assigning samples to only a treatment or control strata [75]. There are also arbitrary requirements for the selection of control sites, which include being far enough away to be unaffected by the potential disturbance, but close enough that the areas are comparable. Some stressors have a large area of potential effect, which makes it difficult to identify suitable control sites with similar ecological characteristics. Differences in variability between sites can also be a problem in statistically detecting impacts [76]. The BACI design is appropriate where there are defined boundaries for the impacted areas, but a gradient design will be more sensitive to change when a contaminant or sound disperses with distance from a point source. A gradient design requires classifying samples according to distance and removes the issue of selecting a control site. It is also more powerful than a randomized Control Impact design at detecting changes due to disturbance [77]. It has recently been demonstrated to be more effective in terms of studying spatial displacement of harbor porpoises in response to pile-driving and detecting how temporal effects differ with distance [48,78]. Furthermore, whilst BACI designs provide opportunities to identify whether or not impacts have occurred, gradient designs can also be used to assess the spatial scale of any impacts, thus informing future spatial planning decisions.

Data collection techniques used for characterizing a site in the planning stages may not be the most appropriate tools for assessing impacts. Visual surveys for both birds and mammals have generally been used to describe their abundance and distribution in planning applications. These techniques are unlikely to have enough power to detect changes in behavior or fine scale spatial or temporal shifts in distribution, since observers can only be in one place at a time and can only reliably survey in calm sea conditions during daylight hours. Our research has shown that acoustic methods for assessing impacts to marine mammals have much greater power to detect change [49,55], and techniques such as GPS tracking, radar, and fixed cameras are likely to provide more useful data for seabirds [64]. GPS tracking has been used for many species and provides high resolution data. In a recent study, it revealed harbor seals foraging around wind turbines in the North Sea [11]. Acoustic telemetry has been a valuable tool for tracking fish [79] and also turtle movements [80], and the technology now exists to use this to examine the long-term, fine-scale movements of aquatic animals [81].

The presence of other disturbance sources unrelated to the wind farm activity may compromise efforts to compare periods before and after construction events. For example, during our study of the impacts of the Beatrice Demonstrator Wind Farm project, we later discovered that hydrographic and seismic surveys had also been conducted in the area during the construction period [49]. Determining the cause of any observed effects is therefore confounded by these additional activities. Communication and coordination amongst planners, regulators, industry and scientists is therefore essential to ensure that impact studies can be properly designed, and any cumulative impacts caused by multiple events during construction or by other human activities can be taken into account [14,82]. Greater involvement of species group specialists during the planning process and engineering design phases may also help to minimize any environmental conflicts at an early stage. Careful spatial planning of wind farms has been identified as a key factor for profitability and environmental protection [83]. Consideration of the increased development of local ports to support construction and maintenance of offshore wind farms and the consequent environmental impacts is also important.

Learning from other industries

There are three existing industries whose experiences have been usefully applied to environmental research surrounding offshore wind energy developments in European waters. These are onshore wind farms, seismic surveys for oil and gas exploration, and floating oil platforms. Our knowledge of bird vulnerability and mortality from wind farms has largely been based on those on land. Direct measurements of mortality from offshore wind farms are much more difficult because of the difficulty of finding corpses at sea. The lack of direct measurements of flight height distributions and avoidance responses for many seabird species means there is still considerable uncertainty in the mortality estimates and the consequent energetic costs of avoidance behaviors for offshore wind farms, but the modeling approaches developed for terrestrial wind farms have provided a robust framework to begin these assessments [84,85].

Airguns used in seismic surveys for oil and gas exploration produce loud multiple pulsed noises with energy mainly below 1 kHz, but also extending to much higher frequencies [55,86]. Pile-driving also produces loud multiple-pulsed sounds, although they tend to be more broadband with the major amplitude at 100–500 Hz compared to a seismic airgun array at 10–120 Hz [87]. In addition, pile-driving has a shorter interval between pulses at about one second [47,88] as opposed to seismic airgun surveys at 10 seconds or more [87]. Changes in distribution and vocal behavior by marine mammals [55,89-92], and diving behavior by loggerhead turtles [93] have been observed to occur in response to seismic surveys. Following environmental concerns about the impact of these explosive sounds,

underwater sound propagation models have been developed to estimate received levels. These are used to determine the distance at which injury or disturbance may occur and to develop mitigation and monitoring plans to reduce noise exposure [94,95]. These approaches have often been adopted during assessments and mitigation of pile-driving activity at offshore wind farms.

Mitigation measures for marine mammals during seismic surveys typically include a soft start or ramp-up to gradually increase the intensity of an airgun array up to full power over a period of 20 minutes or more. This approach is to allow sufficient time for animals to leave the immediate vicinity and avoid harmful noise levels. Similar approaches have been applied to the blow energy intensity during pile-driving. However, whilst this mitigation measure is implemented as a 'common sense' approach, no studies have yet investigated its effectiveness systematically [94]. The form, probability and extent of a marine mammal's response to anthropogenic sound will be affected by a variety of factors. Animals may have different tolerances for increasing sound levels depending on their current behavior, experience, motivation and conditioning [53]. One study observed an avoidance response away from the ramp-up of a 2-D seismic survey by a subgroup of pilot whales (*Globicephala macrorhynchus*) that began when they were 750 m from the airgun array [96], but interpreting the reactions of animals can be difficult because responses can be vertical and/or horizontal. There is a need for further research to assess the efficacy of the ramp-up soft start procedure for mitigating effects on marine mammals.

Another mitigation measure typically used is the monitoring of an exclusion zone. Marine mammal observers are required to visually, and sometimes also acoustically, monitor within a zone in close proximity to the source to ensure the absence of marine mammals (and possibly other protected species such as sea turtles) before beginning piling e.g. [97]. This zone may be a pre-defined fixed distance from the source or based on the expected sound levels. However, there is generally a mismatch between the relatively small area monitored for animals and the potential area of impact, which is likely to be considerably larger [98]. The exclusion zone is aimed at reducing near-field noise exposure and protecting animals from direct physical harm.

Visual observations will be limited during poor visibility conditions and for deep-diving species, such as beaked whales. It is also recognized that this is unlikely to be effective in mitigating behavioral responses over greater distances and that disturbance in the far-field is still likely to occur [55,86,99]. The use of real-time technologies, such as passive acoustic monitoring [100], may be a cost-effective approach to achieve detection coverage over a much larger area for vocalizing animals. Detailed studies to estimate received levels at various distances should be conducted during the planning stages to take into account variations in sound propagation among locations and use this, together with spatiotemporal information on marine mammal occurrence, to identify priority areas for monitoring and mitigation. Current mitigation plans also do not consider the impacts on marine mammal prey species. It should be identified whether any prey species (e.g. fish, squid) are potentially sensitive to noise and disturbance and considered in management plans accordingly to avoid secondary, trophic-level effects, as well as impacts to the fishing industry [18,98]. Efforts are underway to develop technologies to reduce source levels and noise propagation around offshore wind farm sites to help minimize biological impacts e.g. [101].

One measure that could reduce or eliminate the need for pile-driving is the development of floating wind turbine technologies, which are now being considered for deep water (>50 m) sites [2,4,102]. Concerns have been raised over possible entanglement risk in the moorings used to secure the platform to anchors on the seabed. However, the risk would appear to be small as the cables will be under tension and such moorings would be very similar to those widely used for floating oil platforms. Assessments of interactions with wildlife and existing floating oil platforms could therefore inform risk assessments for floating offshore wind turbines and identify what species or groups, if any, may be vulnerable to entanglement.

The future
Emerging technologies
The greatest change that is likely to occur in offshore wind energy is the increased use of floating foundations.

These are designed for deep water areas where the water depth is greater than 50 m (Figure 4). They can currently be used in water depths up to about 300 m but have the potential to reach water depths of up to 700 m, which would greatly increase the potential area for offshore wind energy development [4]. There are many possible designs for floating wind turbines and much more research needs to be done to determine the feasibility of these different options [2]. The first floating wind turbine was installed off Norway in water 220 m deep [4]. Experimental floating turbines have also been installed off Sweden and Portugal, with the latter being a full-scale 2 MW grid connected model [103]. A floating turbine demonstration project of 2 MW off Japan is being followed by a plan for a 1 GW wind farm consisting of up to 143 floating turbines scheduled for start-up in 2018 [104]. There are also currently proposals in the planning system for floating wind turbines off Scotland (http://www.scotland.gov.uk/Topics/marine/Licensing/marine/scoping). The difference in construction of these floating foundations from those that are fixed directly to the seabed means that the potential impact pathways for marine species and habitats may change. Although there may be reduced impacts in terms of noise, our knowledge of the environment and species distributions tends to decrease further offshore and in deeper water.

Data requirements
The environmental assessment process for offshore wind farms in Europe has highlighted the need for more synoptic studies to complement the site-specific surveys

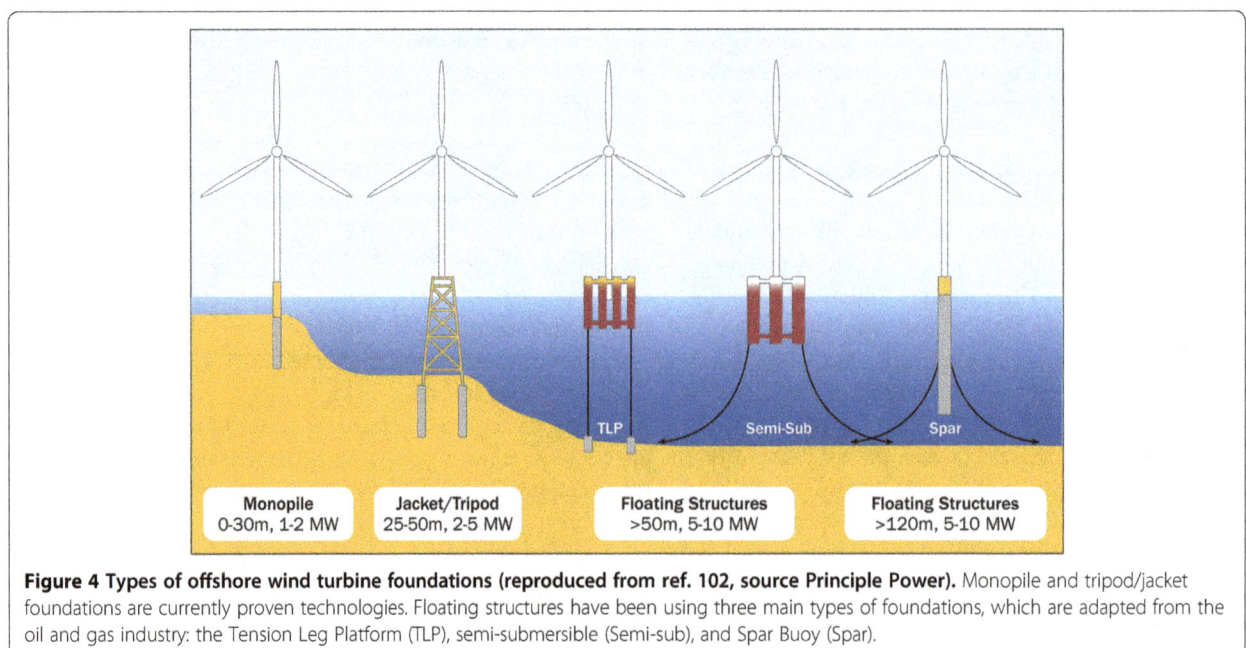

Figure 4 Types of offshore wind turbine foundations (reproduced from ref. 102, source Principle Power). Monopile and tripod/jacket foundations are currently proven technologies. Floating structures have been using three main types of foundations, which are adapted from the oil and gas industry: the Tension Leg Platform (TLP), semi-submersible (Semi-sub), and Spar Buoy (Spar).

and monitoring that may be required around particular developments. Experience in Denmark, Germany and the Netherlands has highlighted the value of having a few key demonstrator sites to study interactions with key receptor species in these shallow North Sea areas e.g. [9,48,78]. Development of a broader suite of demonstrator sites is now required to understand potential interactions with a wider range of species and habitats that will result from the expansion of this industry. Such demonstrator sites could be focused on areas that build on existing research programs or where there are specific species of concern so that parameters of interest can be determined and models for assessing impacts developed and tested. Where regulators are required to consider potential population level effects on protected species, demonstrator sites should be selected to maximize the opportunity for linkage with individual based demographic studies [105]. This approach offers the potential to explore whether individual and colony specific variation in exposure to stressors such as noise or collision risk influences reproduction and survival rates. Critically, individual based studies can also be used to assess how the impacts of particular stressors interact with broader scale variation in environmental conditions e.g. [106] or vary over time e.g. [107]. For example, long-term data collection on bottlenose dolphins and harbor seals in the Moray Firth, Northeast Scotland, means that data on demography and fecundity are available as a baseline and can be used to determine if there are any changes in these vital rates during construction activities [58,108]. Focused studies such as these will be especially important for developing and testing the modeling frameworks that have been used to assess the impacts of construction noise [58] or collision risk [65], supporting a more general understanding of the longer-term population consequence of the short-term interactions recognized in current assessments. Information on these ecological processes that can only be obtained from a few focused studies at demonstrator sites can then be integrated with site specific data on distribution and abundance at proposed wind energy sites. This will provide more robust assessments of the population consequences of future developments.

Conclusions

As offshore wind farms grow in size and number around the world, several changes in the priorities for environmental research and assessments are occurring. Firstly, there are an increasing number of cases where more than one wind farm project may occur within the home range of a population. Consequently, cumulative impact assessments, which should be made at the population level, will become increasingly important when assessing the effect of these activities on marine species and populations. Secondly, for species such as marine mammals,

it is becoming increasingly clear that the most significant consequences of offshore wind farm construction and operation are likely to occur as a result of avoidance of construction noise or structures rather than direct mortality. Hence there needs to be a greater focus on assessing the longer-term impact of any behavioral responses through changes in energetic costs, survival or fecundity. Finally, as offshore wind farms increase in scale, there is a need to put any observed biological impacts into a population context. This requires an understanding of the relative scale of any impacts in relation to existing natural variation and other anthropogenic drivers such as fisheries bycatch or exploitation. Only then can the population consequences be modeled and conservation priorities be identified.

Drewitt and Langston [109] previously recommended a number of best practice measures for reducing the impacts of wind farms on birds. These recommendations included ensuring that key areas of conservation importance and sensitivity are avoided, grouping turbines to avoid alignment perpendicular to main flight paths or migration corridors, timing construction to avoid sensitive periods, and timing and routing maintenance trips to reduce disturbance from boats, helicopters and personnel. In practice, it is unlikely that all of these recommendations can be met given the challenge of balancing the needs of all stakeholders in the marine spatial planning process. In particular, many of the sites suitable for offshore wind energy development, such as offshore sandbanks, are also important habitats for marine species and fisheries. There is therefore a need for careful consideration in finer scale spatial planning and the identification of other mitigation measures to minimize environmental and human user conflicts. Strategic and targeted research around the next generation of offshore wind farm sites is now required to support the regulators' need to achieve a balance between climate change targets and existing environmental legislation.

Competing interests
The authors declare that they have no competing interests.

Authors' contributions
The manuscript was drafted by HB and revised and finalized together with KB and PT. All authors have read and approved the final manuscript.

Acknowledgements
HB is grateful to UMCES faculty and members of the UMCES Offshore Wind focus group for many interesting discussions. KB and PT acknowledge the facilitation of their collaboration through the Marine Collaboration Research Forum (MarCRF) in Aberdeen. We thank the many colleagues in academia, industry and government who informed our understanding of these issues, and Finlay Bennet, Ian Davies and Nancy McLean for comments that greatly improved the manuscript. This is Contribution 4956 of the University of Maryland Center for Environmental Science, Chesapeake Biological Laboratory.

Author details
[1]Chesapeake Biological Laboratory, University of Maryland Center for Environmental Science, 146 Williams Street, Solomons, MD 20688, USA.

[2]Marine Scotland Science, 375 Victoria Road, Aberdeen AB11 9DB, UK.
[3]Institute of Biological and Environmental Sciences, Lighthouse Field Station, University of Aberdeen, George Street, Cromarty, Ross-shire IV11 8YJ, UK.

References

1. Toke D: **The UK offshore wind power programme: a sea-change in UK energy policy?** *Energy Policy* 2011, **39**:526–534.

2. Breton SP, Moe G: **Status, plans and technologies for offshore wind turbines in Europe and North America.** *Renew Energy* 2009, **34**:646–654.

3. European Wind Energy Association: *The European Offshore Wind Industry - Key Trends and Statistics 2013.* Brussels, Belgium: A report by the European Wind Energy Association; 2014.

4. Sun X, Huang D, Wu G: **The current state of offshore wind energy technology development.** *Energy* 2012, **41**:298–312.

5. Inger R, Attrill MJ, Bearhop S, Broderick AC, Grecian WJ, Hodgson DJ, Mills C, Sheehan E, Votier SC, Witt MJ, Godley BJ: **Marine renewable energy: potential benefits to biodiversity? An urgent call for research.** *J Appl Ecol* 2009, **46**:1145–1153.

6. Boehlert GW, Gill AB: **Environmental and ecological effects of ocean renewable energy development: a current synthesis.** *Oceanography* 2010, **23**:68–81.

7. Gill AB: **Offshore renewable energy: ecological implications of generating electricity in the coastal zone.** *J Appl Ecol* 2005, **42**:605–615.

8. Wilhelmsson D, Malm T, Öhman MC: **The influence of offshore windpower on demersal fish.** *ICES J Mar Sci* 2006, **63**:775–784.

9. Lindeboom HJ, Kouwenhoven HJ, Bergman MJN, Bouma S, Brasseur S, Daan R, Fijn RC, De Haan D, Dirksen S, van Hal R, Hille Ris Lambers R, ter Hofstede R, Krijgsveld KL, Leopold M, Scheidat M: **Short-term ecological effects of an offshore wind farm in the Dutch coastal zone; a compilation.** *Environ Res Lett* 2011, **6**:035101.

10. Maar M, Bolding K, Petersen JK, Hansen JLS, Timmermann K: **Local effects of blue mussels around turbine foundations in an ecosystem model of Nysted off-shore wind farm, Denmark.** *J Sea Res* 2009, **62**:159–174.

11. Russell DJF, Brasseur SMJM, Thompson D, Hastie GD, Janik VM, Aarts G, McClintock BT, Matthiopoulos J, Moss SEW, McConnell B: **Marine mammals trace anthropogenic structures at sea.** *Curr Biol* 2014, **24**:R638–R639.

12. Buck BH, Krause G, Rosenthal H: **Extensive open ocean aquaculture development within wind farms in Germany: the prospect of offshore co-management and legal constraints.** *Ocean Coast Manag* 2004, **47**:95–122.

13. Renewable UK: *Cumulative impact assessment guidelines: Guiding principles for cumulative impacts assessment in offshore wind farms.* 2013. Available at: http://www.renewableuk.com/en/publications/index.cfm/cumulative-impact-assessment-guidelines.

14. Masden EA, Fox AD, Furness RW, Bullman R, Haydon DT: **Cumulative impact assessments and bird/wind farm interactions: Developing a conceptual framework.** *Environ Impact Assess Rev* 2010, **30**:1–7.

15. Dolman S, Simmonds M: **Towards best environmental practice for cetacean conservation in developing Scotland's marine renewable energy.** *Mar Policy* 2010, **34**:1021–1027.

16. Madsen PT, Wahlberg M, Tougaard J, Lucke K, Tyack P: **Wind turbine underwater noise and marine mammals: implications of current knowledge and data needs.** *Mar Ecol Prog Ser* 2006, **309**:279–295.

17. Thomsen F, Lüdemann K, Kafemann R, Piper W: *Effects of offshore wind farm noise on marine mammals and fish.* Biola, Hamburg: Germany on behalf of COWRIE Ltd.; 2006.

18. Popper AN, Hastings MC: **The effects of human-generated sound on fish.** *Integr Zool* 2009, **4**:43–52.

19. Gill AB, Bartlett M, Thomsen F: **Potential interactions between diadromous fishes of U.K. conservation importance and the electromagnetic fields and subsea noise from marine renewable energy developments.** *J Fish Biol* 2012, **81**:664–695.

20. Popper AN, Hastings MC: **The effects of anthropogenic sources of sound on fishes.** *J Fish Biol* 2009, **75**:455–489.

21. Tougaard J, Henriksen OD, Miller LA: **Underwater noise from three types of offshore wind turbines: Estimation of impact zones for harbor porpoises and harbor seals.** *J Acoust Soc Am* 2009, **125**:3766–3773.

22. Marmo B, Roberts I, Buckingham MP, King S, Booth C: *Modelling of noise effects of operational offshore wind turbines including noise transmission through various foundation types.* Edinburgh: Scottish Government; 2013.

23. Desholm M, Kahlert J: **Avian collision risk at an offshore wind farm.** *Biol Lett* 2005, **1**:296–298.

24. Furness RW, Wade HM, Masden EA: **Assessing vulnerability of marine bird populations to offshore wind farms.** *J Environ Manag* 2013, **119**:56–66.

25. Normandeau, Exponent, Tricas T, Gill A: *Effects of EMFs from undersea power cables on elasmobranchs and other marine species.* Camarillo, CA: U.S. Dept. of the Interior, Bureau of Ocean Energy Management, Regulation, and Enforcement, Pacific OCS Region; 2011. OCS Study BOEMRE 2011–09.

26. Gill AB, Huang Y, Gloyne-Philips I, Metcalfe J, Quayle V, Spencer J, Wearmouth V: *COWRIE 2.0 Electromagnetic Fields (EMF) Phase 2: EMF-sensitive fish response to EM emmisions from sub-sea electricity cables of the type used by the offshore renewable energy industry.* Thetford, UK: Commissioned by COWRIE Ltd (project reference COWRIE-EMF-1-06); 2009.

27. Westerberg H, Lagenfelt I: **Sub-sea power cables and the migration behaviour of the European eel.** *Fish Manag Ecol* 2008, **15**:369–375.

28. Southall BL, Bowles AE, Ellison WT, Finneran JJ, Gentry RL, Greene CR Jr, Kastak D, Ketten DR, Miler JH, Nachtigall PE, Richardson WJ, Thomas JA, Tyack PL: **Marine mammal noise exposure criteria: initial scientific recommendations.** *Aquat Mamm* 2007, **33**:411–521.

29. National Research Council: *Marine Mammal Populations and Ocean Noise: Determining When Ocean Noise Causes Biologically Significant Effects.* Washington, DC: National Academy Press; 2005.

30. Wahlberg M, Westerberg H: **Hearing in fish and their reactions to sounds from offshore wind farms.** *Mar Ecol Prog Ser* 2005, **288**:295–309.

31. Hawkins AD, Popper AN: **Assessing the impacts of underwater sounds on fishes and other forms of marine life.** *Acoust Today* 2014, **10**:30–41.

32. Casper BM, Halvorsen MB, Matthews F, Carlson TJ, Popper AN: **Recovery of barotrauma injuries resulting from exposure to pile driving sound in two sizes of hybrid striped bass.** *PLoS ONE* 2013, **8**:e73844.

33. Casper BM, Popper AN, Matthews F, Carlson TJ, Halvorsen MB: **Recovery of barotrauma injuries in Chinook salmon, *Oncorhynchus tshawytscha* from exposure to pile driving sound.** *PLoS ONE* 2012, **7**:e39593.

34. Halvorsen MB, Casper BM, Matthews F, Carlson TJ, Popper AN: **Effects of exposure to pile-driving sounds on the lake sturgeon, Nile tilapia and hogchoker.** *Proc R Soc Lond B Biol Sci* 2012, **279**:4705–4714.

35. Bolle LJ, de Jong CAF, Bierman SM, Van Beek PJG, Van Keeken OA, Wessels PW, Van Damme CJG, Winter HV, De Haan D, Dekeling RPA: **Common sole larvae survive high levels of pile-driving sound in controlled exposure experiments.** *PLoS ONE* 2012, **7**:e33052.

36. Thomsen F, Mueller-Blenkle C, Gill A, Metcalfe J, McGregor PK, Bendall V, Andersson MH, Sigray P, Wood D: **Effects of pile driving on the behavior of cod and sole.** In *The Effects of Noise on Aquatic Life*, Advances in Experimental Medicine and Biology, Volume 730. Edited by Popper AN, Hawkins A. New York, USA: Springer; 2012:387–388.

37. Mueller-Blenkle C, Gill AB, McGregor PK, Andersson MH, Sigray P, Bendall V, Metcalfe J, Thomsen F: **A novel field study setup to investigate the behavior of fish related to sound.** In *The Effects of Noise on Aquatic Life*, Advances in Experimental Medicine and Biology, Volume 730. Edited by Popper AN, Hawkins A. New York, USA: Springer; 2012:389–391.

38. Kaiser MJ, Galanidi M, Showler DA, Elliott AJ, Caldow RWG, Rees EIS, Stillman RA, Sutherland WJ: **Distribution and behaviour of Common Scoter Melanitta nigra relative to prey resources and environmental parameters.** *Ibis* 2006, **s1**:110–128.

39. Pelletier SK, Omland K, Watrous KS, Peterson TS: *Information Synthesis on the Potential for Bat Interactions with Offshore Wind Facilities - Final Report.* Herndon, VA: U.S. Department of the Interior, Bureau of Ocean Energy Management, Headquarters; 2013. OCS Study BOEM 2013–01163.

40. Sjollema AL, Gates JE, Hilderbrand RH, Sherwell J: **Offshore activity of bats along the Mid-Atlantic Coast.** *Northeast Nat* 2014, **21**:154–163.

41. Kunz TH, Arnett EB, Erickson WP, Hoar AR, Johnson GD, Larkin RP, Strickland MD, Thresher RW, Tuttle MD: **Ecological impacts of wind energy development on bats: questions, research needs, and hypotheses.** *Front Ecol Environ* 2007, **5**:315–324.

42. Waring GT, Wood SA, Josephson E: *Literature search and data synthesis for marine mammals and sea turtles in the U.S. Atlantic from Maine to the Florida Keys.* New Orleans, LA: U.S Department of the Interior, Bureau of Ocean Energy Management, Gulf of Mexico OCS Region; 2012. OCS Study BOEM 2012–109.

43. Dow Piniak WE, Eckert SA, Harms CA, Stringer EM: *Underwater Hearing Sensitivity of The Leatherback Sea Turtle (Dermochelys coriacea): Assessing The Potential Effect of Anthropogenic Noise*. Herndon, VA: U.S Department of the Interior, Bureau of Ocean Energy Management, Headquarters; 2012. OCS Study BOEM 2012-01156.

44. Rein CG, Lundin AS, Wilson SJK, Kimbrell E: *Offshore Wind Energy Development Site Assessment and Characterization: Evaluation of the Current Status and European Experience*. Herndon, VA: U.S. Department of the Interior, Bureau of Ocean Energy Management, Office of Renewable Energy Programs; 2013. OCS Study BOEM 2013-0010.

45. Carstensen J, Henriksen OD, Teilmann J: **Impacts of offshore wind farm construction on harbour porpoises: acoustic monitoring of echolocation activity using porpoise detectors (T-PODs)**. *Mar Ecol Prog Ser* 2006, **321**:295–308.

46. Tougaard J, Carstensen J, Teilmann J: **Pile driving zone of responsiveness extends beyond 20 km for harbor porpoises (*Phocoena phocoena* (L.))**. *J Acoust Soc Am* 2009, **126**:11–14.

47. Bailey H, Senior B, Simmons D, Rusin J, Picken G, Thompson PM: **Assessing underwater noise levels during pile-driving at an offshore windfarm and its potential impact on marine mammals**. *Mar Pollut Bull* 2010, **60**:888–897.

48. Brandt MJ, Diederichs A, Betke K, Nehls G: **Responses of harbour porpoises to pile driving at the Horns Rev II offshore wind farm in the Danish North Sea**. *Mar Ecol Prog Ser* 2011, **421**:205–216.

49. Thompson PM, Lusseau D, Barton T, Simmons D, Rusin J, Bailey H: **Assessing the responses of coastal cetaceans to the construction of offshore wind turbines**. *Mar Pollut Bull* 2010, **60**:1200–1208.

50. Kastelein RA, Van Heerden D, Gransier R, Hoek L: **Behavioral responses of a harbor porpoise (*Phocoena phocoena*) to playbacks of broadband pile driving sounds**. *Mar Environ Res* 2013, **92**:206–214.

51. Koschinski S, Culik BM, Henriksen OD, Tregenza N, Ellis G, Jansen C, Kathe G: **Behavioural reactions of free-ranging porpoises and seals to the noise of a simulated 2 MW windpower generator**. *Mar Ecol Prog Ser* 2003, **265**:263–273.

52. Degraer S, Brabant R, Rumes B: *Environmental impacts of offshore wind farms in the Belgian part of the North Sea: Learning from the past to optimise future monitoring programmes: 26–28 November 2013*. Brussels, Belgium: Royal Belgian Institute of Natural Sciences; 2013.

53. Ellison WT, Southall BL, Clark CW, Frankel AS: **A new context-based approach to assess marine mammal behavioral responses to anthropogenic sounds**. *Conserv Biol* 2012, **26**:21–28.

54. Diederichs A, Nehls G, Dähne M, Adler S, Koschinski S, Verfuß U: *Methodologies for measuring and assessing potential changes in marine mammal behaviour, abundance or distribution arising from the construction, operation and decommissioning of offshore windfarms*. Germany: BioConsult SH report to COWRIE Ltd; 2008.

55. Thompson PM, Brookes KL, Graham IM, Barton TR, Needham K, Bradbury G, Merchant ND: **Short-term disturbance by a commercial two-dimensional seismic survey does not lead to long-term displacement of harbour porpoises**. *Proc R Soc Lond B Biol Sci* 2013, **280**:20132001.

56. Bailey H, Hammond PS, Thompson PM: **Modelling harbour seal habitat by combining data from multiple tracking systems**. *J Exp Mar Biol Ecol* 2014, **450**:30–39.

57. Bailey H, Clay G, Coates EA, Lusseau D, Senior B, Thompson PM: **Using T-PODs to assess variations in the occurrence of coastal bottlenose dolphins and harbour porpoises**. *Aquat Conserv Mar Freshwat Ecosyst* 2010, **20**:150–158.

58. Thompson PM, Hastie GD, Nedwell J, Barham R, Brookes KL, Cordes LS, Bailey H, McLean N: **Framework for assessing impacts of pile-driving noise from offshore wind farm construction on a harbour seal population**. *Environ Impact Assess Rev* 2013, **43**:73–85.

59. Barrios L, Rodríguez A: **Behavioural and environmental correlates of soaring-bird mortality at on-shore wind turbines**. *J Appl Ecol* 2004, **41**:72–81.

60. Garthe S, Hüppop O: **Scaling possible adverse effects of marine wind farms on seabirds: developing and applying a vulnerability index**. *J Appl Ecol* 2004, **41**:724–734.

61. Camphuysen KCJ, Fox TAD, Leopold MMF, Petersen IK: *Towards standardised seabirds at sea census techniques in connection with environmental impact assessments for offshore wind farms in the U.K*. Report by Royal Netherlands Institute for Sea Research and the Danish National Environmental Research Institute to COWRIE BAM 02–2002. London: Crown Estate Commissioners; 2004.

62. Cook ASCP, Johnston A, Wright LJ, Burton NHK: *A Review of Flight Heights and Avoidance Rates of Birds in Relation to Offshore Wind Farms*. Norfolk, UK: British Trust for Ornithology on behalf of The Crown Estate, Project SOSS-02; 2012. BTO Research Report Number 618.

63. Johnston A, Cook ASCP, Wright LJ, Humphreys EM, Burton NHK: **Modelling flight heights of marine birds to more accurately assess collision risk with offshore wind turbines**. *J Appl Ecol* 2014, **51**:31–41.

64. Plonczkier P, Simms IC: **Radar monitoring of migrating pink-footed geese: behavioural responses to offshore wind farm development**. *J Appl Ecol* 2012, **49**:1187–1194.

65. Chamberlain DE, Rehfisch MR, Fox AD, Desholm M, Anthony SJ: **The effect of avoidance rates on bird mortality predictions made by wind turbine collision risk models**. *Ibis* 2006, **148**:198–202.

66. Masden EA, Haydon DT, Fox AD, Furness RW: **Barriers to movement: Modelling energetic costs of avoiding marine wind farms amongst breeding seabirds**. *Mar Pollut Bull* 2010, **60**:1085–1091.

67. Busch M, Kannen A, Garthe S, Jessopp M: **Consequences of a cumulative perspective on marine environmental impacts: Offshore wind farming and seabirds at North Sea scale in context of the EU Marine Strategy Framework Directive**. *Ocean Coast Manag* 2013, **71**:213–224.

68. McCann J: *Developing Environmental Protocols and Modeling Tools to Support Ocean Renewable Energy and Stewardship*. Herndon, VA: U.S. Department of the Interior, Bureau of Ocean Energy Management, Office of Renewable Energy Programs; 2012. OCS Study BOEM 2012-082.

69. Teilmann J, Carstensen J: **Negative long term effects on harbour porpoises from a large scale offshore wind farm in the Baltic - evidence of slow recovery**. *Environ Res Lett* 2012, **7**:045101.

70. Wade PR: **Calculating limits to the allowable human-caused mortality of cetaceans and pinnipeds**. *Mar Mammal Sci* 1998, **14**:1–37.

71. Butler JRA, Middlemas SJ, McKelvey SA, McMyn I, Leyshon B, Walker I, Thompson PM, Boyd IL, Duck C, Armstrong JD, Graham IM, Baxter JM: **The Moray Firth Seal Management Plan: an adaptive framework for balancing the conservation of seals, salmon, fisheries and wildlife tourism in the UK**. *Aquat Conserv Mar Freshwat Ecosyst* 2008, **18**:1025–1038.

72. New LF, Clark JS, Costa DP, Fleishman E, Hindell MA, Klanjšček T, Lusseau D, Kraus S, McMahon CR, Robinson PW, Schick RS, Schwarz LK, Simmons SE, Thomas L, Tyack P, Harwood J: **Using short-term measures of behaviour to estimate long-term fitness of southern elephant seals**. *Mar Ecol Prog Ser* 2014, **496**:99–108.

73. Harwood J, King S, Schick R, Donovan C, Booth C: **A protocol for implementing the interim population consequences of disturbance (PCoD) approach: Quantifying and assessing the effects of UK offshore renewable energy developmenets on marine mammal populations: Report number SMRUL-TCE-2013-014**. *Scott Mar Freshwater Sci* 2014, **5**:2.

74. Frederiksen M, Wanless S, Harris MP, Rothery P, Wilson LJ: **The role of industrial fisheries and oceanographic change in the decline of North Sea black-legged kittiwakes**. *J Appl Ecol* 2004, **41**:1129–1139.

75. Underwood AJ: **On beyond BACI: Sampling designs that might reliably detect environmental disturbances**. *Ecol Appl* 1994, **4**:3–15.

76. Hewitt JE, Thrush SE, Cummings VJ: **Assessing environmental impacts: Effects of spatial and temporal variability at likely impact scales**. *Ecol Appl* 2001, **11**:1502–1516.

77. Ellis JI, Schneider DC: **Evaluation of a gradient sampling design for environmental impact assessment**. *Environ Monit Assess* 1997, **48**:157–172.

78. Dähne M, Gilles A, Lucke K, Peschko V, Adler S, Krügel K, Sundermeyer J, Siebert U: **Effects of pile-driving on harbour porpoises (*Phocoena phocoena*) at the first offshore wind farm in Germany**. *Environ Res Lett* 2013, **8**:025002.

79. Heupel MR, Semmens JM, Hobday AJ: **Automated acoustic tracking of aquatic animals: scales, design and deployment of listening station arrays**. *Mar Freshw Res* 2006, **57**:1–13.

80. Scales KL, Lewis JA, Lewis JP, Castellanos D, Godley BJ, Graham RT: **Insights into habitat utilisation of the hawksbill turtle, *Eretmochelys imbricata* (Linnaeus, 1766), using acoustic telemetry**. *J Exp Mar Biol Ecol* 2011, **407**:122–129.

81. Espinoza M, Farrugia TJ, Webber DM, Smith F, Lowe CG: **Testing a new acoustic telemetry technique to quantify long-term, fine-scale movements of aquatic animals**. *Fish Res* 2011, **108**:364–371.

82. Maxwell SM, Hazen EL, Bograd SJ, Halpern BS, Breed GA, Nickel B, Teutschel NM, Crowder LB, Benson S, Dutton PH, Bailey H, Kappes MA, Kuhn CE, Weise MJ, Mate B, Shaffer SA, Hassrick JL, Henry RW, Irvine L, McDonald BI, Robinson PW, Block BA, Costa DP: **Cumulative human impacts on marine predators**. *Nat Commun* 2013, **4**:2688.

83. Punt MJ, Groeneveld RA, Van Ierland EC, Stel JH: **Spatial planning of offshore wind farms: A windfall to marine environmental protection?** *Ecol Econ* 2009, **69**:93–103.

84. Band W, Madders M, Whitfield DP: **Developing field and analytical methods to assess avian collision risk at wind farms.** In *Birds and Wind Power.* Edited by De Lucas M, Janss G, Ferrer M. Barcelona, Spain: Lynx Edicions; 2005.

85. Band B, Band B: *Using a collision risk model to assess bird collision risks for offshore windfarms.* Norway: SOSS report for The Crown Estate; 2012.

86. Gordon J, Gillespie D, Potter J, Frantzis A, Simmonds MP, Swift R, Thompson D: **A review of the effects of seismic surveys on marine mammals.** *Mar Technol Soc J* 2003, **37**:16–34.

87. OSPAR: *Overview of the impacts of anthropogenic underwater sound in the marine environment.* North-East Atlantic: OSPAR Convention for the Protection of the Marine Environment of the North-East Atlantic; 2009. www.ospar.org.

88. Nedwell JR, Parvin SJ, Edwards B, Workman R, Brooker AG, Kynoch JE, Nedwell JR, Parvin SJ, Edwards B, Workman R, Brooker AG, Kynoch JE: **Measurement and interpretation of underwater noise during construction and operation of offshore windfarms in UK waters.** In *Subacoustech Report No. 544R0738 to COWRIE Ltd.*; 2007. ISBN: 978-0-9554279-5-4.

89. Di Iorio L, Clark CW: **Exposure to seismic survey alters blue whale acoustic communication.** *Biol Lett* 2010, **6**:51–54.

90. Castellote M, Clark CW, Lammers MO: **Acoustic and behavioural changes by fin whales (*Balaenoptera physalus*) in response to shipping and airgun noise.** *Biol Conserv* 2012, **147**:115–122.

91. Blackwell SB, Nations CS, McDonald TL, Greene CR, Thode AM, Guerra M, Macrander AM: **Effects of airgun sounds on bowhead whale calling rates in the Alaskan Beaufort Sea.** *Mar Mammal Sci* 2013, **29**:E342–E365.

92. Harris RE, Miller GW, Richardson WJ: **Seal responses to airgun sounds during summer seismic surveys in the Alaskan Beaufort Sea.** *Mar Mammal Sci* 2001, **17**:795–812.

93. DeRuiter SL, Doukara KL: **Loggerhead turtles dive in response to airgun sound exposure.** *Endanger Species Res* 2012, **16**:55–63.

94. Rutenko AN, Borisov SV, Gritsenko AV, Jenkerson MR: **Calibrating and monitoring the western gray whale mitigation zone and estimating acoustic transmission during a 3D seismic survey, Sakhalin Island, Russia.** *Environ Monit Assess* 2007, **134**:21–44.

95. Nowacek DP, Vedenev A, Southall BL, Racca R: **Development and implementation of criteria for exposure of western gray whales to oil and gas industry noise.** In *The Effects of Noise on Aquatic Life*, Advances in Experimental Medicine and Biology, Volume 730. Edited by Popper AN, Hawkins A. New York, USA: Springer; 2012:523–528.

96. Weir CR: **Short-finned pilot whales (*Globicephala macrorhynchus*) respond to an airgun ramp-up procedure off Gabon.** *Aquat Mamm* 2008, **34**:349–354.

97. JNCC: *Statutory Nature Conservation Agency Protocol for Minimising the Risk of Injury to Marine Mammals from Piling Noise.* Aberdeen, UK: Joint Nature Conservation Committee; 2010.

98. Parsons ECM, Dolman SJ, Jasny M, Rose NA, Simmonds MP, Wright AJ: **A critique of the UK's JNCC seismic survey guidelines for minimising acoustic disturbance to marine mammals: Best practise?** *Mar Pollut Bull* 2009, **58**:643–651.

99. Miller PJO, Johnson MP, Madsen PT, Biassoni N, Quero M, Tyack PL: **Using at-sea experiments to study the effects of airguns on the foraging behavior of sperm whales in the Gulf of Mexico.** *Deep-Sea Res I* 2009, **56**:1168–1181.

100. Van Parijs SM, Clark CW, Sousa-Lima RS, Parks SE, Rankin S, Risch D, Van Opzeeland IC: **Management and research applications of real-time and archival passive acoustic sensors over varying temporal and spatial scales.** *Mar Ecol Prog Ser* 2009, **395**:21–36.

101. Bellmann MA, Remmers P: **Noise mitigation systems (NMS) for reducing pile driving noise: Experiences with the "big bubble curtain" relating to noise reduction.** *J Acoust Soc Am* 2013, **134**:4059.

102. European Wind Energy Association: *Deep Water: The Next Step for Offshore Wind Energy.* Brussels, Belgium: A report by the European Wind Energy Association; 2013.

103. European Wind Energy Association: *The European Offshore Wind Industry Key 2011 Trends and Statistics.* Brussels, Belgium: A report by the European Wind Energy Association; 2012.

104. Stokes I: **Hotspots: Scotland and Fukushima.** *Renewable Energy Focus* 2013, **14**:10–11.

105. Clutton-Brock T, Sheldon BC: **Individuals and populations: the role of long-term, individual-based studies of animals in ecology and evolutionary biology.** *Trends Ecol Evol* 2010, **25**:562–573.

106. Votier SC, Hatchwell BJ, Beckerman A, McCleery RH, Hunter FM, Pellatt J, Trinder M, Birkhead TR: **Oil pollution and climate have wide-scale impacts on seabird demographics.** *Ecol Lett* 2005, **8**:1157–1164.

107. Véran S, Gimenez O, Flint E, Kendall WL, Doherty PF, Lebreton JD: **Quantifying the impact of longline fisheries on adult survival in the black-footed albatross.** *J Appl Ecol* 2007, **44**:942–952.

108. New LF, Harwood J, Thomas L, Donovan C, Clark JS, Hastie G, Thompson PM, Cheney B, Scott-Hayward L, Lusseau D: **Modelling the biological significance of behavioural change in coastal bottlenose dolphins in response to disturbance.** *Funct Ecol* 2013, **27**:314–322.

109. Drewitt AL, Langston RHW: **Assessing the impacts of wind farms on birds.** *Ibis* 2006, **148**:29–42.

16

Comparative quantitative proteomics of *prochlorococcus* ecotypes to a decrease in environmental phosphate concentrations

Matthew A Fuszard, Phillip C Wright and Catherine A Biggs*

Abstract

Background: The well-lit surface waters of oligotrophic gyres significantly contribute to global primary production. Marine cyanobacteria of the genus *Prochlorococcus* are a major fraction of photosynthetic organisms within these areas. Labile phosphate is considered a limiting nutrient in some oligotrophic regions such as the Caribbean Sea, and as such it is crucial to understand the physiological response of primary producers such as *Prochlorococcus* to fluctuations in the availability of this critical nutrient.

Results: *Prochlorococcus* strains representing both high light (HL) (MIT9312) and low light (LL) (NATL2A and SS120) ecotypes were grown identically in phosphate depleted media (10 μM P_i). The three strains displayed marked differences in cellular protein expression, as determined by high throughput large scale quantitative proteomic analysis. The only strain to demonstrate a significantly different growth rate under reduced phosphate conditions was MIT9312. Additionally, there was a significant increase in phosphate-related proteins such as PhoE (> 15 fold increase) and a depression of the Rubisco protein RbcL abundance in this strain, whereas there appeared to be no significant change within the LL strain SS120.

Conclusions: This differential response between ecotypes highlights the relative importance of phosphate availability to each strain and from these results we draw the conclusion that the expression of phosphate acquisition mechanisms are activated at strain specific phosphate concentrations.

Keywords: *Prochlorococcus*, PstS, PhoA, PhoE, Growth, Phosphate

Background

Within marine oligotrophic systems, such as central subtropical gyres, orthophosphate (P_i) is a crucial macronutrient governing microbial population densities, particularly within the well-lit surface waters of the euphotic zone [1-3]. The principal photosynthetic organism numerically dominating these areas is *Prochlorococcus*, which is estimated to represent about 50% of all photosynthetic activity within them [4,5]. *Prochlorococcus* has been broadly delineated into two clades, or ecotypes, high light (HL) and low light (LL) based upon the ratios of divinylchlorophyll*a* and *b* within their light harvesting apparatuses and as such their assumed depth within the water column [6,7]. Further clade subdivisions have been implemented through phylogenetic analyses of 16S rRNA sequences [8]. As a taxon, *Prochlorococcus* is characterised by its small size (~ 1 μm³), and significantly reduced genomes which ranges from 1.64 Mbps (the HL strain MIT9301) to 2.68 Mbps (the LL strain MIT9303) [9]. This diminished volume and genome is hypothesised to be the result of an accelerated evolutionary process adapting to reduced phosphorus in its environment [10,11]. Indeed, *Prochlorococcus* is known to replace phospholipids in its membranes with sulpholipids, which dramatically reduce its P_i requirements [12].

Given the importance of P_i to *Prochlorococcus*, perhaps it is surprising to find no significant correlation between ecotype distribution and P_i concentration [13]. However, fluxes in P_i transport within these regions are important considerations, which could help to explain

* Correspondence: c.biggs@sheffield.ac.uk
ChELSI Institute, Department of Chemical and Biological Engineering, University of Sheffield, Mappin Street, Sheffield S1 3JD, UK

the discrepancy. Nevertheless the observation of a large number of known P_i acquisition genes in some LL ecotypes (i.e. MIT9313 and NATL2A), and not others (i.e. SS120) [14,15] is confusing. Indeed, P_i acquisition genes are present in some HL strains (i.e. MED4) and not others (i.e. MIT9515) [14]. However it was recently observed that the prevalence of *Prochlorococcus* genes involved in acquisition of phosphate substrates were correlated with areas of low P_i such as the Caribbean Sea and NW Mediterranean [16]. This conflict is likely resolved due to the presence of hypervariable genomic islands within *Prochlorococcus*, allowing for evolutionarily rapid niche adaptation [17]. Given this, it was hypothesised that the presence or absence of these genes could directly affect the protein content of cells when P_i stressed, and as such directly affect the ability of a strain to acclimate to environmental P_i fluctuations [16]. So the question arises, how effective are cells with and without these genes at acclimating to a shift in environmental P_i? Indeed, the levels of mRNA transcripts of two strains, MED4 and MIT9313, which both contain the two component response regulation system *phoBR*, behaved quite differently to P_i starvation [14].

To address this we selected three strains, MIT9312, NATL2A and SS120, each representative of an ecotype and a position within the water column (Table 1). MIT9312 is a HLII strain isolated at depth from the Gulf Stream. NATL2A is a LLI strain isolated from the North Atlantic which contains most of the P_i acquisition genes found in MED4 and MIT9312, and yet is thought to experience both high and low light environments due to storm mixing events. SS120, originally isolated in the Sargasso Sea, does not possess *phoBR*, yet has two copies of the periplasmic phosphate binding protein, PstS. We took these three strains and allowed cells to acclimate to a significant reduction in environmental P_i and investigated their respective protein contents.

Results and discussion
Overview
The experimental growth data for each strain under P_i replete and P_i deplete cultures is shown in Figure 1. Logistic curve fitting and statistical analysis of the experimental growth data reveals no significant differences between the growth rates between P_i replete and P_i deplete cultures, with the exception of MIT9312 growth rates whereby P_i replete growth was significantly greater than P_i deplete growth ($p < 0.05$), as can be seen in Figure 1. It is important to consider the physiological status of the cells at the harvest point when considering protein relative abundances. Importantly, growth analysis shows that both MIT9312 and SS120 were in late exponential/early stationary phase at harvest, whilst NATL2A was in mid exponential phase. As the point of harvest differs for NATL2A, it would be difficult to directly compare the protein complement of NATL2A cells to either MIT9312 or SS120. Given this, the results for NATL2A will be discussed separately.

Thirty eight, 63 and 34 proteins were identified with 2 or more peptides for strains MIT9312, NATL2A and SS120 respectively (Additional file 1: Table S1) with no false positives. An overview of the respective proteomes, through plotting theoretical values of isoelectric points (*pI*) against molecular weights (MW) reveal significant

Table 1 Details of the strains used in this study, as obtained from NCBI and CCMP.

Strain and genome details

Strain	Refseq	Reference	Genome size (Mbp)	Protein coding	%GC	Chl b/a ratio	Ecotypic clade
MIT9312	NC_007577	[18][a]	1.71	1810	31	0.34	HLII
NATL2A	NC_007335	[19]	1.84	2162	35	0.97	LLI
SS120	NC_005042	[20]	1.75	1883	36	1.41	LLII

P acquisition mechanisms

	PhoBR cluster	PtrA cluster	PhoA	PhoE	PstS cluster	ArsA cluster	ArsB cluster	ArsC cluster
MIT9312	√	x	√	√	√	√	x	√
NATL2A	√	√	√	√	√	√	x	√
SS120	x	√	x	x	(2)	x	x	√

Isolation details and culture conditions

	Location	Depth	Isolated by	Date	Culture temp (°C)	Deposited in	Media
MIT9312	Gulf stream	135 m	L. Moore	17/07/1993	22-26	CCMP	Pro99
NATL2A	N. Atlantic	10 m	D. Scanlan	01/04/1990	18-22	CCMP	Pro99
SS120	Sargasso Sea	120 m	S. Frankel & L West-Johnsrud	01/01/1991	18-22	CCMP	Pro99

[a] indicates that the genome sequence has been submitted, yet not cleared. Ticks in '*P acquisition mechanisms*' indicates presence of gene/cluster, and copies are in parentheses

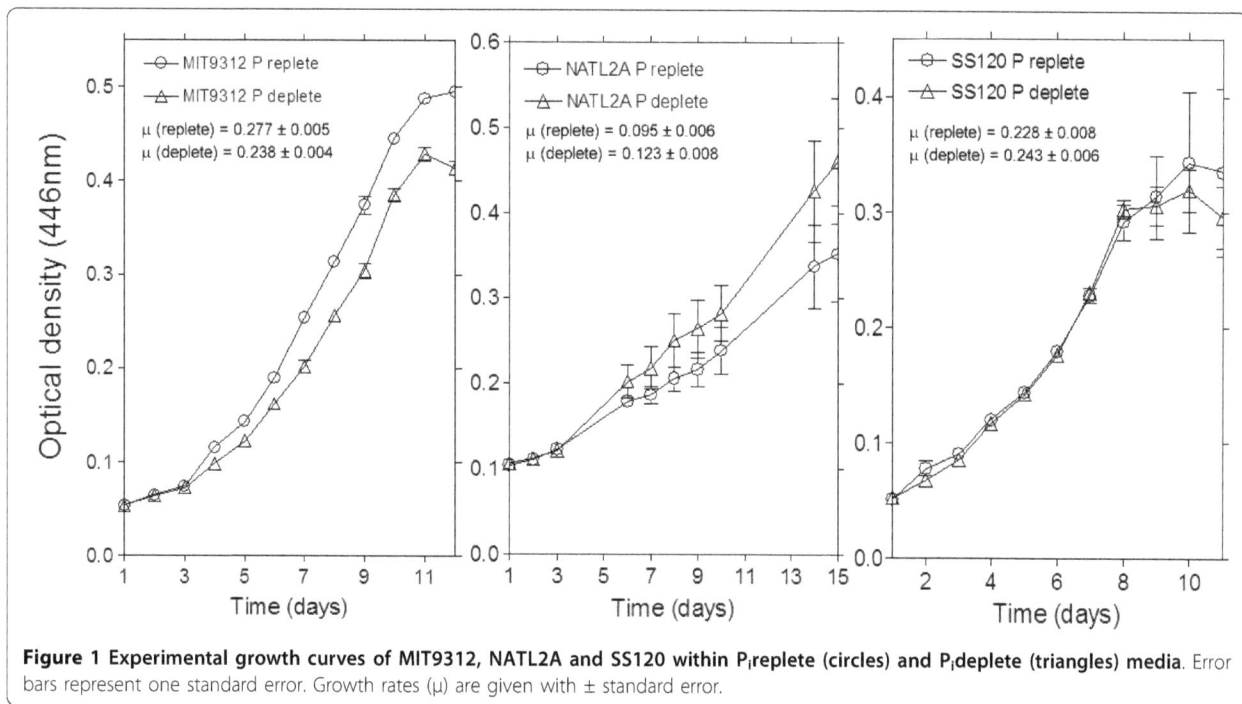

Figure 1 Experimental growth curves of MIT9312, NATL2A and SS120 within P$_i$replete (circles) and P$_i$deplete (triangles) media. Error bars represent one standard error. Growth rates (μ) are given with ± standard error.

bias towards low *pI* values (Additional file 2: Figure S1), with no further correlation to MW, relative protein abundance, nor total peptide hits per protein (data not shown). This bias may be an artefact of the mass spectrometric analysis, where peptides are protonated directly before entry into the MS in order to assist flight and detection. As a consequence, naturally occurring proton-donor peptides may be preferentially selected. However, as there are no observable correlations between *pI* and peptide hits per protein, we can be confident that the intracellular protein abundances reported are directly reflective of the physiological status of the cells. Indeed, when interrogating the proportion of proteins with ≥ 50% of peptide hits, we see similarities between strains, such as the presence of RplL, RbcL and CsoS1 (Additional file 1: Table S1), however all three proteins have *pI* values < 7. Nevertheless, a high *pI* protein, PetH, is present in both MIT 9312 and SS120 samples. Also, identified proteins from all three strains are located evenly across the genomes, and are representative of most major functional groups such as central metabolism, photosynthesis, transcription and translation, biosynthesis and nutrient acquisition (Figure 2A). Of the 105 unique proteins identified, 6 were found in all three strains (Figure 2B). They are the ATP synthase subunits AtpA and AtpD, the PSII protein PsbO, the nitrogen regulatory protein GlnK, rubisco subunit RbcL, and the carboxysome shell protein CsoS1.

Using relative abundance cut-offs of 1.6 and 0.6 fold differences to represent increased or decreased relative

abundances [21,22], 4 proteins were more abundant in MIT9312 and 4 were less abundant than the replete cultures. Within NATL2A, 6 proteins were more abundant and 1 was less abundant than the replete cultures. In SS120, 4 were more abundant and none were lower than the replete cultures (Figure 2A).

Nutrient acquisition
What is immediately apparent from our results is the differential abundance of P$_i$ acquisition proteins exhibited by all three strains to being grown in 10 µM P$_i$. MIT 9312 demonstrates the greatest sensitivity to P$_i$-deplete media, whereby the P$_i$ stress related porinPhoE is > 15-fold more abundant (Figure 3), the putative alkaline phosphatase PhoA appears to be > 9-fold greater, and the periplasmic P$_i$ binding protein PstS > 3 times more than the replete cultures. This result is directly in line with an earlier proteomic assay of P stress in a HL ecotype, MED4 [21], and closely reflective of microarray analyses of both MED4 and MIT9313 [14], *Synechococcucs* WH8102 [23], measured alkaline phosphatase activity of MIT9312 [15] and in line with observed responses within earlier P$_i$ depletion studies of other cyanobacteria [15,24-26].

Within NATL2A, PstS abundance is significantly greater within P$_i$-deplete conditions, though with greater uncertainty (Additional file 1: Table S1). However neither PhoA nor PhoE was observed with mass spectrometry here, which is surprising as we showed previously that both PhoA and PhoE are greater in

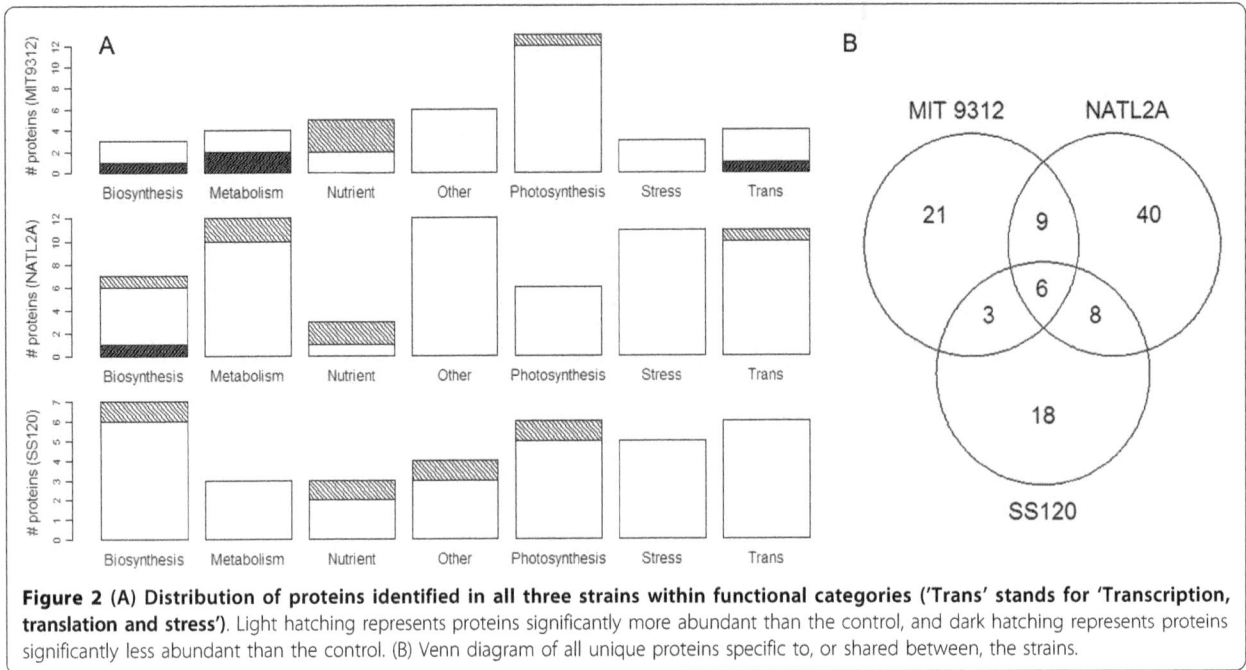

Figure 2 (A) Distribution of proteins identified in all three strains within functional categories ('Trans' stands for 'Transcription, translation and stress'). Light hatching represents proteins significantly more abundant than the control, and dark hatching represents proteins significantly less abundant than the control. (B) Venn diagram of all unique proteins specific to, or shared between, the strains.

abundance alongside PstS in the high light ecotype MED4 [21], as is true with MIT9312 in this study. However, considering that NATL2A cells are in mid-exponential phase as opposed to early stationary phase this may indicate a progressive strategy of protein expression within the cells, however more work is needed to clarify this.

What was also unexpected, was the absence of any P_i acquisition mechanisms (as reflected in observed peptide identifications) within SS120 cells (Additional file 1: Table S1), allied with no significant difference in growth rates between P_i-replete and P_i-deplete cultures (p > 0.05). SS120 is deficient in most P_i acquisition genes [14,15], however it does have two copies of PstS, neither of which were present in our assay. At first glance, this

result appears counter-intuitive, as a 'very' LL strain typically present *in vivo* within P_i-replete environments would be expected to be adversely affected by a substantial decrease in P_i. However, the absence of a *phoBR* regulon suggests that the strain is incapable of regulating a response to shifts in environmental concentrations of P_i that are not immediately starvation inducing [27]. Curiously, this also infers that activation of the *phoBR* response mechanisms within MIT9312 and NATL2A were directly due to the mechanism's innate sensitivity to changing external P_i concentrations. This suggests that the intensity of response is directly proportional to external P_i concentration, coincidentally specific to each strain, and may be reflective of each strain's environmental niche and/or obligate cellular requirements.

Photosynthesis, biosynthesis and central metabolism

The exposure of all three strains to lower P_i concentrations appears to have had little effect upon the photosynthetic machinery (Figure 4A and Additional file 1: Table S1). This is unusual, as P_i depleted conditions have been previously noted to directly affect both photosystems in cyanobacteria [21,23,28]. In contrast, it is interesting to note that, for MIT9312, both Rubisco subunits (RbcL and RbcS) are noticeably lower in abundance (Figure 3B). This suggests that there is a progressive strategy within the cell when acclimating to lowered P_i, whereby photosynthesis is initially dissociated from glycolysis, to then strategically break down the photosynthetic apparatus. This is a reasonable conclusion, considering a P_i-induced organised break down

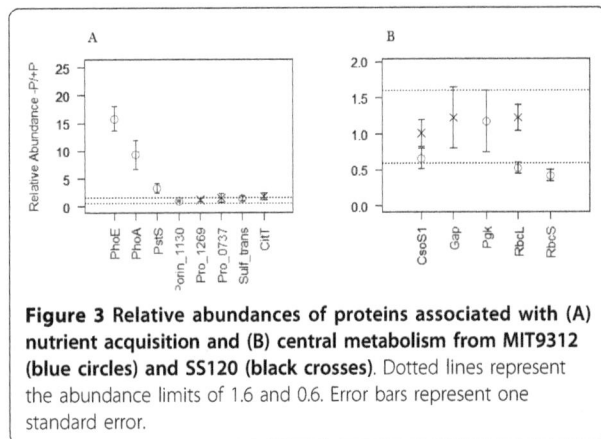

Figure 3 Relative abundances of proteins associated with (A) nutrient acquisition and (B) central metabolism from MIT9312 (blue circles) and SS120 (black crosses). Dotted lines represent the abundance limits of 1.6 and 0.6. Error bars represent one standard error.

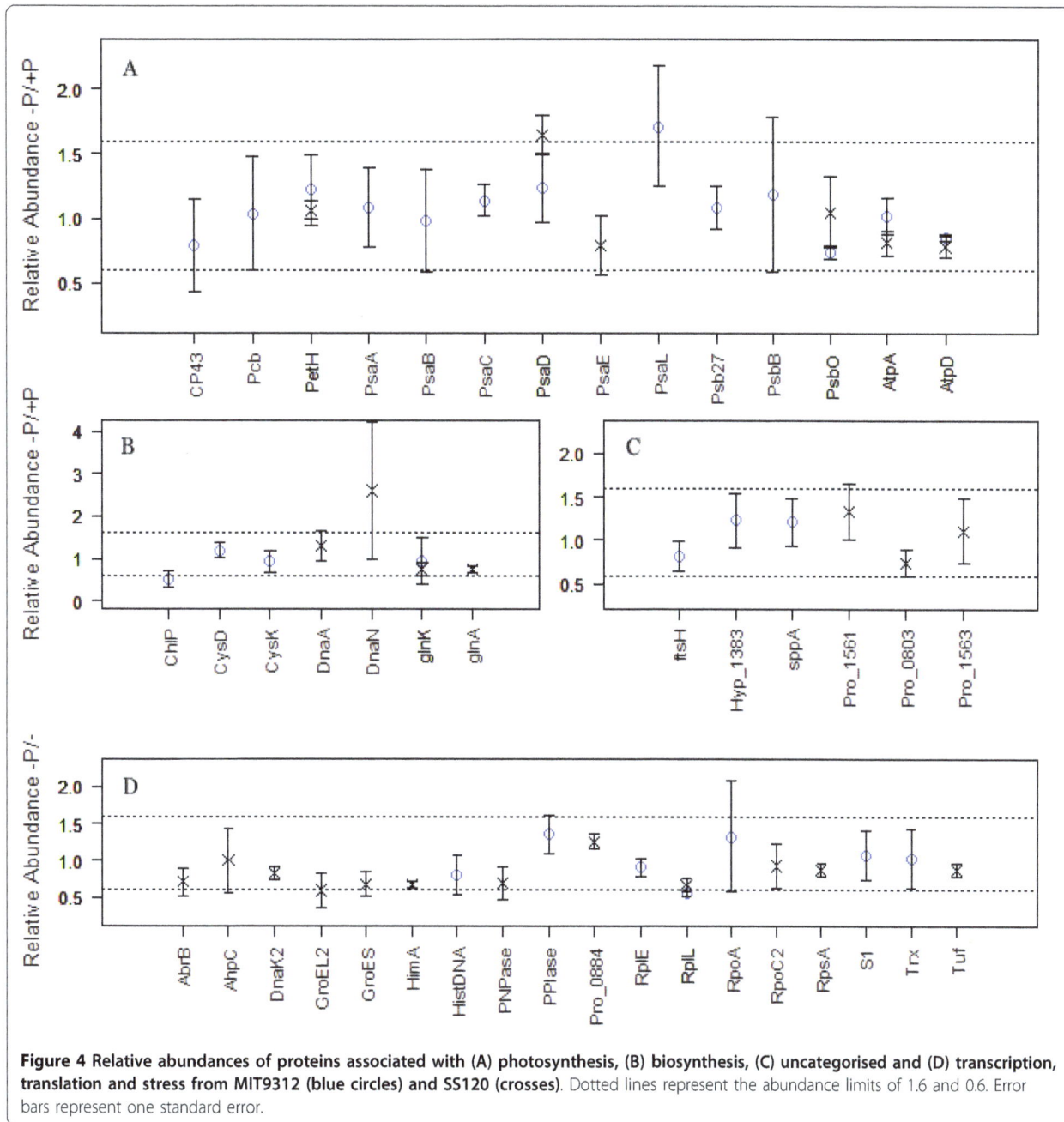

Figure 4 Relative abundances of proteins associated with (A) photosynthesis, (B) biosynthesis, (C) uncategorised and (D) transcription, translation and stress from MIT9312 (blue circles) and SS120 (crosses). Dotted lines represent the abundance limits of 1.6 and 0.6. Error bars represent one standard error.

of phycobilisomes has been previously observed in *Synechococcus* sp. PCC 7942 [29], chlorosis has been observed in thermophillic*Synechococcus* under P_i-stress [28], and a strategic approach to a reduction in photosynthetic function has been hypothesised in MED4 [21]. Indeed, within WH8102 it appears that PSII was degraded before PSI, allowing continued cyclic photophosphorylation-based ATP generation to continue [23]. In this context, this could explain why an essential chlorophyll biosynthetic protein (ChlP) appears to be less abundant within P_i-deplete MIT9312 cells (Figure

4B). However, it would be parsimonious to also expect a concurrent reduction in the light harvesting protein (Pcb) within P_i-deplete MIT9312, which was noticed in MED4 [21], but there is no change. The reason for this is not clear.

When considering NATL2A solely, there appear to be a few subtle discrepancies in protein abundances between stressed and non-stressed cultures. Fumerase (FumC) is an enzyme associated with both the tricarboxylic acid (TCA) cycle and arginine/proline biosynthesis, and appears to be more abundant within NATL2A cells when P_i-deplete (Additional file 1: Table S1). As NATL2A has an incomplete TCA cycle, it is safe to assume that its function within the cell is within arginine and proline metabolism. Also, the acyl carrier protein (AcpP) is an essential component of fatty acid biosynthesis, and is more abundant in P_i-deplete NATL2A cells (Additional file 1: Table S1). Fatty acids are for the most part used within either fuel storage or membrane manufacture. However it may be misleading to arrive at the conclusion that this is a specific cellular response to lower P_i concentrations. It is possibly a function of apparently slightly elevated (albeit not significant) growth within NATL2A P_i-deplete cultures, and as such could reflect comparatively greater metabolic activity. Nevertheless, this explanation cannot immediately address the lower abundance of CobJ, a Precorrin-3B C17-methyltransferase region-containing protein (Additional file 1: Table S1), part of the aerobic vitamin B12 biosynthesis pathway within P_i-stressed cells. However, B12 synthesis is a sub pathway offshoot from the main chlorophyll biosynthetic pathway, and as such may reflect a metabolic preference for chlorophyll production that, again, may be representative of faster growing populations.

An interesting observation is the abundance of CitT within P_i-stressed SS120 cells (Figure 3A). This protein functions as a di/tricarboxylate transporter, which implies that the cells are scavenging lysed cellular material from the environment. That stressed SS120 cells appear to be preferentially acquiring tricarboxylic acid intermediates when growing in P_i-deplete conditions, and not upregulatingPstS, is puzzling. However, it may indicate that this strain may be supplementing an affected glycolysis pathway through acquiring external carbon sources, and that this is more evidence that the cells response to an environmental stress is an iterative, evolving process. SS120 may simply have not initiated transcription of PstS in sufficiently detectable quantities. Indeed, even in starvation experiments *pstS*experession is far from an immediate response [14,23].

Other proteins
An interesting observation is the presence of LuxR, the response regulatory family protein involved in quorum sensing within bacteria, in NATL2A cells (Additional file 1: Table S1). To our knowledge, this is the first instance of observing proteins putatively indicated in quorum sensing capability in any marine cyanobacteria. However, we were unable to locate any LuxI homologues, an essential protein required for effective quorum sensing, within NATL2A (data not shown). However LuxR is known to be a transcriptional regulator activated when cell concentrations of a particular trigger compound (usually N-(3-oxohexanoyl)-$_L$-homoserine

lactone, which is generated through the enzymatic functioning of LuxI) reach particular levels. As such, we speculate that the protein acts as a density-dependant transcriptional regulator, but for an unknown function, and through another trigger compound.

Conclusions
Prochlorococcus are now widely considered to be evolutionarily adept at environmental niche domination, particularly within nutrient poor oligotrophic waters. The genus is typified by genomes characterised by hypervariable genomic islands [17], which are thought to contain genes obtained through phage-mediated horizontal gene transfer, and infer niche-specific advantages such as nutrient acquisition and phage resistance. Our results reinforce previous results concerning the importance of phosphate concentrations to specific strains, but also highlight the possibility of the cells employing a progressive acclimation strategy. It appears that *Prochlorococcus* strains evolutionarily adapted to life in a P_i-deplete environment respond to phosphate fluctuations through a succession of cellular processes, such as the upregulation of P_i acquisition mechanisms, a dissociation of photosynthesis from central metabolic pathways, and a staggered breakdown of the photosystems allowing prolonged photophosphorylated ATP generation. This progressive response allows the cell to react quickly to any subsequent increases in ambient P_i concentrations. It is our hypothesis that HL strains are also particularly sensitive to changes in P_i, and that ambient phosphate concentrations initiate a strong response regardless of being predominantly growth limited elsewhere.

We also note that our results strongly infer that the induction of P_i acquisition mechanisms are concentration specific between strains, particularly considering the absence of any stress response of the LL strain SS120 compared to MIT9312 when grown from identical initial concentration levels.

Methods
For a complete description of the Materials and Methods used please refer to the (Additional file 3: Material and Methods). In brief, however, biological triplicates of all three strains (MIT9312, NATL2A and SS120 (CCMP, Maine)) were grown under 2 separate conditions: P_i replete (Pro99 media with 50 μM NaH_2PO_4 [30]) and P_i deplete (Pro99 media with 10 μM NaH_2PO_4), and moderate white light intensities (30, 10 and 20 μE m^{-2} s^{-1} respectively), in a 13:11 h light:dark regime at 23°C.

For the proteomic analysis, the cells were harvested once measured optical densities reached 0.4 (after which populations had been observed to crash), and proteins were extracted from the three biological replicates for each phenotype [31]. 100 µg of protein from each replicate was then reduced, alkylated, digested and labelled with 8-plex iTRAQ reagents according to the manufacturer's (ABSciex, Framingham, MA) protocol. The labelled replicates were then pooled before primary strong cation exchange (SCX) fractionation [21]. Mass spectrometric analysis of the SCX fractions was performed with a QStar XL Hybrid ESI Quadrupole time-of-flight tandem mass spectrometer, ESI-qQ-TOF-MS/MS (Applied Biosystems; MDS Sciex, Concord, Ontario, Canada), coupled with an online capillary liquid chromatography system (Ultimate 3000, Dionex/LC Packings, The Netherlands) [21,22]. Preliminary data analysis, protein identification and quantitation were carried out using the PHENYX [Geneva Bioinformatics (GeneBio), Geneva, Switzerland] software platform.

Additional material

> **Additional file 1: Table S1**. Full list of identified proteins and peptides for all 3 strains used in this study.
>
> **Additional file 2: Figure S1**. Virtual 2D gel representations of proteins identified from MIT9312 (top left), NATL2A (top right), and SS120 (bottom left).
>
> **Additional file 3: Materials and methods** [32-34].

Acknowledgements
The authors wish to acknowledge the provision of an EPSRC studentship, Advanced Research Fellowship for CAB (EP/E053556/01) and further EPSRC funding (GR/S84347/01 and EP/E036252/1). We also acknowledge the Provasoli-Guillard National Center for Culture of Marine Phytoplankton for the kind provision of cells. Thanks also to Adam Martiny for additional input.

Authors' contributions
MAF designed the study, carried out the proteomics, analysed the data and drafted the manuscript. PCW and CAB conceived of the study and participated in its design. All authors read and approved the final manuscript.

Competing interests
The authors declare that they have no competing interests.

References
1. Ammerman JW, Hood RR, Case DA, Cotner JB: **Phosphorus deficiency in the Atlantic: An emerging paradigm in oceanography.** *Eos Trans Am Geophys Union* 2003, **84**:165-170.
2. Thingstad TF, Krom MD, Mantoura RF, *et al*: **Nature of phosphorus limitation in the ultraoligotrophic eastern Mediterranean.** *Science* 2005, **309**:1068-1071.
3. Thingstad TF, Zweifel UL, Rassoulzadegan F: **P limitation of heterotrophic bacteria and phytoplankton in the northwest Mediterranean.** *Limnol Oceanogr* 1998, **43**:88-94.
4. Chisholm SW, Olson RJ, Zettler ER, Goericke R, Waterbury JB, Welschmeyer NA: **A novel free-living prochlorophyte abundant in the oceanic euphotic zone.** *Nature* 1988, **334**:340-343.
5. Partensky F, Hess WR, Vaulot D: *Prochlorococcus*, **a marine photosynthetic prokaryote of global significance.** *Microbiol Mol Biol Rev* 1999, **63**:106-127.
6. Moore LR, Chisholm SW: **Photophysiology of the marine cyanobacterium** *Prochlorococcus*: **ecotypic differences among cultured isolates.** *Limnol Oceanogr* 1999, **44**:628-638.
7. Moore LR, Rocap G, Chisholm SW: **Physiology and molecular phylogeny of coexisting** *Prochlorococcus* **ecotypes.** *Nature* 1998, **393**:464-467.
8. Rocap G, Distel DL, Waterbury JB, Chisholm SW: **Resolution of** *Prochlorococcus* **and** *Synechococcus* **ecotypes by using 16S-23S ribosomal DNA internal transcribed spacer sequences.** *Appl Environ Microbiol* 2002, **68**:1180-1191.
9. Partensky F, Garczarek L: *Prochlorococcus*: **Advantages and Limits of Minimalism.** *Ann Rev Marine Sci* 2009, **2**:305-331.
10. Coleman ML, Chisholm SW: **Ecosystem-specific selection pressures revealed through comparative population genomics.** *ProcNatlAcadSci* 2010, **107**:18634-18639.
11. Dufresne A, Garczarek L, Partensky F: **Accelerated evolution associated with genome reduction in a free-living prokaryote.** *Genome Biol* 2005, **6**: R14.
12. Van Mooy BA, Rocap G, Fredricks HF, Evans CT, Devol AH: **Sulfolipids dramatically decrease phosphorus demand by picocyanobacteria in oligotrophic marine environments.** *Proc Natl Acad Sci USA* 2006, **103**:8607-8612.
13. Johnson ZI, Zinser ER, Coe A, McNulty NP, Woodward EM, Chisholm SW: **Niche partitioning among** *Prochlorococcus* **ecotypes along ocean-scale environmental gradients.** *Science* 2006, **311**:1737-1740.
14. Martiny AC, Coleman ML, Chisholm SW: **Phosphate acquisition genes in** *Prochlorococcus* **ecotypes: evidence for genome-wide adaptation.** *Proc Natl Acad Sci USA* 2006, **103**:12552-12557.
15. Moore LR, Ostrowski M, Scanlan DJ, Feren K, Sweetsir T: **Ecotypic variation in phosphorus-acquisition mechanisms within marine picocyanobacteria.** *AquatMicrobEcol* 2005, **39**:257-269.
16. Martiny AC, Huang Y, Li W: **Occurrence of phosphate acquisition genes in** *Prochlorococcus* **cells from different ocean regions.** *Environ Microbiol* 2009, **11**:1340-1347.
17. Coleman ML, Sullivan MB, Martiny AC, Steglich C, Barry K, Delong EF, Chisholm SW: **Genomic islands and the ecology and evolution of** *Prochlorococcus*. *Science* 2006, **311**:1768-1770.
18. Copeland A, Lucas S, Lapidus A, Barry K, Detter JC, Hammon N, Israni S, Pitluck S, Thiel J, Schmutz J, Larimer F, Land M, Kyrpides N, Lykidis A, Richardson P: **Complete sequence of** *Prochlorococcus marinus* **str.** *MIT* 2005, 9312, Unpublished.
19. Kettler GC, Martiny AC, Huang K, Zucker J, Coleman ML, Rodrigue S, Chen F, Lapidus A, Ferriera S, Johnson J, *et al*: **Patterns and implications of gene gain and loss in the evolution of** *Prochlorococcus*. *PLoS Genet* 2007, **3**: e231.
20. Dufresne A, Salanoubat M, Partensky F, Artiguenave F, Axmann IM, Barbe V, Duprat S, Galperin MY, Koonin EV, Le Gall F, Makarova KS, Ostrowski M, Oztas S, Robert C, Rogozin IB, Scanlan DJ, Tandeau de Marsac N, Weissenbach J, Wincker P, Wolf YI, Hess WR: **Genome sequence of the cyanobacterium** *Prochlorococcus marinus* **SS120, a nearly minimal oxyphototrophic genome.** *Proc Natl Acad Sci USA* 2003, **100(17)**:10020-10025.
21. Fuszard MA, Wright PC, Biggs CA: **Cellular acclimation strategies of a minimal picocyanobacterium to phosphate stress.** *FEMS MicrobiolLett* 2010, **306**:127-134.
22. Pandhal J, Wright PC, Biggs CA: **A quantitative proteomic analysis of light adaptation in a globally significant marine cyanobacterium** *Prochlorococcus marinus* **MED4.** *J Proteome Res* 2007, **6**:996-1005.
23. Tetu SG, Brahamsha B, Johnson DA, Tai V, Phillippy K, Palenik B, Paulsen IT: **Microarray analysis of phosphate regulation in the marine cyanobacterium** *Synechococcus* **sp. WH8102.** *ISME J* 2009, **3**:835-849.
24. Huber AL, Hamel KS: **Phosphatase activities in relation to phosphorus nutrition in** *Nodularia spumigen* **(Cyanobacteriaceae).** *Hydrobiologia* 1985, **123**:81-88.
25. Natesan R, Shanmugasundaram S: **Extracellular phosphate solubilization by the cyanobacterium** *Anabaena* **ARM310.** *J Biosci* 1989, **14**:203-208.
26. Scanlan DJ, Mann NH, Carr NG: **The response of the picoplanktonic marine cyanobacterium** *Synechococcus* **species WH7803 to phosphate starvation involves a protein homologous to the periplasmic phosphate-binding protein of** *Escherichia coli*. *Mol Microbiol* 1993, **10**:181-191.

27. Scanlan DJ, West NJ: **Molecular ecology of the marine cyanobacterial genera** *Prochlorococcus* **and** *Synechococcus*. *FEMS Microbiol Ecol* 2002, **40**:1-12.

28. Adams MM, Gomez-Garcia MR, Grossman AR, Bhaya D: **Phosphorus Deprivation Responses and Phosphonate Utilization in a Thermophilic** *Synechococcus* **sp. from Microbial Mats.** *J Bacteriol* 2008, **190**:8171-8184.

29. Collier JL, Grossman AR: **Chlorosis induced by nutrient deprivation in** *Synechococcus* **sp. strain PCC 7942: not all bleaching is the same.** *J Bacteriol* 1992, **174**:4718-4726.

30. Moore LR, Post AF, Rocap G, Chisholm SW: **Utilization of Different Nitrogen Sources by the Marine Cyanobacteria** *Prochlorococcus* **and** *Synechococcus*. *LimnolOceanogr* 2002, **47**:989-996.

31. Meijer EA, Wijffels RH: **Development of a Fast, Reproducible and Effective Method for the Extraction and Quantification of Proteins of Micro-algae.** *Biotechnol Tech* 1998, **12**:353-358.

32. Moore LR, Goericke R, Chisholm SW: **Comparative physiology of** *Synechococcus* **and** *Prochlorococcus*: **Influence of light and temperature on growth, pigments, fluorescence and absorptive properties.** *Mar Ecol Prog Ser* 1995, **116**:259-276.

33. Ow SY, Noirel J, Cardona T, Taton A, Lindblad P, Stensjo K, Wright PC: **Quantitative Overview of N2 Fixation in** *Nostoc punctiforme* **ATCC 29133 through Cellular Enrichments and iTRAQ Shotgun Proteomics.** *Journal Proteome Research* 2009, **8**:187-198.

34. Team RDC: *R: A Language and Environment for Statistical Computing* R Foundation for Statistical Computing: Vienna, Austria; 2011.

Permissions

The contributors of this book come from diverse backgrounds, making this book a truly international effort. This book will bring forth new frontiers with its revolutionizing research information and detailed analysis of the nascent developments around the world.

We would like to thank all the contributing authors for lending their expertise to make the book truly unique. They have played a crucial role in the development of this book. Without their invaluable contributions this book wouldn't have been possible. They have made vital efforts to compile up to date information on the varied aspects of this subject to make this book a valuable addition to the collection of many professionals and students.

This book was conceptualized with the vision of imparting up-to-date information and advanced data in this field. To ensure the same, a matchless editorial board was set up. Every individual on the board went through rigorous rounds of assessment to prove their worth. After which they invested a large part of their time researching and compiling the most relevant data for our readers.

The editorial board has been involved in producing this book since its inception. They have spent rigorous hours researching and exploring the diverse topics which have resulted in the successful publishing of this book. They have passed on their knowledge of decades through this book. To expedite this challenging task, the publisher supported the team at every step. A small team of assistant editors was also appointed to further simplify the editing procedure and attain best results for the readers.

Apart from the editorial board, the designing team has also invested a significant amount of their time in understanding the subject and creating the most relevant covers. They scrutinized every image to scout for the most suitable representation of the subject and create an appropriate cover for the book.

The publishing team has been an ardent support to the editorial, designing and production team. Their endless efforts to recruit the best for this project, has resulted in the accomplishment of this book. They are a veteran in the field of academics and their pool of knowledge is as vast as their experience in printing. Their expertise and guidance has proved useful at every step. Their uncompromising quality standards have made this book an exceptional effort. Their encouragement from time to time has been an inspiration for everyone.

The publisher and the editorial board hope that this book will prove to be a valuable piece of knowledge for researchers, students, practitioners and scholars across the globe.

List of Contributors

Steven H Ferguson
Fisheries and Oceans Canada, Central and Arctic Region,
501 University Crescent, Winnipeg, Manitoba R3T 2N6
Canada
Department of Environment and Geography, University of
Manitoba, 501 University Crescent, Winnipeg, Manitoba
R3T 2N6 Canada

Jeff W Higdon
Higdon Wildlife Consulting, 45 Pilgrim Avenue, Winnipeg,
Manitoba R2M 0L3 Canada

Kristin H Westdal
Oceans North Canada, 515-70 Arthur Street, Winnipeg,
Manitoba R3B 1G7 Canada

Daniel Guzmán
Centro de Biotecnología, Facultad de Ciencias y Tecnología,
Universidad Mayor de San Simón, Cochabamba, Bolivia
Department of Biotechnology, Lund University, P.O. Box
124, SE-221 00 Lund, Sweden

Andrea Balderrama-Subieta
Centro de Biotecnología, Facultad de Ciencias y Tecnología,
Universidad Mayor de San Simón, Cochabamba, Bolivia

Carla Cardona-Ortuño
Centro de Biotecnología, Facultad de Ciencias y Tecnología,
Universidad Mayor de San Simón, Cochabamba, Bolivia

Mónica Guevara-Martínez
Centro de Biotecnología, Facultad de Ciencias y Tecnología,
Universidad Mayor de San Simón, Cochabamba, Bolivia

Nataly Callisaya-Quispe
Centro de Biotecnología, Facultad de Ciencias y Tecnología,
Universidad Mayor de San Simón, Cochabamba, Bolivia

Jorge Quillaguamán
Centro de Biotecnología, Facultad de Ciencias y Tecnología,
Universidad Mayor de San Simón, Cochabamba, Bolivia

Kabilan Mani
Department of Biological Sciences, BITS PILANI, K K
Birla Goa Campus, NH 17 B, Zuarinagar, Sancoale, Goa
403 726, India

Bhakti B Salgaonkar
Department of Biological Sciences, BITS PILANI, K K
Birla Goa Campus, NH 17 B, Zuarinagar, Sancoale, Goa
403 726, India

Judith M Braganca
Department of Biological Sciences, BITS PILANI, K K
Birla Goa Campus, NH 17 B, Zuarinagar, Sancoale, Goa
403 726, India

Renato De Medeiros Rocha
Departamento de Geografia, Universidade Federal do Rio
Grande do Norte, Campus de Caicó, Joaquim Gregório,
s/n, Penedo 59.300-000, Caicó-RN, Brasil

Diógenes FS Costa
Departamento de Geografia, Universidade Federal do Rio
Grande do Norte, Campus de Caicó, Joaquim Gregório,
s/n, Penedo 59.300-000, Caicó-RN, Brasil
Departamento de Biologia, Universidade de Aveiro,
Campus de Santiago, 3810-193 Aveiro, Portugal

Milton A Lucena-Filho
Departamento de Geografia, Universidade Federal do Rio
Grande do Norte, Campus de Caicó, Joaquim Gregório,
s/n, Penedo 59.300-000, Caicó-RN, Brasil

Rodolfo M Bezerra
Departamento de Geografia, Universidade Federal do Rio
Grande do Norte, Campus de Caicó, Joaquim Gregório,
s/n, Penedo 59.300-000, Caicó-RN, Brasil

David HM Medeiros
Departamento de Geografia, Universidade Federal do Rio
Grande do Norte, Campus de Caicó, Joaquim Gregório,
s/n, Penedo 59.300-000, Caicó-RN, Brasil

Antonio M Azevedo-Silva
Departamento de Geografia, Universidade Federal do Rio
Grande do Norte, Campus de Caicó, Joaquim Gregório,
s/n, Penedo 59.300-000, Caicó-RN, Brasil

Cristian N Araújo
Departamento de Geografia, Universidade Federal do Rio
Grande do Norte, Campus de Caicó, Joaquim Gregório,
s/n, Penedo 59.300-000, Caicó-RN, Brasil

Lauro Xavier-Filho
Instituto de Tecnologia e Pesquisa, Universidade
Tiradentes, Av. Murilo Dantas, 300. Bairro Farolândia,
49032-490 Aracaju, Sergipe, Brasil

Emmanuelle Apper-Bossard
Tereos Syral, Z.I. Portuaire, 67 390, Marckolsheim, France

Aurélien Feneuil
Tereos Syral, Z.I. Portuaire, 67 390, Marckolsheim, France

Anne Wagner
Tereos Syral, Z.I. Portuaire, 67 390, Marckolsheim, France

Frédérique Respondek
Tereos Syral, Z.I. Portuaire, 67 390, Marckolsheim, France

Nardi Cribb
School of Biological Sciences, Flinders University, Box 2100, Adelaide, SA 2001, Australia

Cara Miller
School of Biological Sciences, Flinders University, Box 2100, Adelaide, SA 2001, Australia
Faculty of Science, Technology and the Environment, University of the South Pacific, Laucala campus, Suva, Fiji
The Whale and Dolphin Conservation Society International, Chippenham, UK

Laurent Seuront
School of Biological Sciences, Flinders University, Box 2100, Adelaide, SA 2001, Australia
South Australian Research and Development Institute, Aquatic Science, West Beach, SA 5022, Australia
Centre National de la Recherche Scientifique, Laboratoire d'Océanologie et de Géosciences, UMR LOG 8187, Université des Sciences and Technologies de Lille, Station Marine, Wimereux 62930, France

Kabilan Mani
Department of Biological Sciences, BITS PILANI, K K Birla Goa Campus, Zuarinagar, Goa 403 726, India

Bhakti B Salgaonkar
Department of Biological Sciences, BITS PILANI, K K Birla Goa Campus, Zuarinagar, Goa 403 726, India

Deepthi Das
Department of Biological Sciences, BITS PILANI, K K Birla Goa Campus, Zuarinagar, Goa 403 726, India

Judith M Bragança
Department of Biological Sciences, BITS PILANI, K K Birla Goa Campus, Zuarinagar, Goa 403 726, India

Jean-Lou Justine
UMR 7138 Systématique, Adaptation, Évolution, Muséum National d'Histoire Naturelle, Case postale 51, 55, rue Buffon, 75231 Paris cedex 05, France

Ian Beveridge
Department of Veterinary Science, University of Melbourne, Veterinary Clinical Centre, Werribee 3030, Victoria, Australia

Geoffrey A Boxshall
Department of Zoology, Natural History Museum, Cromwell Road, London SW7 5BD, UK

Rodney A Bray
Department of Zoology, Natural History Museum, Cromwell Road, London SW7 5BD, UK

Terrence L Miller
Biodiversity Program, Queensland Museum, PO Box 3300, South Brisbane, Queensland 4101, Australia

František Moravec
Institute of Parasitology, Biology Centre, Academy of Sciences of the Czech Republic, Branišovská 31 370 05, České Budějovice, Czech Republic

Jean-Paul Trilles
Équipe Adaptation écophysiologique et Ontogenèse, UMR 5119 (CNRS-UM2-IRD-UM1-IFREMER), Université Montpellier 2, Place Eugène Bataillon, 34095 Montpellier cedex 05, France

Ian D Whittington
Monogenean Research Laboratory, The South Australian Museum, Adelaide 5000, & Marine Parasitology Laboratory, & Australian Centre for Evolutionary Biology and Biodiversity, The University of Adelaide, North Terrace, Adelaide 5005, South Australia, Australia

Fulvio Cerfolli
Department of Ecological and Biological Sciences (DEB), Ichthyogenic Experimental Marine Centre (CISMAR), Tuscia University, Borgo Le Saline, 01016, Tarquinia, VT, Italy

Bruno Bellisario
Department of Ecological and Biological Sciences (DEB), Ichthyogenic Experimental Marine Centre (CISMAR), Tuscia University, Borgo Le Saline, 01016, Tarquinia, VT, Italy

Corrado Battisti
'Torre Flavia' LTER (Long Term Environmental Research) Station, Environmental Service, Provincia di Roma, via Tiburtina, 691, 00159, Rome, Italy

Brandon J Sansom
Department of Biology, Washington and Jefferson College, Washington, PA, USA
Oklahoma Biological Survey and Department of Biology, University of Oklahoma, 111 East Chesapeake St, Norman, OK, 73019, USA

Daniel J Hornbach
Department of Environmental Studies, Macalester College, St. Paul, MN, USA
Department of Biology, Macalester College, St. Paul, MN, USA

Mark C Hove
Department of Biology, Macalester College, St. Paul, MN, USA

Jason S Kilgore
Department of Biology, Washington and Jefferson College, Washington, PA, USA

Terry J McGenity
School of Biological Sciences, University of Essex, Wivenhoe Park, Colchester CO4 3SQ, UK

Benjamin D Folwell
School of Biological Sciences, University of Essex, Wivenhoe Park, Colchester CO4 3SQ, UK

Boyd A McKew
School of Biological Sciences, University of Essex, Wivenhoe Park, Colchester CO4 3SQ, UK

Gbemisola O Sanni
School of Biological Sciences, University of Essex, Wivenhoe Park, Colchester CO4 3SQ, UK

M Belal Hossain
Environmental and Life Sciences, Faculty of Science, Universiti Brunei Darussalam, BE1410 Jalan Tunkgu Link, Brunei Darussalam
Department of Fisheries and Marine Science, Noakhali Science and Technology University, Sonapur 3814, Bangladesh

David J Marshall
Environmental and Life Sciences, Faculty of Science, Universiti Brunei Darussalam, BE1410 Jalan Tunkgu Link, Brunei Darussalam

Maíra Duarte Cardoso
Programa de Pós-Graduação em Saúde Pública e Meio Ambiente, ENSP/ Fiocruz, Rua Leopoldo Bulhões, 1480, Manguinhos, Rio de Janeiro 21041-210, RJ, Brasil

Jailson Fulgencio de Moura
Systems Ecology, Leibniz Center for Tropical Marine Ecology (ZMT), Fahrenheitstrasse 6, 28359 Bremen, Germany

Davi C Tavares
Departamento de Endemias Samuel Pessoa – DENSP & Grupo de Estudos de Mamíferos Marinhos da Região dos Lagos – GEMM-Lagos, Escola Nacional de Saúde Pública /FICORUZ, Rua Leopoldo Bulhões, 1.480, 6° andar, Sala 611, Manguinhos, Rio de Janeiro 21041-210, RJ, Brasil

Rodrigo A Gonçalves
Departamento de Química, Pontifícia Universidade Católica do Rio de Janeiro, Rua Marquês de São Vicente, 225, Gávea, Rio de Janeiro 22453-900, RJ, Brasil

Fernanda I Colabuono
Universidade de São Paulo, Instituto Oceanográfico, Praça do Oceanográfico 191, Cidade Universitária, São Paulo 05508-120, SP, Brasil

Emily M Roges
Instituto Oswaldo Cruz/FIOCRUZ, Laboratório de Referência Nacional de Enteroinfecções Bacterianas, Av. Brasil, 4365, Manguinhos, Rio de Janeiro 21040-360, RJ, Brasil

Roberta Laine de Souza
Instituto Oswaldo Cruz/FIOCRUZ, Laboratório de Referência Nacional de Enteroinfecções Bacterianas, Av. Brasil, 4365, Manguinhos, Rio de Janeiro 21040-360, RJ, Brasil

Dalia Dos Prazeres Rodrigues
Instituto Oswaldo Cruz/FIOCRUZ, Laboratório de Referência Nacional de Enteroinfecções Bacterianas, Av. Brasil, 4365, Manguinhos, Rio de Janeiro 21040-360, RJ, Brasil

Rosalinda C Montone
Universidade de São Paulo, Instituto Oceanográfico, Praça do Oceanográfico 191, Cidade Universitária, São Paulo 05508-120, SP, Brasil

Salvatore Siciliano
Departamento de Endemias Samuel Pessoa – DENSP & Grupo de Estudos de Mamíferos Marinhos da Região dos Lagos – GEMM-Lagos, Escola Nacional de Saúde Pública /FICORUZ, Rua Leopoldo Bulhões, 1.480, 6° andar, Sala 611, Manguinhos, Rio de Janeiro 21041-210, RJ, Brasil

Zheng Liu
Institute of Marine and Environmental Technology, University of Maryland Center for Environmental Science, 701 E Pratt St, Baltimore, MD 21202, USA

Fan Zhang
Institute of Marine and Environmental Technology, University of Maryland Center for Environmental Science, 701 E Pratt St, Baltimore, MD 21202, USA

Feng Chen
Institute of Marine and Environmental Technology, University of Maryland Center for Environmental Science, 701 E Pratt St, Baltimore, MD 21202, USA

Helen Bailey
Chesapeake Biological Laboratory, University of Maryland Center for Environmental Science, 146 Williams Street, Solomons, MD 20688, USA

Kate L Brookes
Marine Scotland Science, 375 Victoria Road, Aberdeen AB11 9DB, UK

Paul M Thompson
Institute of Biological and Environmental Sciences, Lighthouse Field Station, University of Aberdeen, George Street, Cromarty, Ross-shire IV11 8YJ, UK

Matthew A Fuszard
ChELSI Institute, Department of Chemical and Biological Engineering, University of Sheffield, Mappin Street, Sheffield S1 3JD, UK

Phillip C Wright
ChELSI Institute, Department of Chemical and Biological Engineering, University of Sheffield, Mappin Street, Sheffield S1 3JD, UK

Catherine A Biggs
ChELSI Institute, Department of Chemical and Biological Engineering, University of Sheffield, Mappin Street, Sheffield S1 3JD, UK

www.ingramcontent.com/pod-product-compliance
Lightning Source LLC
Chambersburg PA
CBHW050451200326
41458CB00014B/5147